Algorithms and Computation in Mathematics · Volume 19

Editors

Arjeh M. Cohen   Henri Cohen
David Eisenbud   Bernd Sturmfels

Wieb Bosma
John Cannon (Eds.)

Discovering Mathematics with Magma

Reducing the Abstract to the Concrete

With 9 Figures and 16 Tables

 Springer

Editors

Dr. Wieb Bosma

Katholieke Univ. Nijmegen
Vakgroep Wiskunde
Postbus 9010, 6500 GL
Nijmegen, Netherlands
E-mail: bosma@math.kun.nl

Professor Dr. John Cannon

University Sydney
Department Pure Mathematics F07
2006 Sydney, Australia
E-mail: john@maths.usyd.edu.au

Mathematics Subject Classification (2000): 68W30, 11Yxx, 12Y05, 13Pxx, 14Qxx, 16Z05

ISSN 1431-1550
ISBN 978-3-642-07231-4
e-ISBN 978-3-540-37634-7

Springer is a part of Springer Science+Business Media
springer.com
© Springer-Verlag Berlin Heidelberg 2006
Softcover reprint of the hardcover 1st edition 2006

Cover design: *design & production* GmbH, Heidelberg

Preface

The appearance of this volume celebrates the first decade of Magma, a new computer algebra system launched at the First Magma Conference on Computational Algebra held at Queen Mary and Westfield College, London, August 1993.

This book introduces the reader to the role Magma plays in advanced mathematical research. Each paper examines how the computer can be used to gain insight into either a single problem or a small group of closely related problems. The intention is to present sufficient detail so that a reader can (a), gain insight into the mathematical questions that are the origin of the problems, and (b), develop an understanding as to how such computations are specified in Magma. It is hoped that the reader will come to a realisation of the important role that computational algebra can play in mathematical research. Readers not primarily interested in using Magma will easily acquire the skills needed to undertake basic programming in Magma, while experienced Magma users can learn both mathematics and advanced computational methods in areas related to their own.

The core of the volume comprises 14 papers. The authors were invited to submit articles on designated topics and these articles were then reviewed by referees. Although by no means exhaustive, the topics range over a considerable part of Magma's coverage of algorithmic algebra: from number theory and algebraic geometry, via representation theory and computational group theory to some branches of discrete mathematics and graph theory. The papers are preceded by an outline of the Magma project, a brief summary of the papers and some instructions on reading the Magma code. A basic introduction to the Magma language is given in an appendix.

The editors express their gratitude to the contributors to this volume, both for the work put into producing the papers and for their patience.

John Cannon, Sydney

Wieb Bosma, Nijmegen

April 2006

Magma: The project

Computer Algebra

Computers are transforming the manner in which mathematics is discovered, communicated and applied. Thus, computers are used to determine the properties of, and relations between, mathematical objects, thereby enabling the mathematician to formulate new conjectures or disprove old conjectures. The application of mathematics often requires extensive calculations which are practical only if performed by a computer. Increasingly, computers are used to generate or verify formal mathematical proofs, sometimes in combination with extensive computations. Finally, computers allow mathematicians to typeset their own papers and books, which can then be placed in a web archive for instant access by colleagues in every part of the world.

The non-trivial use of computers in the creation and application of mathematics typically requires sophisticated software and it is here that a major problem appears. Even experiments in elementary number theory require multi-precision integer arithmetic, something which is found in very few general scientific programming languages. If the software necessary for performing an algebraic calculation in an abstract area of mathematics has to be implemented in a standard scientific programming language, it is likely to be at a degree of sophistication and complexity far beyond the ability or time-budget of most mathematicians.

General systems

To address this problem, a number of so-called *general symbolic* Computer Algebra systems have emerged. These packages are designed for the easy manipulation of symbolic expressions and have, as their central motivation, the solution of differential equations. The packages MACSYMA and Reduce, whose development commenced in the 1960's, are two early examples. Currently, Maple and Mathematica are the two most popular general symbolic systems.

The developers of these systems claim that they are suitable for all areas of mathematics. However, while their symbolic-expression, transformation-rule (SETR) model of computation with mathematical objects works well for the large number of applied and theoretical problems that use tools such as calculus and differential equations, it is inappropriate for much of mathematics. For example, implementing code to determine the structure of a finite group in either Maple or Mathematica is unnatural, difficult and results in extremely inefficient code.

Specific area systems

The failure of general systems to meet the needs of many areas led to the proliferation of specialised libraries and packages that focus on a fairly narrow area. In particular, such systems can provide functions that are highly efficient (compared to a general symbolic system) and which also provide specialised coverage of the area. Furthermore, being comparatively small means that one person or a small group can write and maintain the package. We will refer to this class of packages as the *specific area* systems. The computational model is usually quite different to the SETR model and it is often not explicit. An important example of a specific area system is the number theory package Pari, which provides a C library of functions for higher-order objects (such as number fields and elliptic curves). Pari also provides a simple user interface (GP) so that for many applications it is unnecessary to resort to writing C code in order to use the package.

The narrow focus of specific area systems has a major downside in that the user soon reaches the point where tools outside the scope of the package are needed. For example, advanced methods for computing the Galois group of an equation (a number theory problem) require the use of sophisticated tools from computational group theory. This results in the user having to acquire and learn a number of different packages and then somehow enable the packages to communicate with one another. Generally speaking, mathematicians are reluctant to invest the considerable time needed to gain a basic fluency with more than one or two packages such as Pari.

The Magma project

A good computational model properly implemented should make it possible to code calculations at a high degree of abstraction in a manner which is both concise and natural to the subject matter. Because of the quite different flavours of different parts of mathematics, it seems very unlikely that a single model will work effectively across most of mathematics. A key aim of the Magma project is the development of an effective model for a broad area sharing a strong commonality of approach. We concentrate on topics where the concepts of structure and morphism are central, areas where the

SETR model provides an inadequate description of the underlying data and its operations. Specifically, we chose to concentrate on topics that make heavy use of the axiomatic approach to modern algebra as exemplified by van der Waerden. This is taken to include algebra, number theory, algebraic geometry, algebraic topology and algebraic combinatorics. Indeed, modern algebra provides a natural setting for the application of the Object-Oriented (OO) programming paradigm, which is also the basis of the Axiom system.

Some general characteristics of Magma include:

- Concepts from universal algebra and category theory are used to provide a formal basis for the computational model.
- The language design attempts to approximate the usual algebraic modes of thought and notation. In particular, the principal constructs in the user language are set, algebraic structure and morphism.
- The system is designed to automatically remember and utilise relationships between objects.
- An important goal is to provide highly-tuned implementations of the most efficient algorithms known for computing with the various classes of structure supported. The objective is to provide performance similar to, or better than, specialised stand-alone programs.

The computational model

Each object definable in Magma is considered to belong to a unique mathematical structure (*magma*), called its *parent*, and this parent must be present at the time the object is created. For instance, Magma does not have the concept of a matrix as a primitive object; rather, a matrix may only exist as an element of some particular ring or bimodule. Moreover, Magma employs *evaluation semantics*, so that an expression may only be evaluated if every identifier appearing in it has a value. Thus, notions such as "indeterminates" or "unknowns" must be defined as elements of the appropriate magma, such as a univariate polynomial ring over a field, before any use is made of them. This philosophy makes programming a great deal easier, since there is never any ambiguity as to the current interpretation of an object. In programming language parlance, Magma is a strongly typed language.

Magmas are organised hierarchically. At the top level are *varieties*, which are families of magmas sharing the same basic operations and axioms. Typical varieties include groups, commutative rings and modules. A variety has attached to it functions (or methods in OO parlance) that do not depend upon properties of a specific representation of a magma. When computing with a structure A it is necessary that A be represented in some concrete form; here we come to the level of *categories*. For example, within the variety of groups there are categories for finitely-presented groups, permutation groups and matrix groups (among others). The category to which a magma belongs determines not only how the magma and its elements are represented, but also the permitted operations. Note that operations independent of the representation are inherited from the parent variety while representation-dependent operations are attached to the category. Categories of magma that are specialised subcategories of an existing category C may, in turn, inherit operations from C, in addition to having further operations that are specific to the subcategory.

The general model for constructing a magma is to start with a 'free' magma and then successively form submagmas, quotients and extensions until the desired magma is obtained. This approach, which has its foundations in Universal Algebra, provides a powerful and general mechanism for constructing a vast range of algebraic and geometric structures. Since most commonly occurring structures in computational algebra are finitely generated, in general, a magma M is given in terms of a finite generating set. Special handling applies to the rare cases of infinitely generated magmas, such as $\mathbb{R}$, $\mathbb{C}$, and Laurent series rings.

The user language

The Magma user language is an interpreted high-level language with dynamic typing. The data structures are designed around the ideas of set, sequence, mapping and structure. While the language formally belongs to the class of imperative languages there is also a strong functional flavour. Powerful constructors allow complex sets and magmas to be constructed in a single statement. Typically, a magma is created from another magma M by means of *constructor* expressions, such as the submagma and quotient constructors:

> **sub**$< M \mid$ *generators* $>$
> **quo**$< M \mid$ *normal subgroup* $>$

These constructors not only return the new submagma **S** or quotient **Q**, but also return a morphism relating the new magma to the existing magma **M** given in the left part of the constructor; here the morphisms are the inclusion monomorphism $S \to M$ and the natural epimorphism $M \to Q$. Other mappings from magma M_1 to M_2 may be directly created by means of the mapping or the homomorphism constructors:

> **map**$< M_1 \to M_2 \mid x \mapsto$ *expression* **in** $x >$
> **hom**$< M_1 \to M_2 \mid$ *images of generators of* $M_1 >$

Magma has a powerful collection of set and sequence constructors whose syntax is based on traditional mathematical set notation and was influenced by SETL [1]. For example, $\{1..10\}$ is the set of integers $\{x \in \mathbb{Z} \mid 1 \le x \le 10\}$ and

> $\{ i^2 : i$ **in** $\{1..10\} \mid$ **not** IsPrime(i) $\}$;
>
> > $\{ 1, 16, 36, 64, 81, 100 \}$

is the set of squares of the non-primes in this set.

The language has a functional subset, providing functions and procedures as first-class objects. Functions are allowed multiple return values. Value arguments and parameters are supported both for functions and procedures while reference arguments are supported for procedures only. As first-class objects, functions and procedures are created as the values of expressions, rather than by means of special statements.

For a more complete description, the reader should consult the concise introduction in the appendix on page 331, or [2] and [3] for a formal overview, or [4] and [5] for full user documentation.

Acknowledgements

The core of the Magma system has been developed by the Computational Algebra Group at the University of Sydney. In addition, Magma includes algorithms and code generously contributed by many mathematicians from around the world. In many ways the Magma project represents a major cooperative effort by mathematicians to develop a system properly tuned to their needs. A full list of contributors may be found at the front of Volume 1 of [5].

References

1. J. T. Schwartz, R. B. K. Dewar, E. Dubinsky, E. Schonberg, *Programming With Sets: An Introduction to SETL*, New York: Springer, 1986.
2. Wieb Bosma, John Cannon, Catherine Playoust, *The Magma algebra system I: The user language*, J. Symbolic Comput. **24** (1997), 235–265.
   See also the Magma home page at http://magma.maths.usyd.edu.au/magma/.
3. Wieb Bosma, John Cannon, Graham Matthews, *Programming with algebraic structures: design of the Magma language*, pp. 52–57 in: M. Giesbrecht (ed), *Proceedings of International Symposium on Symbolic and Algebraic Computation, Oxford (1994)*, Association for Computing Machinery, 1994.
4. Wieb Bosma, John Cannon, Catherine Playoust, *Algebraic programming with Magma*, to appear.
5. John Cannon, Wieb Bosma (eds.), *Handbook of Magma Functions*, Version 2.13, Volumes **1–10**, Sydney, July 2006, 4320 pages.

Discovering mathematics: About this volume

Introduction

A major motivation behind the design and implementation of Magma has been the desire to make advanced areas of algebra and geometry much more accessible to effective computation. We believe that the system represents a major step towards achieving these aims. Since its first public release in 1994, Magma has been widely adopted and has been used for applications in a wide variety of fields both inside and outside of mathematics. Magma has been used by many mathematicians to develop packages for new areas. Examples include algebraic geometry via schemes (Gavin Brown), cohomological rings of group modules (Jon Carlson), quantum groups (Willem de Graaf), Kronecker's algorithm for solving systems of polynomial equations (Gregoire Lecerf), Aschbacher analysis of large degree matrix groups (Eamonn O'Brien), hyperelliptic curves (Michael Stoll) and modular forms and modular abelian varieties (William Stein). Magma has been cited in more than 2000 books, papers and theses.

This volume is a collection of case studies showing how Magma is applied to problems across a range of different areas. In most instances the work led to the publication of one or more theoretical papers. The corresponding paper in this volume presents the Magma computations behind the theoretical results. The collection of papers can be regarded as a series of tutorials on the application of Magma to advanced research problems. The authors were asked to provide some mathematical background on their problem(s) and then present the Magma code with an extended commentary. It is hoped that readers will gain insight into the possible applications of Magma and also detailed examples of its use in various contexts. For the convenience of readers not particularly familiar with the Magma language, a brief introduction written by Geoff Bailey is included as an appendix to this book. In addition, a reader meeting Magma for the first time may also wish to consult Volume 1 of the *Handbook of Magma Functions* (shipped in electronic form as part of the Magma distribution) in order to get a detailed description of the language and its intrinsic functions.

Number theory

The paper by Wieb Bosma, *Some computational experiments in number theory*, opens with a number of examples of short Magma programs which are applied to (open) questions in elementary number theory. In particular, readers new to the system will gain insight into how some of the basic language features may be used to construct simple but effective programs. The material includes set covering problems, with applications to primality tests, and properties of Euler's ϕ function and its inverse image. The second part of the paper reproduces computations carried out jointly with Bart de Smit as part of a study of relations between class numbers in the subfields of Galois extensions of $\mathbb{Q}$ with a fixed Galois group. The work involves calculations in the ring of integers of a number field as well as the determination of its ideal class group. Access to the machinery in Magma for working with small permutation groups, and especially their characters is essential.

Claus Fieker, in his paper *Applications of the class field theory of global fields*, begins by introducing the reader to the basics of computing with number fields and global function fields, stressing their similarity. Maps are used to move between ideal class groups and their structure as abstract abelian groups. The correspondence between subfields of ray class fields and quotients of the ray class group is made explicit. The class field theory for number fields is then used to construct tables of number fields with given Galois group. As an application of class fields of function fields, curves of small genus with many rational points are constructed. These curves give rise to linear codes with good error-correcting properties.

Algebraic geometry

Applying methods from arithmetic algebraic geometry to Diophantine equations in number theory, Nils Bruin succeeded in determining all solutions in coprime integers of $x^n + y^n = Dz^2$ for small values of n and D. In *Some ternary Diophantine equations of signature* $(n, n, 2)$ he describes the computational details for the method. It involves the construction of subcovers of (hyper)elliptic curves over number fields, deciding their local solvability, and using Selmer groups and Chabauty-like methods for the determining Mordell-Weil groups of elliptic curves.

In *Studying the Birch and Swinnerton-Dyer conjecture for modular abelian varieties using Magma*, William Stein describes how the Magma-package written by him can be used to compute the quantities conjecturally related by the Birch and Swinnerton-Dyer conjecture. He does this for abelian varieties (generalizing from elliptic curves) that have some extra structure given in terms of modular forms.

Curves of genus 2 are the subject of Paul van Wamelen's paper *Computing with the analytic Jacobian of a genus 2 curve*. The author gives solutions in Magma for two problems. First, given a CM field, construct the equation for a genus 2 curve over $\mathbb{Q}$ with complex multiplication by the maximal order of this field. The construction uses computations in the number field, with real and complex numbers, and with integral matrices. The second problem is to find rational isogenies between the Jacobians of two genus 2 curves. One of the tools developed by the author constructs maps between the algebraic description of the Jacobian (in terms of divisors) and the analytic version of it (as a complex torus).

Gavin Brown, in his paper *Finding special K3 surfaces with Magma*, introduces his Magma database of K3 surfaces. He then applies graded ring methods to construct 27 families of K3 surfaces that appear as degenerate cases of surfaces in the standard lists. Before doing so, the reader is shown how to set up varieties in Magma as projective schemes, how to manipulate power series representing the Hilbert series, and how to compute the dimension of Riemann-Roch spaces at certain divisors.

Algebras and representation theory

The paper of Don Taylor, *Constructing the split octonions* presents Magma code for constructing the algebra of split octonions $\mathbb{O}(R)$ over a ring R, firstly in terms of structure constants and secondly using Lie algebras. The latter construction is then used to give an explicit representation of the algebraic group of type $G_2(F)$, F a field, as the automorphism group of the split octonions $\mathbb{O}(F)$. The calculations involve the use of standard machinery for working with Lie algebras, root systems, groups of Lie type, and matrix representations. In addition, numerous examples of user-defined functions illustrate many useful features of the language.

In the first of two papers, *Support varieties for modules*, Jon Carlson presents a probabilistic algorithm for computing the support variety of a module defined over a group algebra. (A support variety is an affine variety that encodes many of the homological properties of the module). The construction of the support variety of a module over a group algebra can be reduced to considering the support varieties of the restriction of the module to elementary abelian subgroups. In this case the support variety is isomorphic (as a set) to a rank variety which may be computed by finding the ranks of certain matrices that arise from the action of group elements on the module. Carlson describes the theory and presents Magma code for the construction of the support variety of a module over the group ring of an elementary abelian p-group. In addition to kG-modules and matrix algebras, the calculation involves working with Gröbner bases for ideals of multivariate polynomial rings.

Jon Carlson, in his second paper, *When is projectivity detected on subalgebras?* is concerned with understanding when the projectivity of a module over an algebra can be determined by considering its restriction to certain subalgebras. Using a standard Magma package he developed for computing the cohomology ring of a module, Carlson first constructs an example where projectivity is detected on subalgebras and then an example where it is not.

The constructions involve performing cohomological calculations with modules over split basic algebras corresponding to extraspecial p-groups. The examples presented in this paper played an important role in the classification of the torsion endotrivial modules for a finite p-group by Jon Carlson and Jacques Thévenaz.

Group theory

In the paper *Cohomology and extensions of groups*, Derek Holt describes algorithms for computing the first and second cohomology of a finite group which reduce the problem to the calculation of the nullspace of a matrix. The algorithm for the second cohomology group has not been previously published. As an application, Holt considers the cohomology group $\mathrm{H}^2(G, M)$, where G is $\Omega^-(12, 2)$ and M is the natural module over $\mathrm{GF}(2)$, and shows that it has dimension 1. Until recently this was an open question. In the second part of the paper, Holt uses the cohomology as the basis of an algorithm for enumerating all non-isomorphic extensions of a soluble group by a finite group. This machinery plays a key role in Claus Fieker's work on constructive class field theory.

Colva Roney-Dougal and Bill Unger in their paper *Classifying primitive groups of degree up to 1000*, describe how Magma was used to determine the primitive affine subgroups of $\mathrm{AGL}(n, p)$ for $p^n < 1000$. This reduces to the problem of finding the distinct conjugacy classes of irreducible subgroups of $\mathrm{GL}(n, p)$, p a prime. The paper focuses on the development of an efficient method for determining conjugacy of two subgroups in $\mathrm{GL}(n, q)$ based on Aschbacher's theorem on the subgroups of classical groups. Taken together with previously published work, the work of Roney-Dougal and Unger completed the classification of primitive groups of degree up to 1000. (Since this paper was written, Roney-Dougal has extended the classification up to degree 2500).

Volker Gebhardt in his paper *Computer-aided discovery of a fast algorithm for solving the conjugacy problem in braid groups* describes Magma experiments that led him to discover an algorithm much faster than any previously known for testing conjugacy of elements in braid groups. Gebhardt begins by taking the reader through a simple implementation of the Birman, Ko and Lee algorithm (BKL algorithm) for conjugacy testing. The BKL algorithm is based on the concept of the *super-summit set* invariant for a conjugacy class. Gebhardt then shows how an extensive study of the behaviour of the BKL algorithm led him to discover a much more powerful conjugacy class invariant, the *ultra-summit set* and consequently, a much faster conjugacy algorithm. This work has had major ramifications for braid group cryptography. The paper concludes with an example of constructing and breaking a simple braid group cryptosystem.

Discrete mathematics

A. E. Brouwer maintains an on-line table giving upper and lower bounds on the minimum distance of linear codes over small fields. However, explicit constructions for many of the codes in this table are not available. In the paper *Searching for linear codes with large minimum distance*, Markus Grassl describes some of the techniques that are being employed to find and construct codes that meet or exceed the Brouwer lower-bounds. In particular, some improvements by Markus Grassl and Greg White to the Brouwer-Zimmermann algorithm for computing the minimum distance of a code are outlined. The Magma implementation (not given) is illustrated by using it to search for certain ternary codes with high minimum weights.

In the paper *Colouring planar graphs*, Paulette Lieby investigates the problem of finding an efficient algorithm for constructing a 4-colouring for a planar graph. Her approach is based on a theorem which states that a planar graph G has a k-colouring if and only if the dual of G has a nowhere-zero k-flow. Magma is used to explore possible strategies for finding nowhere-zero k-flows. After showing how to write Magma code to produce the dual graph G^* of the given graph G, Lieby then uses a depth-first search to construct a graph G^{**} that is a totally cyclic orientation of G^*. Finally, she turns G^{**} into a network and uses a maximal flow algorithm to determine the existence of a nowhere-zero k-flow.

How to read the Magma code

The authors of the papers in this book were asked to provide Magma code to illustrate their constructions and computations. As the reader will notice, rather than using the obligatory `verbatim` style in LATEX we have used Don Taylor's sophisticated MagmaTEX style package. The code fragments appear in a much more structured way, as MagmaTEX parses the Magma code and chooses fonts according to the function of the tokens of the language. Moreover, a limited amount of sub- and superscripting is done by MagmaTEX, a Greek font is used for Greek identifier names, loop control structures appear in boldface, etc. In our opinion, this enables the reader to determine the effect of a program statement at a glance and results in code that is much more pleasing to the eye.

Some authors initially expressed concern about the difficulty readers would have in reproducing the program code from these typeset pages. We believe that this concern is not justified if the reader will keep the following points in mind.

Firstly, all of the code appearing in the book is available from a website whose location is noted in a footnote* appearing in each paper.

Secondly, in some cases it will be necessary to use additional code that has been omitted from the published text for reasons of length. The additional material is available, in a form suitable for immediate execution, as part of the code available at the web site. This will also make it possible to update the code slightly if necessary to comply with changes in future releases of Magma.

Thirdly, it is necessary to note only a small number of conventions used throughout this book. Readers will have little trouble recognizing and applying the most important rules when trying to reproduce Magma sessions:

- Font styles (*italics*, **bold face** and the like) should be ignored completely for reproduction purposes:

---

*See the Preface to this volume for style conventions regarding Magma code; code appearing in this book is available at `http://magma.maths.usyd.edu.au/magma/`.

for x **in** $[1..10]$ **do**

can simply be typed into Magma as

```
for x in [1..10] do
```

It is useful to know that identifier names are typeset in a slanted, sans serif font, like *myVar*, unless they refer to a function (intrinsic or user-defined), in which case there is no slant on the sans serif, as in MyFunction.

- Greek letters simply result from Greek letter names as identifiers; thus

$$\gamma := \Gamma(3);$$

is the typeset version of the statement

```
gamma := Gamma(3);
```

in Magma.

- Subscripts appear in the code only when trailing digits are used on identifier names (they do not need an underscore character). Thus, the lines of actual Magma code

```
x10 := 11;
y_11 := 12;
z_A := b;
Z2F3 := func< n | GF(3) ! n >;
```

are typeset in this book as

$x_{10} := 11;$
$y_{-11} := 12;$
$z\_A := b;$
$Z2F_3 := \textbf{func}< n \mid GF(3) \ ! \ n >;$

Superscripts arise when a caret ^ is used:

```
v3 := v_1^3;
y := x^Conjugate(g);
```

will appear in this book as

$v_3 := v_{-1}{}^3;$
$y := x^{\text{Conjugate}(g)};$

In particular, the caret is not printed.

- There are a very small number of composite characters such as the symbol $\rightarrow$ which is composed of the two keyboard characters - (dash) and >:

$h := \textbf{hom}< P \ \rightarrow \ Q \mid x >;$

```
h := hom< P -> Q | x >;
```

Original Magma code	Appearance in this book
x^n	x^n
P1, P2, ...	$P_1, P_2, \ldots$
A -> B	$A \rightarrow B$
x :-> y	$x \mapsto y$
alpha, beta, ...	$\alpha, \beta, \ldots$

Finally, we mention the use of the prompt

> $x := 3$;

It is shown on some parts of the code, but not on others. The general rule is that the prompt is shown on fragments of interactive sessions but not on stand-alone declarations of functions and constructions (that may or may not be applied later on).

A glance at the index of this book will give a general impression of the vocabulary of the Magma language. Besides ordinary text entries from all papers, many keywords, intrinsic names, operators, and constructors are included. Of course not every occurrence is recorded; often a reference is given either to the typical use of intrinsics or to a more or less random place where frequently used constructors can be found. The keywords appear with the same font conventions as in the main text.

Contents

Some computational experiments in number theory

Wieb Bosma

Mathematisch Instituut
Radboud University
Nijmegen, the Netherlands
bosma@math.ru.nl

Summary. The Magma code and some computational results of experiments in number theory are given. The experiments concern covering systems with applications to explicit primality tests, the inverse of Euler's totient function, and class number relations in Galois extensions of $\mathbb{Q}$. Some evidence for various conjectures and open problems is given.

1 Introduction

In the course of 10 years of working with and for Magma [6], I have conducted a large number of computational experiments in number theory. Many of them were meant, at least initially, as tests for new algorithms or implementations. In this paper I have collected results from, and code for, a few of those experiments.

Three main themes can be recognized in the material below: *covering systems*, the *Euler ϕ function*, and *class number relations*.

In section 3 it is shown how covering systems can be used, with cubic reciprocity, to produce a simple criterion for the primality of $n = h \cdot 3^k + 1$ in terms of a cubic recurrence modulo n; the starting value depends only on the residue class of k modulo some covering modulus M. These primality tests generalize the Lucas–Lehmer type tests for numbers of the form $h \cdot 2^k + 1$. They lead to a question about values of h for which $h \cdot 3^k + 1$ is composite for every k, and a generalization of a problem of Sierpiński. We found a 12-digit number h with this property — a candidate analogue for the number 78577, which is most likely the smallest h with the property that $h \cdot 2^k + 1$ is composite for every k. As a simpler application of our methods to produce covering systems in section 2, we improve slightly on the known results for a problem of Erdős; this problem asks for a finite set of congruence classes with distinct moduli, each at least c, that cover all integers.

There is also a connection between Sierpiński's problem and the image of Euler's totient function ϕ; this is explained in section 4, which is devoted to various questions about the ϕ function, its image, its inverse image and its iterates. We describe our implementation of the function that computes all n with $\phi(n) = m$ for a given m; as a test we produced some statistics on the size of the inverse image for the first 327 million even integers. We also experimented extensively with iterates of the composite function $\phi \circ \sigma$, found a new candidate smallest starting value for which this function may not reach a cycle, and recorded many cycles and the frequency with which they occur. We searched for (and found many) new fixed points n, for which $\phi \circ \sigma(n) = n$. For many of the experiments in sections 2–4 the problems and the references collected by Richard K. Guy in [20] proved very valuable.

In section 5 we present some details of computations done with Bart de Smit on relations between class numbers in the subfields of Galois extensions of $\mathbb{Q}$ with some fixed Galois group. This requires computations with transitive permutation groups of small degree and all of their subgroups, and the characters of these groups. Focusing on pairs of fields with the same zeta-function, it is shown how Magma can now deal routinely with questions about the class number quotients for such pairs; in particular, we use resultant computations on polynomial rings over rational function fields to obtain symbolically the explicit defining relations for a family of equivalent number fields in degree 7.

2 Covering systems

A collection of residue classes a_i mod m_i is called a *covering system* if every integer n satisfies at least one of the congruences $n \equiv a_i$ mod m_i. Several constraints in various combinations are possible: for example, one may require the system to be *finite*, to consist of *distinct moduli*, or of *odd moduli* only, or to be *disjoint*. For a finite covering system we will call the least common multiple $M = \text{lcm}\{m_i\}$ of the moduli the *covering modulus*.

A problem of Erdős

Erdős considered the question of whether for all c there exists a covering system with finitely many distinct moduli satisfying $c = m_1 < m_2 < \ldots m_k$ (for some k) to be 'Perhaps my favorite problem of all', and offered \$1000 for a solution [16].

A simple search for solutions for small values of c can be conducted in Magma as follows[*]

```
for c := 2 to 10 do
    D := 0;
```

---

[*]See the Preface to this volume for style conventions regarding Magma code; code appearing in this book is available at http://magma.maths.usyd.edu.au/magma/.

```
    done := false;
  repeat
    D +:= 4;
    S := [ x : x in Divisors(D) | x ge c ];
    if &+[ Integers() | D div s : s in S ] ge D then
      done, F := try2cover(S, D);
    end if;
  until done;
    print < Min([ f[2] : f in F ]), D, F>;
  end for;
```

In this loop one attempts to find solutions with covering modulus D. Only D with many divisors are useful, and this search takes only $D \equiv 0 \bmod 4$ into account. Only if there are enough divisors of D exceeding c to make a covering by residue classes feasible is a call to the function **try2cover** made. The test uses the summation over D **div** s for this, where the specification [Integers() | ...] is used to ensure that the integer 0 is returned when the sum is taken over an empty sequence.

Here is the function **try2cover**:

```
  try2cover := function(S, D)
    Z := Integers();
    for tries := 1 to 50 do
      T := [ ];
      Q := [ 1 : i in [1..D] ];
      for i in [1..#S] do
        addm(~Q, ~T, S[i]);
        if &+[ Z | D div s : s in S[i+1..#S] ] lt &+Q then
          if &+Q/D gt 0.1 then
            break tries;
          end if;
          break;
        elif &+Q eq 0 then
          return true, T;
        end if;
      end for;
    end for;
    return false, _;
  end function;
```

For at most 50 times (a value that could be modified, but which worked well in our experiments) an attempt is made to add one residue class for each modulus (which is a divisor of D) stored in S; this residue class is added to a list (initially empty) that is kept in T. The sequence Q of length D stores 0 in position i precisely when the residue system in T covers the residue class $i-1 \bmod D$ and 1 otherwise. If not enough divisors are left to have any hope of completing the system, this try is aborted and a new one attempted, unless the

50 tries have been completed or the fraction covered in this aborted attempt is so low that the search is abandoned prematurely because it seems hopeless altogether. An attempt is aborted if the cover could not even be completed if all remaining residue classes were disjoint, and this implementation considers the case hopeless (and no more attempt is made) if an aborted attempt occurs when still more than 10% is uncovered.

The procedure **addm** is the most interesting part of this simple search.

```
addm := procedure(~Q, ~T, m)
  if &+Q eq #Q then
    addr(~Q, ~T, 1, m);
  else
    mx := [ &+Q[[i..#Q by m]] : i in [1..m] ];
    mm := Max(mx);
    im := Random([ i : i in [1..#mx] | mx[i] eq mm ]);
    addr(~Q, ~T, im−1, m);
  end if;
end procedure;
```

In it, an attempt is made to find a good residue r for the given modulus m to add to the system T; 'goodness' is measured in terms of the number of previously uncovered classes modulo D that will be covered when adding $r \bmod m$. This success rate is computed for all possible choices (unless no residue class is covered yet, in which case we may and will just choose the class of one) and among the best a random choice is made. This random aspect of the otherwise 'greedy' algorithm makes it useful to have several attempts in **try2cover**.

The function **addr** (that is called but will not be listed here) simply adjusts the sequences Q and T according to the choice made: newly covered classes $i \bmod D$ get their entry in Q replaced by a 0 and the pair (r, m) is appended to T.

Example 2.1. The little program described is fairly successful for small values for c. Of course it immediately finds a version of the well-known 'smallest' covering with modulus 12 for $c = 2$, such as

$$1 \bmod 2, \quad 2 \bmod 3, \quad 2 \bmod 4, \quad 4 \bmod 6, \quad 0 \bmod 12$$

and for $c = 3$ one obtains a covering using divisors of 120. Already for $c = 4$ this algorithm does better than the deterministic algorithm given in the (very early) work of Churchhouse [14]. He gives a solution with moduli that are divisors of 720, whereas our solution

1 mod 4,	1 mod 5,	4 mod 6,	3 mod 8,	8 mod 10,
0 mod 12,	14 mod 15,	15 mod 16,	2 mod 20,	18 mod 24,
20 mod 30,	7 mod 32,	30 mod 40,	7 mod 48,	32 mod 60,
55 mod 80,	87 mod 96,	54 mod 120,	87 mod 160,	23 mod 240,

uses divisors of 480. The other values found are listed in the table below, and compared to those obtained by Churchhouse:

2	3	4	5	6	7	8	9	10
12	120	480	2520	5040	20160	60480	151200	1663200
12	120	720	2520	10080	30240	75600	604800	–

For $c = 10$ we found a covering modulus 1663200; we did not find any cover for this case elsewhere in the literature. Note that the values are not necessarily smallest possible, although we tried fairly hard to beat them.

At this stage the method of storing an indicator for every residue class becomes cumbersome. Some larger examples were constructed by Choi [13] and Morikawa [29].

The conjecture of de Polignac

Covering systems were introduced by Erdős, and used by him to give a disproof of a conjecture made in [33] (and quickly retracted; cf. [19]) by de Polignac, namely that every odd $n > 1$ can be written as $2^k + p$ for some k and some prime number p. De Polignac himself had already found small counterexamples; $k = 127, 149, 251$ are the smallest, and there are 14 others below 1000. But the disproof by Erdős exhibits an arithmetic progression of counterexamples, by noting that covering systems $\{r_i \bmod m_i\}$ like

$$0 \bmod 2, \quad 0 \bmod 3, \quad 1 \bmod 4, \quad 3 \bmod 8, \quad 7 \bmod 12, \quad 23 \bmod 24$$

can be used for this purpose. Indeed, observing that the moduli m_i here are equal to the multiplicative order e_i of 2 modulo p_i, where p_i is $3, 7, 5, 17, 13$ or 241 respectively, we see that if we choose simultaneously

$$N \equiv 2^{r_i} \bmod p_i$$

for all classes $r_i \bmod m_i$ in the cover, then for every integer k there exists at least one i for which $k \equiv r_i \bmod m_i$ and hence $N - 2^k \equiv 0 \bmod p_i$; that is, $N - 2^k$ is divisible by p_i. As it is easy to see (by working modulo 31, for example) that $N - 2^k$ cannot be *equal* to p_i, we see that this N gives rise to a counterexample to the conjecture, as does every odd integer in the arithmetic progression of N with modulus $Q = \prod_i p_i = 3 \cdot 5 \cdot 7 \cdot 13 \cdot 17 \cdot 241$. The covering system above yields $N = 2036812$ as the smallest non-negative solution (by application of the Chinese Remainder Theorem) and 7629217 as the smallest counterexample:

```
>   m := [ 0, 0, 1, 3, 7, 23 ];
>   P := [ 3, 7, 5, 17, 13, 241 ];
>   N := CRT([ 2^j : j in m ], P); N;
        2036812
```

The additional condition that m is odd produces the smallest counterexample.

```
>     N + &*P;
       7629217
```

The problems of Sierpiński and Riesel

The fact that $N-2^k$ is always divisible by one of the primes $3, 5, 7, 13, 17, 241$ is closely related to the solution of another problem, concerning the existence of integers H such that $H \cdot 2^k + 1$ is composite for every $k \geq 0$. Sierpiński showed that there is an infinitude of such H. Indeed, if we let $Q = 3 \cdot 5 \cdot 7 \cdot 13 \cdot 17 \cdot 241$ again, then

$$N - 2^k \equiv 0 \bmod p_i \quad \Longleftrightarrow \quad H \cdot 2^k + 1 \equiv 0 \bmod p_i,$$

if we let H be such that $H \equiv -N^{-1} \bmod Q$.

In the current example, we let H be the smallest positive element in this residue class and find:

```
>     Q := &*P; Q;
       5592405

>     H := InverseMod(−N, Q); H;
       1624097
```

and the output of

```
>     for k := 0 to 100 do
>        print k, [ p : p in P | (H*2^k+1) mod p eq 0 ];
>     end for;
```

will demonstrate that $H \cdot 2^k + 1$ is divisible by 3 when $k \equiv 0 \bmod 2$, divisible by 7 when $k \equiv 0 \bmod 3$, by 5 when $k \equiv 1 \bmod 4$, etc.

Since there are, in general, several covering systems for a fixed covering modulus, there will be several pairs of solutions N, H as above. The smallest H with covering modulus 24 is $H = 271129$.

However, there exists a smaller integer H with the property that $H \cdot 2^k + 1$ is divisible by at least one prime in a fixed, finite collection, but it comes from a covering system with covering modulus 36: the covering system

$$0 \bmod 2, \ 2 \bmod 3, \ 3 \bmod 4, \ 1 \bmod 9, \ 9 \bmod 12, \ 13 \bmod 18, \ 25 \bmod 36$$

yields $N = 20512783$, and $H = 314228$, the odd part of which is 78557.

The problem of determining the smallest H such that $H \cdot 2^k + 1$ is always composite is sometimes referred to as *Sierpiński's problem*. The number $H = 78557$ is the most likely candidate for Sierpiński's problem; see [42] for progress on the remaining work on proving that for every smaller h there is a prime of the form $h \cdot 2^k + 1$. It has been conjectured (but never been proven, as far as

I know) that every H such that $H \cdot 2^k + 1$ is always composite arises from a finite covering system.

The similar problem of determining the smallest H such that $H \cdot 2^k - 1$ is always composite is sometimes referred to as *Riesel's problem*; Riesel [36] first showed the existence of infinitely many such H. The most likely candidate is $H = 509203$; see [43].

Note that a Sierpiński number H with covering modulus D also provides a Riesel number $D - H$.

3 Covering systems and explicit primality tests

One way in which the problems of Sierpiński and Riesel (and a generalization) arose naturally for me occurred in [3], and [4]. In these papers the well-known Lucas–Lehmer type tests for $2^n \pm 1$ were generalized to numbers of the form $h \cdot 2^n \pm 1$ and $h \cdot 3^n \pm 1$ using covering systems. We will first explain the connection, and then return to the problems of Sierpiński and Riesel and their generalization.

Non-residue covers

In [3], covering systems were used to solve the following problem: for fixed h find a finite *quadratic non-residue cover* of elements $c_1, c_2, \ldots, c_m \in \mathbb{Z}^*$ satisfying

$$\left(\frac{c_r}{h \cdot 2^k + 1} \right)_2 \neq 1, \quad \text{when} \quad 2 \leq k \equiv r \bmod m.$$

Here the symbol $(\text{-})_2$ on the left is the Jacobi symbol. It turns out that such a finite cover can usually be found, unless h is of the form $h = 4^s - 1$. The reason it is of interest to find such a finite cover is the following result.

Theorem 3.1. *If $c_1, \ldots, c_m$ forms a quadratic non-residue cover and $2^k > h$ then, with $k \equiv r \bmod m$:*

$$n = h \cdot 2^k + 1 \quad \text{is prime} \quad \Longleftrightarrow \quad c_r^{(n-1)/2} \equiv -1 \bmod n.$$

So a quadratic non-residue cover for h provides a nice, very explicit, primality test for the family $h \cdot 2^k + 1$ (for fixed h). The classical example for this is the case where $h = 1$, since in this case $m = 1$ and $c_1 = 3$ work: we get a well-known test for Fermat numbers

$$n = 2^k + 1 \quad \text{is prime} \quad \Longleftrightarrow \quad 3^{(n-1)/2} \equiv -1 \bmod n.$$

In fact, this cover works for any h not divisible by 3.

Similar, but slightly more complicated tests based on covering systems can be derived for families $h \cdot 2^k - 1$, again for h not of the form $4^s - 1$. The extra complication amounts to the following: The cover does not consist of integers

c_r but of pairs (D_r, α_r), $r = 1, \ldots, m$, where D_r is an integer discriminant $0 < D_r \equiv 0, 1 \bmod 4$ and α_r is an element of the quadratic field $\mathbb{Q}(\sqrt{D_r})$. These pairs have the property

$$\left(\frac{D_r}{h \cdot 2^k - 1} \right)_2 \neq 1 \quad \text{and} \quad \left(\frac{\alpha_r \bar{\alpha}_r}{h \cdot 2^k - 1} \right)_2 \neq 1 \quad \text{whenever} \quad 2 \leq k \equiv r \bmod m;$$

here $\bar{}$ is the non-trivial $\mathbb{Q}$-automorphism of the quadratic field. Again, this provides explicit primality tests for families $h \cdot 2^k - 1$, which can be formulated either in terms resembling Theorem 3.1 above

$$n = h \cdot 2^k - 1 \quad \text{is prime} \quad \Longleftrightarrow \quad \left(\frac{\alpha_r}{\bar{\alpha}_r} \right)^{(n+1)/2} \equiv -1 \bmod n,$$

or in terms of a recurrence relation

$$n = h \cdot 2^k - 1 \quad \text{is prime} \quad \Longleftrightarrow \quad e_{k-2} \equiv 0 \bmod n,$$

where $e_{j+1} = e_j^2 - 2$ for $j \geq 0$, and the starting value e_0 is determined by α_r. The classical Lucas–Lehmer case (for Mersenne numbers) is $h = 1$, where again the length of the cover is $m = 1$ and the single pair $(12, 2 + \sqrt{12})$ works; in this case the starting value e_0 equals -4.

The connection with ordinary congruence covers is as follows. Since the Jacobi symbol is multiplicative (in the top argument), it is not a restriction to assume that the c_r are prime. Then I claim that for prime c

$$\left(\frac{c}{h \cdot 2^k + 1} \right)_2 \neq 1 \quad \Longrightarrow \quad \left(\frac{c}{h \cdot 2^j + 1} \right)_2 \neq 1 \quad \text{for every} \quad j \equiv k \bmod d,$$

where d is the multiplicative order of 2 mod c. This is clear from quadratic reciprocity and the fact that $h \cdot 2^k + 1 \equiv h \cdot 2^{k+d} + 1 \bmod c$. So any good pair (c, k) provides a solution for the whole residue class $k \bmod d$. The aim in our search for a quadratic non-residue cover then simply becomes that of finding a congruence cover using such residue classes $k \bmod d$. The dependence of the modulus d on (the prime) c is that c is a primitive divisor of $2^d - 1$; that is, c divides $2^d - 1$ but not $2^i - 1$ for any value of $i < d$. In other words, d is the multiplicative order of 2 modulo c. The way k and c are related depends on h. For prime $c > 2$ precisely $(c + 1)/2$ residue classes modulo c consist of non-residues, so those values $k \bmod d$ can be used for which $h \cdot 2^k + 1 \bmod c$ is such quadratic non-residue class.

What we would like to show here is how Magma was used [4] to generalize these results to integers of the form $h \cdot 3^k + 1$ using cubic non-residue covers.

Cubic reciprocity

Let $\zeta = \zeta_3$ be a primitive third root of unity. For prime $\pi \in \mathbb{Z}[\zeta]$ with $n = $ Norm $\pi \neq 3$, we let $\left(\frac{\alpha}{\pi} \right)_3$ be the element of $\{0, 1, \zeta, \zeta^2\} \subset \mathbb{Z}[\zeta]$ defined as follows. If π divides α then the value is 0, in all other cases it is the element ζ^i satisfying $\alpha^{\frac{n-1}{3}} \equiv \zeta^i \bmod \pi$.

As a consequence, for $\alpha \in \mathbb{Z}[\zeta]$ and prime $\pi \in \mathbb{Z}[\zeta]$ of norm $n > 3$:

$$\alpha^{\frac{n-1}{3}} \not\equiv 1 \bmod \pi \quad \Longleftrightarrow \quad \forall\, x \neq 0 : \ x^3 \not\equiv \alpha \bmod \pi \quad \Longleftrightarrow \quad \left(\frac{\alpha}{\pi}\right)_3 \neq 1.$$

Next, one extends the definition by multiplicativity: for $\alpha, \beta \in \mathbb{Z}[\zeta]$ with Norm β not divisible by 3 we define

$$\left(\frac{\alpha}{\beta}\right)_3 = \left(\frac{\alpha}{\pi_1}\right)_3 \left(\frac{\alpha}{\pi_2}\right)_3 \cdots \left(\frac{\alpha}{\pi_k}\right)_3,$$

where $\pi_i \in \mathbb{Z}[\zeta]$ is prime and $\beta = \pi_1 \pi_2 \cdots \pi_k$.

Other important properties of the cubic residue symbol are its multiplicativity (in the top argument) and periodicity (in the top argument modulo the bottom argument).

An element $\alpha \in \mathbb{Z}[\zeta]$ is *primary* if and only if $\alpha \equiv 2 \bmod 3$. The primary prime elements of $\mathbb{Z}[\zeta]$ are precisely the positive rational primes $q \equiv 2 \bmod 3$ and the elements $\pi = a + b\zeta$ with $a \equiv 2 \bmod 3$ and $b \equiv 0 \bmod 3$ for which Norm $\pi = a^2 - ab + b^2 = p \equiv 1 \bmod 3$ is prime. Among the associates of any $\beta \in \mathbb{Z}[\zeta]$ of norm not divisible by 3 exactly one is primary, and if β is primary it can be written uniquely (up to order) as a product of primary prime elements and a power of the primary unit -1.

The following theorem summarizes the results of the cubic reciprocity law, its supplementary law, and a result on units. For proofs see [21], [2].

Theorem 3.2. *If $\alpha, \beta \in \mathbb{Z}[\zeta]$ are primary elements of norm not divisible by 3 then:*

$$\left(\frac{\alpha}{\beta}\right)_3 = \left(\frac{\beta}{\alpha}\right)_3.$$

If $\beta \in \mathbb{Z}[\zeta]$ is a primary prime element, $\beta = (3m-1) + b\zeta$, with $b \equiv 0 \bmod 3$ then:

$$\left(\frac{1-\zeta}{\beta}\right)_3 = \zeta^{2m}.$$

If $\pi \in \mathbb{Z}[\zeta]$ is a prime element of norm not equal to 3 then

$$\left(\frac{-1}{\pi}\right)_3 = \left(\frac{1}{\pi}\right)_3 = 1 \quad and \quad \left(\frac{\zeta}{\pi}\right)_3 = \begin{cases} 1 & if \ \ \text{Norm } \pi \equiv 1 \bmod 9, \\ \zeta & if \ \ \text{Norm } \pi \equiv 4 \bmod 9, \\ \zeta^2 & if \ \ \text{Norm } \pi \equiv 7 \bmod 9. \end{cases}$$

Explicit primality tests

If, for fixed even h and $k = 1, 2, \ldots$

$$\left(\frac{\alpha_r}{h \cdot 3^k + 1}\right)_3 \neq 1, \quad \text{when} \quad 2 \leq k \equiv r \bmod m,$$

we call $\alpha_1, \alpha_2, \ldots, \alpha_m \in \mathbb{Z}[\zeta]^*$ a *cubic non-residue cover* for h. We obtain the following analogue of Theorem 3.1; by $\bar{}$ we denote the automorphism of $\mathbb{Q}(\zeta)$ sending ζ to ζ^2.

Theorem 3.3. *If $\alpha_1, \alpha_2, \ldots, \alpha_m \in \mathbb{Z}[\zeta]^*$ forms a cubic non-residue cover for h and $3^k > h$ then*

$$N = h \cdot 3^k + 1 \text{ is a prime number} \quad \Longleftrightarrow \quad w_{k-1} \equiv \pm 1 \bmod N,$$

where, with $r \equiv k \bmod m$,

$$w_0 = \mathrm{Tr}\left(\frac{\alpha_r}{\bar{\alpha}_r}\right)^{\frac{h}{2}} \quad and \quad w_{j+1} = w_j(w_j^2 - 3), \quad for \ j \ge 0.$$

The same result holds for the family $h \cdot 3^k - 1$. So we see that we find an explicit primality criterion for these families, if we can solve the following problem.

Problem 3.4. Given an even positive integer h not divisible by 3, find a finite set $\mathcal{S}_h^+ = \{(r, m, \alpha)_j : j = 1, \ldots, t\}$ of tuples (r, m, α) consisting of residue classes $r \bmod m$ that form a finite covering system such that for integers k with $3^k > h$ and $k \equiv r \bmod m$ it holds that

$$\left(\frac{\alpha}{h \cdot 3^k + 1}\right)_3 \ne 1.$$

Similarly, for the set $\mathcal{S}_h^-$, we require

$$\left(\frac{\alpha}{h \cdot 3^k - 1}\right)_3 \ne 1.$$

Here is the Magma code with which we solved the problem for all $h < 10^5$ (except for h of the form $h = 27^s - 1$ for which there are no solutions). We give the case $h \cdot 3^k - 1$ below; for the similar function **plusfind** replace the appropriate $-$ sign by a $+$ in the computation of N.

```
minfind := function(h, bound, PX)
  i := 0;
  K := [1];
  repeat
    i +:= 1;
    N := h*3^i - 1;
    _, I := get(N, PX);
    if IsEmpty(I) then return 0; end if;
    J := cut(I);
    K := Sort([ Lcm(k,j) : k in K, j in J | Lcm(k,j) lt bound ]);
    if IsEmpty(K) then
      return 0;
```

```
    else
        K := cut(K);
    end if;
    until i in K;
    return i;
end function;
```

The function minfind calls a function get, where most of the work is done, as well as the function cut that simply removes from a sequence of positive integers I all entries that are divisible by an entry with smaller index.

The main function get (which will work both for minfind and for plusfind) does the following. For given N (which will be of the form $h \cdot 3^k \pm 1$), a list P of primes p is found for which the cubic residue symbol for p over N is not equal to 1; the smallest e such that p divides $3^e - 1$ is also stored. For each prime p a prime $\pi \in \mathbb{Z}[\zeta]$ lying over p is found and put in PX, that will then consist of a sequence of sequences of prime divisors π of $3^i - 1$ in position i. The function cubicsymbol is a straightforward implementation of the cubic residue symbol (code not reproduced here).

```
get := function(N, PX)
    S := [ Parent(ζ) | ];
    OS := [ ];
    for i in [1..#PX] do
        if IsEmpty(PX[i]) then
            continue;
        end if;
        for x in PX[i] do
            if cubicsymbol(x[1], x[2], N, 0) ne 1 then
                if x notin S then
                    Append(~S, x);
                    Append(~OS, i);
                end if;
            end if;
        end for;
    end for;
    return S, OS;
end function;
```

In minfind (or plusfind) the information from get is recorded for $N = h \cdot 3^i - 1$ for $i = 1, 2, \ldots$ until a value $i = k$ is reached with the property that for all i with $1 \le i \le k$ at least one of the prime divisors of $3^k - 1$ has cubic symbol not equal to 1. The value for *bound* is an upper bound for the solution that will be found; a small value gives quicker results, but it may be that no solution k less than this value exists, and a retry with larger bound will be necessary.

Example 3.5. We attempt to find an explicit primality test for integers of the form $1900 \cdot 3^k - 1$. To this end we run minfind with $h = 1900$, and a bound of 25. This means that we will make use only of prime divisors of $3^e - 1$ for e up to 25 (Magma's **Cunningham** facility will happily supply such divisors for e up to 400 or more). In **PX** both the inert primes (those that are 2 mod 3) and the primes in $\mathbb{Z}[\zeta]$ lying over rational primes that are 1 mod 3 are collected.

In the first round **minfind** will find in **PX** primes that give cubic symbol not equal to 1 with $N_1 = 1900 \cdot 3^1 - 1 = 5699$. It finds the element $29 + 36\zeta$ of norm 1093 (which divides $3^7 - 1 = 2186$), the element $5 + 9\zeta$ of norm 61 (dividing $3^{10} - 1$), $8 + 9\zeta$ (dividing $3^{12} - 1$), and elements dividing $3^{13} - 1$, $3^{14} - 1$, $3^{15} - 1$, $3^{16} - 1$, $3^{19} - 1$, $3^{20} - 1$, and $3^{22} - 1$. For example

$$\left(\frac{29 + 36\zeta}{5699}\right)_3 = \left(\frac{5 + 9\zeta}{5699}\right)_3 = \zeta, \quad \left(\frac{8 + 9\zeta}{5699}\right)_3 = \zeta^2.$$

The sequence **K** will then consist of the integers 7, 10, 12, 13, 15, 16, 19, 22, indicating that these (and their multiples) will have a chance left to act as covering modulus.

It then turns to $N_2 = 1900 \cdot 3^2 - 1 = 17099$; this time elements dividing $3^e - 1$ with the required cubic symbol are found in **PX** for $e = 3$ and all its multiples, for $e = 10$ (and 20), as well as for $e = 14, 16, 19$, and 22, but *not* for $e = 7$ or $e = 13$. This leaves $10, 12, 14, 15, 16, 19, 21, 22$ as possible primitive solutions in **K**.

However, the possibilities $e = 15$ and $e = 16$ disappear when we consider $N_3 = 1900 \cdot 3^3 - 1$, and the possibility $e = 10$ (as well as 20) vanishes when looking at $N_6 = 1900 \cdot 3^6 - 1$, as does $e = 19$. Then $e = 12$ is not good for N_7 (but $e = 24$ is fine), $e = 21$ disappears with N_9 and $e = 22$ with N_{10}. The remaining possibilities are then $e = 14$ and $e = 24$.

It turns out that they also furnish suitable elements for N_{11}, N_{12}, N_{13} and N_{14}. But at that stage we are finished because we know that among the prime divisors of $3^{14} - 1$ we can find suitable elements for N_i with $1 \le i \le 14$; if such an element works for N_i it will also work for N_{i+14}, etc. In other words, we have completed the cover! Indeed, for every $k \ge 1$ at least one of

$$\left(\frac{-13 - 27\zeta}{1900 \cdot 3^k - 1}\right)_3 \in \{\zeta, \zeta^2\}, \quad \left(\frac{29 + 36\zeta}{1900 \cdot 3^k - 1}\right)_3 \in \{\zeta, \zeta^2\}$$

holds, which gives an explicit test by Theorem 3.3. The first holds for all $k > 0$ except the residue classes 4 mod 7 and 5 mod 14, the other for all $k > 0$ in residue classes $0, 1, 4, 5, 6$ mod 7. To make the test completely explicit we would have to compute the trace of the 950th power of $(-13 - 27\zeta)/(-13 - 27\zeta^2)$ and of $(29 + 36\zeta)/(29 + 36\zeta^2)$. Numerator and denominator of the first w_0 have over 2600 decimal digits, however. Of course w_0 will be reduced modulo N; for example, with $N_{69} = 15853318198290530141665289245210376699$ we find $w_{68} = -1$, hence N_{69} is prime.

Sierpiński's problem revisited

It is not hard to see from cubic reciprocity that the rational primes q stored in **PX** will never satisfy

$$\left(\frac{q}{h \cdot 3^k \pm 1}\right)_3 \in \{\zeta, \zeta^2\};$$

what can happen and would be useful for Theorem 3.3, however, is that the symbol becomes 0, implying that for k in a certain residue class $h \cdot 3^k \pm 1$ will be divisible by q.

This also gives the link with our earlier problem: If we adapt our function get in such a way that we look for a special cubic non-residue cover consisting only of elements with cubic symbol equal to 0 (rather than not equal to 1), we would detect values for h with the property that $h \cdot 3^k - 1$ or $h \cdot 3^k + 1$ is always divisible by one of a finite set of primes. Conducting this search in the comparable but easier case (using quadratic reciprocity) for numbers of the form $h \cdot 2^k \pm 1$ for h less than 10^6 immediately yields the known examples of Sierpiński and Riesel numbers ([22, 23, 24]). To test divisibility only, there is no need at all to use quadratic or cubic reciprocity, and the test in get could simply be replaced by a test of the type **if** N **mod** p **eq** 0 **then** for p running over the relevant set of primes. Up to 10^7, however, no h with this property for $h \cdot 3^k \pm 1$ was found.

This led us to attempt to *construct* a 'small' solution in another way (cf. [41, 40]). Just as before, we will find generalized Sierpiński (or Riesel) numbers when we find a finite covering system $\{a_i \bmod m_i\}$ for the exponents k provided that for each modulus m_i we find a prime p_i such that the order of c modulo p_i is a divisor of m_i (that is, p_i divides $c^{m_i} - 1$). We use a table P such that its i-th element contains the primitive prime divisors of $3^i - 1$. Now we wish to construct a covering system for the exponents, but contrary to the situation in the problem of Erdős we will not insist that the moduli are all distinct; however, we will only be able to use the modulus m_i with multiplicity k if there are k different primes in $P[i]$. We just apply the function try2cover defined before, with a sequence of moduli satisfying these requirements. As we explained, it is easy to find H (as $-N^{-1} \bmod \prod_i p_i$) once we know the cover. Since we are now interested in the *smallest possible* value for H, we want to generate all possible covering systems with the same (multi)set of moduli.

Example 3.6. First let $c = 2$ again, the case of Sierpiński numbers. For a very small covering modulus it may be possible to enumerate all covering systems; here is a simple way to do it in Magma.

```
>    S := [ 2, 3, 4, 9, 12, 18, 36 ]; CS := [ ];
>    K := CartesianProduct([ Integers(i) : i in S ]);
>    for x in K do
>      C := [ [ Integers() ! x[i], S[i] ]: i in [1..#S] ];
>      if check(C) then Append(~CS, C); end if;
>    end for;
```

The function check returns true if and only if a given system of residue classes forms a cover (which is tested by simply checking every residue).

Out of the $\#K = 1679616$ possibilities, we find 144 different covers with $S = \{m_1, m_2, \ldots, m_7\}$.

We use the intrinsic 'Chinese Remainder Theorem' function CRT to find, for each of the covers found, the unique H with the property that $H \equiv -2^{-x_i} \bmod p_i$ for all residue classes $x_i \bmod m_i$ in the cover, with p_i a primitive divisor of $2^{m_i} - 1$:

```
>    P := [ 3, 7, 5, 73, 13, 19, 109 ];
>    H := CRT([ −Modexp(2, −C[i][1], P[i]) : i in [1..#C] ], P);
```

It turns out that 72 distinct (odd) Sierpiński numbers are generated this way, the smallest being 934909, and the largest 202876561.

In the above we made one particular choice, $p_7 = 109$, for a primitive prime divisor of $2^{36} - 1$, where an alternative $p_7' = 37$ was available. Applying the same call CRT to the same set of covering systems

```
>    P := [ 3, 7, 5, 73, 13, 19, 37 ];
>    H := CRT([ −Modexp(2, −C[i][1], P[i]) : i in [1..#C] ], P);
```

we find 75 Sierpiński numbers (they are all listed in [40]), the smallest this time being 78557, the largest 68496137. Curiously, three numbers appear in both lists: 12151397, 45181667, and 68468753.

For larger covering systems such a complete enumeration will no longer be feasible. To find analogues of the Sierpiński numbers for $h \cdot 3^k + 1$ we had to use a probabilistic approach again.

Example 3.7. Let $c = 3$. Suppose we know that 48 can be used as a covering modulus. We could then use try2cover to obtain a cover (or several covers), and combine the information using the Chinese Remainder Theorem as before to construct the number H. We should be careful, however, only to use residue classes $x_i \bmod m_i$ in our covering system for which there exist primitive prime divisors of $3^{m_i} - 1$.

For example, here is a list of the sets of odd primitive prime divisors of $3^m - 1$ for divisors m of 48:

1	2	3	4	6	8	12	16	24	48
$\emptyset$	$\emptyset$	$\{13\}$	$\{5\}$	$\{7\}$	$\{41\}$	$\{73\}$	$\{17, 193\}$	$\{6481\}$	$\{97, 577, 769\}$

This tells us that we cannot use a residue class with modulus 2 in the cover; also, we are allowed 3 different residue classes modulo 48, and 2 modulo 16.

Feeding [3, 4, 6, 8, 12, 16, 16, 24, 48, 48, 48] to try2cover produced as one covering system

$$1 \bmod 3, \quad 1 \bmod 4, \quad 2 \bmod 6, \quad 3 \bmod 8, \quad 0 \bmod 12,$$
$$7 \bmod 16, \quad 15 \bmod 16, \quad 18 \bmod 24, \quad 30 \bmod 48, \quad 6 \bmod 48.$$

This leads to the solution $H = 41552862226126268$ from

```
>    C := [[1,3],[2,4],[2,6],[3,8],[0,12],[7,16],[5,16],[18,24],[30,48],[6,48]];
```

```
>    P := [ 13, 5, 7, 41, 73, 17, 193, 6481, 97, 577 ];
>    H := CRT([ −Modexp(3, −C[i][1], P[i]) : i in [1..#C] ], P);
```

Note that a different choice for the primes of order 48 produces a different answer, and so does a change in the order in which 17 and 193 are listed.

The smallest number H that we found in our experiments with covering modulus up to 250 with the property that $H \cdot 3^k + 1$ is composite for all $k \geq 1$ is the number 125050976086, which occurs for the covering system

$$2 \bmod 3, \quad 2 \bmod 4, \quad 3 \bmod 6, \quad 0 \bmod 8, \quad 7 \bmod 9,$$
$$4 \bmod 16, \quad 12 \bmod 16, \quad 1 \bmod 18, \quad 13 \bmod 18,$$

with P equal to $[13, 5, 7, 41, 757, 17, 193, 19, 37]$.

Again, a generalized Sierpiński number furnishes an associated generalized Riesel number for $h \cdot 3^k - 1$.

4 The totient function

In this section we consider various questions about the image of ϕ, Euler's totient function. By definition, $\phi(n) = \#\{x : 1 \leq x \leq n \mid \gcd(x, n) = 1\}$. Obviously, $\phi(p^k) = (p - 1) \cdot p^{k-1}$ for any prime p and every $k \geq 1$. Also, $\phi(s \cdot t) = \phi(s) \cdot \phi(t)$ if $\gcd(s, t) = 1$. It follows immediately that $\phi(n)$ is even when $n > 2$, and since $\phi(1) = \phi(2) = 1$ no odd $m > 1$ is in the image of ϕ. But there are also even m that are not in the image of ϕ; these are called non-totients. The smallest non-totient is $m = 14$. There also exist integers divisible by 4 that are non-totients; the smallest is $4 \cdot 17$. In fact (cf. [30]), for every $\alpha \geq 1$ there exists an odd h such that $2^\alpha \cdot h$ is a non-totient; the smallest such h we denote by h_α. There is a connection with Sierpiński numbers, as follows. If $h \cdot 2^n + 1$, for some $n \geq 1$, is a prime number, then $\phi(h \cdot 2^n + 1) = 2^n \cdot h$, and, more generally, $\phi(2^r \cdot (h \cdot 2^n + 1)) = 2^{r-1} \cdot 2^n \cdot h$, so $\phi(x) = 2^k \cdot h$ has solutions for any $k \geq n$. If $h = 2^s + 1$ and $2^s + 1$ is a prime number, then $\phi(x) = 2^s \cdot h$ has solution $x = h^2$, and more generally $2^r \cdot h^2$ is a solution to $\phi(x) = 2^{r-1} \cdot 2^s \cdot h$, so $\phi(x) = 2^k \cdot h$ has solutions for all $k \geq s$. But if h is an odd prime, h is not of the form $2^s + 1$ and $h \cdot 2^n + 1$ is composite for any $n \geq 1$, then there will exist no k for which $\phi(x) = 2^k \cdot h$. Thus any Sierpiński number h that is prime but not a Fermat prime has the property that $\phi(x) = 2^k \cdot h$ has no solution for any $k \geq 0$. The smallest known prime Sierpiński number is $h = 271129$. So, neither h nor any power of 2 times h is in the image of ϕ for $h = 271129$, and $h_\alpha \leq 271129$ for every $\alpha \geq 1$.

The inverse of the Euler ϕ function

We will now describe a function, available in Magma as EulerPhiInverse, that determines $\phi^{-1}(m)$ for any $m \geq 1$.

When solving the equation $\phi(x) = m$ we first note that there will be no solution for odd m exceeding 1. For even m we store the powers of 2 dividing m in an indexed set (for efficient look-up).

```
inv := function(m)
    mfact := Factorization(m);
    if IsEven(m) then
        twopows := {@ 2^i : i in [0..mfact[1][2]] @};
    else
        if m gt 1 then return [ ]; end if;
        twopows := {@ 1 @};
    end if;
```

Any odd prime p dividing x must have the property that $p-1$ divides m and that p^2 can only divide x for such p if p also divides m.

The idea of the algorithm is to build up integers x from primes p for which $\phi(p) = p-1$ divides m. We keep a list of pairs of partially built up integers a and remainder integers $m/\phi(a)$, and have found a solution whenever the remainder becomes 1.

We start by putting the odd primes p such that $m \equiv 0 \bmod p-1$ in P; we deal with the prime 2 separately at the end.

```
    D := Divisors(mfact);
    P := [ ];
    for d in D do
        if d eq 1 then
            continue;
        end if;
        if IsPrime(d+1) then
            Append(~P, d+1);
        end if;
    end for;
```

In S we will store pairs (a, b) such that a is odd (kept in factored form) and $\phi(a) = m/b$ with b even or 1; clearly, when $b = 1$ we have found a solution $n = a$ to our equation, and $2 \cdot a$ is another solution. More generally, when $b = 2^k$ is a power of 2 we always have a solution $n = 2 \cdot b \cdot a$.

Initially we put $(1, m)$ in S, and then loop through the primes p in P, checking for every pair (a, b) already in S whether b is divisible by $p-1$; if so, we append a pair $(a \cdot p, b/(p-1))$ to S, and also a pair $(a \cdot p^2, b/((p-1) \cdot p))$ if p divides b, and so on for higher powers of p, except when the second value is odd and greater than 1.

In this algorithm it is most restrictive, and hence efficient, to treat the primes in P in *descending* order.

```
    S := [ <SeqFact([ ]), m> ];
    for p in Reverse(P) do
        for s in S do
```

```
if s[2] eq 1 then continue; end if;
k := 1;
d, mmod := Quotrem(s[2], p−1);
while mmod eq 0 do
   if IsEven(d) or d eq 1 then
      Append(~S, <SeqFact([<p, k>])*s[1], d>);
   end if;
   k +:= 1;
   d, mmod := Quotrem(d, p);
end while;
   end for;
end for;
```

The last prime $p = 2$ is dealt with in a special way, since at the end of this loop only those pairs (a, b) in S will be of use for which b is a power of 2. Every such pair contributes a solution as we indicated above, or even two in case $b = 1$. On the other hand it is also easy to see that we find all possible solutions this way, and hence all that remains is to assemble these solutions, and to sort and return them.

```
R := { };
for s in S do
   j := Index(twopows, s[2]);
   if j gt 0 then
      Include(~R, SeqFact([<2, j>] cat s[1]));
      if j eq 1 then
         Include(~R, s[1]);
      end if;
   end if;
end for;
return Sort([ Facint(nf) : nf in R]);
end function;
```

Example 4.1. Looking at the equation $\phi(n) = m = 1012$, we find first that $P=[2, 3, 5, 23, 47, 1013]$, and in the loop for $p = 1013$ only the pair $(1013, 1)$ is added to the list S consisting initially of $(1, 1012)$. For $p=47$ we see that the second value of the pair $(1, 1012)$ is divisible by $\phi(47) = 46$, and we add a pair $(47, 22)$ to S. For $p = 23$ we add a pair $(23, 46)$ and a pair $(23^2, 2)$, and also a pair $(23 \cdot 47, 1)$ because $(47, 22)$ was in S. For $p = 5$ nothing happens, but with $p = 3$ we add $(3, 506)$ and also $(3 \cdot 23^2, 1)$ because $(23^2, 2)$ was in S. That means that when we start considering the last prime $p = 2$ in P, S contains the useful pairs $(1013, 1)$, $(47, 22)$, $(23, 46)$, $(23^2, 2)$, $(23 \cdot 47, 1)$, $(3, 506)$, and $(3 \cdot 23^2, 1)$. This furnishes the solutions 1013, and $1081 = 23 \cdot 47$, and $1587 = 3 \cdot 23^2$, as well as twice these numbers. Finally, the pair $(23^2, 2)$ implies that also $2116 = 2^2 \cdot 23^2$ is a solution. Thus $\phi^{-1}(12) = \{1013, 1081, 1587, 2026, 2116, 2162, 3174\}$.

Carmichael's conjecture

One of the striking properties of the inverse Euler-ϕ function is that when n ranges over the natural numbers, the size $\#\phi^{-1}(n)$ of the set of inverse images of n seems to assume every possible natural number — except 1. *Carmichael's conjecture* states that for no n can there be exactly one solution to the equation $\phi(x) = n$. Carmichael thought he had a proof [11], but it was erroneous; it was replaced by an argument showing that any solution would have to be very large [12], an argument that was refined later [25, 37] to show that any solution will have at least 10^7 decimal digits (see also [34]).

We recorded #EulerPhiInverse(m) while executing a simple loop over even m. The results given here concerned the computations for the 327 million even integers up to 654000000. The table lists for some values of k how many m in the range given were found such that #EulerPhiInverse(m) equals k, as well as the smallest n for which #EulerPhiInverse(m) equals k.

0	234369438	14
1	0	–
2	34885680	10
5	3936195	8
10	1964797	24
50	74409	1680
100	18425	34272
500	603	2363904
1000	129	1360800
2500	12	36408960
5000	3	107520000
63255	1	638668800

The last line in the table shows the maximum size that was found: there are 63255 integers x with $\phi(x) = 638668800$.

The smallest value k for which no m was encountered with $\#\phi^{-1}(m) = k$ was $k = 4077$. It is an open conjecture that every $k > 1$ will occur eventually.

Erdős proved [17] that if there exists an integer m for which $\#\phi^{-1}(m) = k$, then there exist infinitely many such m. This was done by a fairly complicated analytic argument, showing that there are very many primes p such that $\#\phi^{-1}((p-1)m) = \#\phi^{-1}(m) = k$.

Iteration of $\phi \circ \sigma$

Another conjecture about ϕ concerns the iteration of the composite $\phi \circ \sigma$ of ϕ and the divisor-σ function, which assigns to n the sum of its divisors $\sigma(n) = \sum_{d|n} d$. The conjecture, formulated by Poulet in [35] as 'loi empirique', states that this function will ultimately cycle for every input n. Meade and Nicol [28] found that for the starting value $n_1 = 455536928 = 2^5 \cdot 7^6 \cdot 11^2$ no cycle had occurred yet when they had computed iterates of $\phi \circ \sigma$ up to

50 digits long, and they state that 'In independent studies Sid Graham has observed that this appears to be the smallest number which does not cycle'. One part of this claim we can prove incorrect here: If the function does not cycle for n_1, this is certainly not the smallest such starting value. The reason is that the sequence of iterates for n_1 merges with the sequence for the starting value $n_0 = 254731536 = 2^4 \cdot 3^2 \cdot 17^2 \cdot 6121$ after a few steps. As a matter of fact, there are almost 400 starting values smaller than n_1 leading to the same sequence, and n_0 is the smallest. After 29781 iterations on n_0 we reached the 179 digit number

$$2^{106} \cdot 3^{70} \cdot 5^{40} \cdot 7^{18} \cdot 11^{11} \cdot 13^4 \cdot 17^2 \cdot 19^3 \cdot 23^3 \cdot 31^2 \cdot 37 \cdot 41 \cdot 59 \cdot$$
$$61^2 \cdot 67 \cdot 229 \cdot 271 \cdot 347 \cdot 733 \cdot 5569 \cdot 18211 \cdot 33791 \cdot 83151337.$$

We stopped at this point for no particular reason.

Here are some more statistics about what happens up to starting value $255 \cdot 10^6$. All but one of the sequences, the one starting with n_0, end in one of 46 different cycles. Of these cycles, 20 are of length 1, namely (listing the number of occurrences in parentheses):

1	(1),	$712800 = 2^5 \cdot 3^4 \cdot 5^2 \cdot 11$	(7741)
2	(3),	$1140480 = 2^8 \cdot 3^4 \cdot 5 \cdot 11$	(44858)
$8 = 2^3$	(6),	$1190400 = 2^9 \cdot 3 \cdot 5^2 \cdot 31$	(1833)
$12 = 2^2 \cdot 3$	(7),	$3345408 = 2^{10} \cdot 3^3 \cdot 11^2$	(73649)
$128 = 2^7$	(37),	$3571200 = 2^9 \cdot 3^2 \cdot 5^2 \cdot 31$	(128258)
$240 = 2^4 \cdot 3 \cdot 5$	(43),	$5702400 = 2^8 \cdot 3^4 \cdot 5^2 \cdot 11$	(1149102)
$720 = 2^4 \cdot 3^2 \cdot 5$	(151),	$14859936 = 2^5 \cdot 3^6 \cdot 7^2 \cdot 13$	(48306)
$6912 = 2^8 \cdot 3^3$	(1919),	$29719872 = 2^6 \cdot 3^6 \cdot 7^2 \cdot 13$	(44113)
$32768 = 2^{15}$	(160),	$50319360 = 2^{12} \cdot 3^3 \cdot 5 \cdot 7 \cdot 13$	(1135829)
$142560 = 2^5 \cdot 3^4 \cdot 5 \cdot 11$	(1374),	$118879488 = 2^8 \cdot 3^6 \cdot 7^2 \cdot 13$	(290673)

We found 11 cycles of length 2, 5 cycles of length 3, 3 cycles of length 4 and 2 cycles of length 6, as well as single cycles of lengths 9, 11, 12, 15, and 18. The following table lists some of them (again, the number of occurrences in brackets), cf. [5].

length 2:

$[4, 6]$	(7),
$[3852635996160, 4702924800000]$	(123),

length 3:

$[16, 30, 24]$	(35),
$[272160, 290304, 290400]$	(413972),

length 4:

$[2142720000, 2935296000, 3311642880, 3185049600]$	(16),

length 9:
[113218560, 124895232, 163296000, 181149696, 170698752,
    125798400, 116121600, 139708800, 136857600]   (7682341),
length 11:
[326592, 550368, 435456, 580608, 851840, 552960,
    524160, 442368, 432000, 368640, 381024]   (1263032),
length 12:
[5033164800, 6808043520, 6291456000, 6220800000,
    5761082880, 5225472000, 6657251328, 5283532800,
    5837832000, 7608287232, 7429968000, 6521389056]   (18458283),
length 15:
[40255488, 48384000, 43130880, 41912640, 47029248,
    70253568, 91998720, 82944000, 83825280, 71663616,
    52428800, 79221120, 70778880, 57600000, 42456960]   (128378949),
length 18:
[150493593600, 152374763520, 202491394560, 167215104000,
    219847799808, 161864220672, 247328774784, 191102976000,
    207622711296, 178362777600, 283740364800, 233003796480,
    221908377600, 204838502400, 214695936000, 237283098624,
    185794560000, 178886400000]   (82683195).

It is surprising that so many sequences end in so few cycles. One should not get the impression, however, that it is difficult to find other cycles. Starting values $n = 2^\ell$ for example, frequently lead to new ones. Indeed, for $\ell = 33$ we find a cycle of length 21, and for $\ell = 40$ we find a cycle of length 22. We list a few more values below.

ℓ	length	minimal entry
33	21	12227604480
41	3	4672651788288000
45	8	140005324800000
52	34	19937391280128000
54	9	140145643808410278297600000
79	5	663450926905517305076121600
88	56	42313405772261648007954432000
89	23	562218111097315629465600000

For larger values of ℓ the sequence of iterates seems to keep growing for a long time. All of this hardly provides evidence for or against the conjecture that every starting value eventually leads to a cycle.

Fixed points

A related question concerns fixed points under $\phi \circ \sigma$: solutions in positive integers to $\phi \circ \sigma(n) = n$. According to Guy (Problem B42 in [20]) Selfridge, Hoffman and Schroeppel found all but the final value $2^8 \cdot 3^6 \cdot 7^2 \cdot 13$ mentioned in the table of the previous section, and in addition

$$2147483648 = 2^{31}$$
$$4389396480 = 2^{13} \cdot 3^7 \cdot 5 \cdot 7^2$$
$$21946982400 = 2^{13} \cdot 3^7 \cdot 5^2 \cdot 7^2$$
$$11681629470720 = 2^{21} \cdot 3^3 \cdot 5 \cdot 11^3 \cdot 31$$
$$58408147353600 = 2^{21} \cdot 3^3 \cdot 5^2 \cdot 11^3 \cdot 31$$

We tried the following code in Magma to generate some more solutions, using various values for A to produce a list of primes P and maximal exponents E:

```
>    A := 35;
>    P := [ n : n in [2..A] | IsPrime(n) ];
>    E := [ Floor(A/p) : p in P ];
>    C := CartesianProduct([ [ e..0 by −1] : e in E ]);
>    for c in C do
>      nfn := SeqFact([ <P[i], c[i]> : i in [1..#P] | c[i] ne 0 ]);
>      if EulerPhi(DivisorSigma(1, nfn)) eq Facint(nfn) then
>        print nfn;
>      end if;
>    end for;
```

Here are some of the 25 new solutions we found (cf. [5]):

$$118879488 = 2^8 \cdot 3^6 \cdot 7^2 \cdot 13$$
$$3889036800 = 2^9 \cdot 3^4 \cdot 5^2 \cdot 11^2 \cdot 31$$
$$1168272833817083904000000 = 2^{25} \cdot 3^{11} \cdot 5^6 \cdot 7^4 \cdot 13^2 \cdot 31$$
$$14877199606392592421126804111136 \cdot 10^7 = 2^{35} \cdot 3^{21} \cdot 5^7 \cdot 7^2 \cdot 11^4 \cdot 13^2 \cdot 19 \cdot 23$$

5 Class number relations

The final examples concern the art of computing with character relations.

A *character relation* for a finite group G consists of a sequence of integers a_H, one for every subgroup H of G, such that $\sum a_H 1_H^G = 0$, when we sum over all subgroups. Here $\chi = 1_H^G$ is the *permutation character* of the subgroup H, so $\chi(g)$ is the integer counting the number of cosets of H left invariant by the action of g. The number theoretic significance of character relations follows from a theorem of Brauer [9],

$$\#\{ \prod_{H<G} h(N^H)^{a_H} : \ \mathrm{Gal}(N/\mathbb{Q}) = G \} < \infty,$$

expressing that the class number products with multiplicities according to a character relation for G assume finitely many different rational values when N ranges over all normal number fields with Galois group G. Here $h(N^H)$ is the ideal class number of the ring of integers of the subfield of N fixed by the elements of the subgroup H of G. To prevent trivial cases we will assume that in the character sums (and the related class number products) the sum (and product) ranges over non-conjugate subgroups only.

In Magma the permutation characters for all subgroups of a given permutation group G can be generated, as a matrix with the characters as rows, by this function:

```
permcharmat := function(G)
  nc := #ConjugacyClasses(G);
  subs := Subgroups(G);
  M := Hom(RSpace(Integers(), #subs), RSpace(Integers(), nc));
  return M ! &cat[
    [ PermutationCharacter(G, s`subgroup)[i] : i in [1..nc] ]
          : s in subs ];
end function;
```

all conjugacy classes of subgroups of a permutation group as a sequence of *records*, one for each class. Each record contains the representative of the class, which can be obtained via the *attribute* 'subgroup', here in the form s`subgroup, where s is one of the records in the sequence **subs**. Other attributes that can be used on this record are s`order for the order of the subgroup and s`length for the number of different subgroups that are in the class.

Here is the result for the alternating group on 4 letters:

```
>    permcharmat( Alt(4) );
          [12  0  0  0]
          [ 6  2  0  0]
          [ 4  0  1  1]
          [ 3  3  0  0]
          [ 1  1  1  1]
```

The character relations are the non-trivial relations between the rows of this matrix, and they can simply be generated as its kernel:

```
>    relations := func< G | Kernel( permcharmat(G) ) >;
>    relations( Alt(4) );
          RSpace of degree 5, dimension 2 over Integer Ring
          Echelonized basis:
          ( 1  0 -3 -1  3)
          ( 0  1 -1 -1  1)
```

Thus, for A_4, all character relations can be derived from the basis pair given here. According to Brauer's theorem the class number products corresponding to these relations

$$\frac{h(N) \cdot h(\mathbb{Q})^3}{h(N_4)^3 \cdot h(N_3)}, \qquad \frac{h(N_6) \cdot h(\mathbb{Q})}{h(N_4) \cdot h(N_3)}$$

take on finitely many values as N ranges over all Galois extensions of $\mathbb{Q}$ with Galois group A_4. Here we used the notation N_d for the degree d subfield of N invariant under the index d subgroup of A_4; of course $h(N_1) = h(\mathbb{Q}) = 1$ in this notation. In [7], Example 5.3, it is shown that the set of rationals that will occur is included in $\{\frac{1}{8}, \frac{1}{4}, \frac{1}{2}, 1, 2\}$.

Arithmetically equivalent fields

The simplest non-trivial character relation occurs when G has a pair H, H' of non-conjugate subgroups with the same permutation character. The corresponding invariant subfields $N^H, N^{H'}$ of the normal field N with Galois group G will then be non-isomorphic but they share many properties: they will have the same zeta-function [32]. Such fields are called *arithmetically equivalent*.

The existence of arithmetically equivalent number fields was shown by Gassmann [18], who exhibited in 1926 two non-conjugate subgroups of the symmetric group on 6 elements (both isomorphic to V_4) with the same permutation character:

```
>    G := Sym(6);
>    U := sub< G | (1,2)(3,4), (1,3)(2,4), (1,4)(2,3) >;
>    V := sub< G | (1,2)(3,4), (1,2)(5,6), (3,4)(5,6) >;
>    PermutationCharacter(G, U);

        ( 180, 0, 0, 12, 0, 0, 0, 0, 0, 0, 0 )

>    PermutationCharacter(G, V);

        ( 180, 0, 0, 12, 0, 0, 0, 0, 0, 0, 0 )

>    Induction(PrincipalCharacter(U), G) eq PermutationCharacter(G, U);
        true
```

The last line is included here by way of explanation for the notation 1_U^G for the permutation character: it is the character on G induced by the principal character on U. In this case it is easy to see that $1_U^G = 1_V^G$ if one uses the equivalent property that $C \cap U = C \cap V$ for all conjugacy classes C of G; the latter is obvious as conjugacy classes in S_n coincide with cycle types, and U and V are clearly the same in this respect. Since U fixes the points $5, 6$ and V is fix-point free, U and V are non-conjugate in G.

The degree of the equivalent number fields in this case is 180 (being the index of U in G, which equals the first character value). Since S_6 is the generic group for an irreducible polynomial of degree 6, the construction will furnish infinitely many pairs of arithmetically equivalent fields.

To search for examples of small degree n in Magma, one uses a simple double loop over all transitive subgroups G of S_n. Since only subgroups of index n are relevant, we set the parameter OrderEqual on the intrinsic Subgroups equal to $\#G/n$, and search for pairs U, V of subgroups isomorphic to a point stabilizer but not conjugate in S_n:

```
>    for n := 1 to 12 do
>      for k := 1 to NumberOfTransitiveGroups(n) do
>        G := TransitiveGroup(n, k);
>        U := Stabilizer(G, 1);
>        χ := PermutationCharacter(G, U);
>        S := Subgroups(G : OrderEqual := Order(G) div n);
>        if exists(i){ i : i in [1..#S] | PermutationCharacter(G, V) eq χ
>                and IsEmpty(Fix(V)) where V := S[i]`subgroup } then
>          < n, k, Order(G)>;
>        end if;
>      end for;
>    end for;
```

```
            <7,  5,  168>
            <8,  15,  32>
            <8,  23,  48>
            <11, 5,  660>
            <12, 26, 48>
            <12, 38, 72>
            <12, 49, 96>
            <12, 57, 96>
            <12, 104, 192>
            <12, 124, 240>
```

This computation reproduces part of the table given in [8], see also [26], and proves that there exist precisely 10 different configurations of pairs of arithmetically equivalent fields in degrees up to 12, namely one in degree 7, two in degree 8, one in degree 11, and six in degree 12. This non-trivial computation can only be done efficiently (in a matter of minutes) because of the availability of a database of transitive groups and a fast subgroup algorithm [10].

This computation confirms the theoretical proof of Perlis [31] that no non-trivial character relations exist for permutation groups of degree less than 7.

An arithmetically equivalent family in degree 7

A family of arithmetically equivalent pairs of number fields consists of a parametrized pair of polynomials that generically generate subfields of a Galois extension invariant under the pair of subgroups of a given configuration (as in the previous section). In this section we show how this can be done for the configuration in the smallest possible degree 7.

If we replace the line that produces output in the previous code fragment by

```
>     < n, k, G, U, S[i]`subgroup >;
```

it would output this for the degree 7 case:

```
<7, 5,
Permutation group G acting on a set of cardinality 7
Order = 168 = 2^3 * 3 * 7
    (1, 2, 3, 4, 5, 6, 7)
    (1, 2)(3, 6),
Permutation group U acting on a set of cardinality 7
Order = 24 = 2^3 * 3
    (2, 3)(4, 7)
    (2, 7, 5)(3, 6, 4)
    (3, 7)(5, 6)
    (3, 6)(5, 7),
Permutation group acting on a set of cardinality 7
Order = 24 = 2^3 * 3
    (1, 6)(4, 7)
    (1, 6, 5)(2, 3, 7)
    (2, 4)(3, 7)
    (2, 3)(4, 7)>
```

In [8] it is shown how the two subgroups U, V of the simple group $G \cong \mathrm{GL}_3(\mathbb{F}_2)$ of 168 elements can be related to each other geometrically. If N is Galois with group G and the invariant field N^U is generated by the irreducible degree 7 polynomial f, then V leaves 'collinear' triples of roots of f invariant, when we identify the 7 roots of f with the points of the projective plane over $\mathbb{F}_2$; so N^V is generated by a polynomial of degree 7 having sums of collinear roots of f as its roots.

The following notation will be used for the particular family of arithmetically equivalent fields that will be considered here. For $s, t \in \mathbb{Q}$ let $f_{s,t}$ be defined as

$$x^7 + (-6t + 2)x^6 + (8t^2 + 4t - 3)x^5 + (-s - 14t^2 + 6t - 2)x^4$$
$$+(s + 6t^2 - 8t^3 - 4t + 2)x^3 + (8t^3 + 16t^2)x^2 + (8t^3 - 12t^2)x - 8t^3.$$

If $f_{s,t}$ is irreducible over $\mathbb{Q}$ then the number field obtained by adjoining a root of $f_{s,t}$ to $\mathbb{Q}$ will be denoted by K, and the field defined by $f_{-s,t}$ will be denoted by K'. The Galois closure of K will be denoted by N as usual.

Magma can be used in the proof of the following proposition, cf. [8].

Proposition 5.1. *If $f_{s,t}$ is irreducible in $\mathbb{Q}[x]$, then so is $f_{-s,t}$; the Galois group of $f_{s,t}$ is a subgroup of $\mathrm{GL}_3(\mathbb{F}_2)$, and when it equals $\mathrm{GL}_3(\mathbb{F}_2)$ then K and K' are arithmetically equivalent.*

LaMacchia [27] already showed that the Galois group $\mathrm{Gal}(N/\mathbb{Q})$ of $f_{s,t}$ is generically $\mathrm{GL}_3(\mathbb{F}_2)$. The remarks above imply that we can identify a polynomial generating K' as an irreducible factor g of degree 7 of the polynomial P of degree 35 that has all sums of three roots of $f_{s,t}$ as roots. We determine this polynomial here symbolically; for the paper [8] a modular approach was used.

```
>    F<s, t> := FunctionField(Rationals(), 2);
```

```
>    Q<x> := PolynomialRing(F);
>    f := x^7 + (−6*t+2)*x^6 + (8*t^2+4*t−3)*x^5 +
>      (−s−14*t^2+6*t−2)*x^4 + (s+6*t^2−8*t^3−4*t+2)*x^3 +
>      (8*t^3+16*t^2)*x^2 + (8*t^3 − 12*t^2)*x − 8*t^3;
```

We determine the polynomial q_1 having as roots all sums of pairs of distinct roots of $f_{s,t}$. For this, observe that the resultant of $f(x - y)$ and $f(y)$ with respect to y is a polynomial in x that consists of the product of all differences of the roots $x - \alpha_i$ of $f(x - y)$ and α_j of f:

$$r = \mathrm{Res}_y(f(x - y), f(y)) = - \prod_{1 \leq i,j \leq 7} (x - \alpha_i - \alpha_j) = -q_1^2 \cdot h_2,$$

where h_2 is the monic polynomial that has the sums of two equal roots of f as its root. To work with monic polynomials throughout we replace f by the monic version h_1 of $f(-x)$.

```
>    h_1 := (−1)^Degree(f)*Evaluate(f, −x);
>    Y<y> := PolynomialRing(Q);
>    r := Resultant(Evaluate(h_1, y−x), Evaluate(f, y));
>    h_2 := 2^7*Evaluate(f, x/2);
>    q_1 := SquareRoot(r div h_2);
```

The resulting polynomial q_1 of degree 21 is

$$q_1 = x^{21} + (-36t + 12)x^{20} + \cdots (-147456t^{13} + \cdots - 8s^4t^3 + \cdots - 384t^5).$$

To find P, repeat the resultant trick. Determine the polynomial q having as roots the sums of three roots of f, at least two of them equal, and also q_2, having as roots the sum of one root and twice another root of f:

```
>    q := Resultant(Evaluate(h_1, y−x), Evaluate(h_2, y));
>    h_3 := 3^7*Evaluate(f, x/3);
>    q_2 := q div h_3;
```

then the polynomial P having sums of three distinct roots of f as its roots is easily obtained:

```
>    R := Resultant(Evaluate(h_1, y−x), Evaluate(q_1, y));
>    P := Root(R div q_2, 3);
```

From this polynomial P of degree 35, which has 2668 non-zero terms, we obtain the desired polynomial g by factorization:

```
>    fP := Factorization(P);
>    g := fP[1][1]; g;
```

```
        x^7 + (−18*t + 6)*x^6 + (124*t^2 − 64*t + 6)*x^5 + (−408*t^3 +
          208*t^2 − 4*t − 16)*x^4 + (6*s*t − 6*s + 640*t^4 − 156*t^3 −
          116*t^2 + 84*t − 27)*x^3 + (−36*s*t^2 + 36*s*t − 12*s −
          384*t^5 − 152*t^4 + 120*t^3 + 88*t^2 − 34*t − 6)*x^2 +
          (−s^2 + 48*s*t^3 − 20*s*t^2 − 2*s*t − 2*s − 64*t^5 − 84*t^4 +
```

```
    52*t^3 - 8*t^2 - 12*t)*x - 8*s*t^3 - 4*s*t^2 + 384*t^6 +
    80*t^5 - 88*t^4 - 24*t^3
```

The bottlenecks in this computation are the root extraction $P = \sqrt[3]{R/q_2}$ and the factorization of P. The polynomial R has degree 147 and

```
>    &+[ #Terms(Integers(F) ! c) : c in Coefficients(R) ];

        165555
```

non-zero terms.

Finally we show that $f_{-s,t}$ and g generate the same number field. Instead of literally pasting in the definition of $f_{-s,t}$ we obtain it by applying the homomorphism h of $F(s,t)[x]$ sending s to $-s$:

```
>    fh := hom< F → F | −s, t >;
>    h := hom< Q → Q | fh, x >;
>    fminus := f @ h;
```

When we apply a particular rational transformation to g the result is divisible by *fminus*,

```
>    gnew := Q ! (x^7*Evaluate(g, (x−1)*(1+2*t/x)));
>    gnew mod fminus;

        0
```

and the proposition follows.

Class number quotients

Arithmetically equivalent number fields have the same zeta-function; the zeta-function encodes a lot of information but it is not true that number fields are characterized (up to isomorphism) by their zeta-function. Equality of zeta-functions for two fields forces many invariants of the fields to be equal, but not necessarily their ideal class numbers. However, the product of class number and *regulator* for arithmetically equivalent fields will be the same.

The first example of number fields with the same zeta-function but different class numbers was published by Perlis and de Smit [39]. It consists of a pair of arithmetically equivalent fields in degree 8 of the form $\mathbb{Q}(\sqrt[8]{a}), \mathbb{Q}(\sqrt[8]{16a})$.

Work by de Smit has produced bounds on the possible class number quotients that appear in the finite set $h(N^H)/h(N^{H'})$ for a fixed triple (G, H, H') that produces arithmetically equivalent number fields N^H and $N^{H'}$ with $G = \mathrm{Gal}(N/\mathbb{Q})$ of small degree $[G : H]$. These bound are fairly tight in the sense that most remaining quotients do occur. For our example in degree 7 the bounds imply that the set is contained in $\{1, \frac{1}{2}, \frac{1}{4}, \frac{1}{8}\}$ and their reciprocals, where $\frac{1}{8}$ would only be possible for a totally real field N (the only other possibility in this configuration is that N has precisely 2 pairs of complex embeddings).

To generate examples the following code could be used. It is useful in practice to search for examples with relatively small discriminant only. Continuing our previous examples with $F = \mathbb{Q}(s,t)$ and $Q = F[x]$

```
>   U<u> := PolynomialRing(Rationals());
>   evalst := func< j, k | hom< Q  →  U | C, u>
>       where C is hom< F  →  Rationals() | j, k > >;
```

the function evalst(j, k) can, for rational values of j, k, be applied to $f_{s,t}$ to cast it into an element of $U = \mathbb{Q}[u]$ by evaluating $s = j$ and $t = k$. Here we use this for some selected values for j and k (obtained from a search):

```
>   for p in [ [1,2], [7,1], [6,−7], [5,4], [1,4], [19,5] ] do
>     j, k := Explode(p);
>     N₁ := NumberField( evalst(j,k)(f) );
>     fD := Factorization(Discriminant(Integers(N₁)));
>     h₁ := ClassNumber(N₁: Bound := 300);
>     N₂ := NumberField( evalst(−j,k)(f) );
>     h₂ := ClassNumber(N₂: Bound := 300);
>     print <p, Min([ h₁/h₂, h₂/h₁ ]), fD, Signature(N₁)>;
>   end for;
```

```
        <[ 1,  2 ], 1,   [ <27277, 2> ], 3>
        <[ 7,  1 ], 1/2, [ <222107, 2> ], 3>
        <[ 6, -7 ], 1/4, [ <2, 4>, <13, 2>, <1728655121887, 2> ], 3>
        <[ 5,  4 ], 1,   [ <8488225021, 2> ], 7>
        <[ 1,  4 ], 1/2, [ <3347, 2>, <2602463, 2> ], 7>
        <[ 19, 5 ], 1/4, [ <270982714837, 2> ], 7>
```

The ideal class numbers for the pair of arithmetically equivalent number fields N_1 and N_2 are calculated, and their quotient is displayed here, together with the field discriminant (in factored form) and the number of real embeddings. The bound of 300, given as a parameter here, speeds up the computation (it puts a bound on the norms of the ideals used as generators for ideal classes), but some additional work is required to obtained guaranteed results. We did not find an example where the quotient equals $\frac{1}{8}$ (or 8).

References

1. R. Baillie, G. Cormack, H. C. Williams, *The problem of Sierpiński concerning $k \cdot 2^n + 1$*, Math. Comp. **37** (1981), 229–231.
2. Bruce C. Berndt, Ronald J. Evans, Kenneth S. Williams, *Gauss and Jacobi sums*, Canad. Math. Soc. series of monographs and advanced texts **21**, New York: John Wiley and Sons, 1997.
3. Wieb Bosma, *Explicit primality criteria for $h \cdot 2^k \pm 1$*, Math. Comp. **61** (1993), 97–109.
4. Wieb Bosma, *Cubic reciprocity and explicit primality tests for $h \cdot 3^k \pm 1$*, pp. 77–89 in: Alf van der Poorten, Andreas Stein (eds.), *High primes and misdemeanours: lectures in honour of the 60th birthday of Hugh Cowie Williams*, Fields Inst. Commun. **41**, Providence: Amer. Math. Soc. 2004.
5. Wieb Bosma, *Some computational experiments in elementary number theory*, report **05-02** Mathematical Institute, Radboud University Nijmegen, 2005.

6. Wieb Bosma, John Cannon, Catherine Playoust, *The Magma algebra system I: The user language*, J. Symbolic Comput. **24** (1997), 235–265. See also the Magma home page at http://magma.maths.usyd.edu.au/magma/.

7. Wieb Bosma, Bart de Smit, *Class number relations from a computational point of view*, J. Symbolic Comput. **31** (2001), 97–112.

8. Wieb Bosma, Bart de Smit, *On arithmetically equivalent number fields of small degree*, pp. 67–79 in: C. Fieker, D.R. Kohel (eds.), *Algorithmic Number Theory Symposium, Sydney, 2002*, Lecture Notes in Computer Science **2369**, Berlin, Heidelberg: Springer, 2002.

9. R. Brauer, *Beziehungen zwischen Klassenzahlen von Teilkörpern eines galoisschen Körpers*, Math. Nachrichten **4** (1951), 158–174.

10. J.J. Cannon, D.F. Holt, *Computing maximal subgroups of a finite group*, J. Symbolic Comput. **37** (2004), 589–609.

11. R.D. Carmichael, *On Euler's φ-function*, Bull. Amer. Math. Soc. **13** (1907), 241–243; Errata: Bull. Amer. Math. Soc. **54** (1948), 1192.

12. R.D. Carmichael, *Note on Euler's φ-function*, Bull. Amer. Math. Soc. **28** (1922), 109–110; Errata: Bull. Amer. Math. Soc. **55** (1949), 212.

13. S.L.G. Choi, *Covering the set of integers by congruence classes of distinct moduli*, Math. Comp. **25** (1971), 885–895.

14. R.F. Churchhouse, *Covering sets and systems of congruences*, pp. 20–36 in: R.F. Churchhouse, J.-C. Herz (eds.), *Computers in mathematical research*, Amsterdam: North-Holland, 1968.

15. D.W. Erbach, J. Fischer, J. McKay, *Polynomials with PSL(2,7) as Galois group*, J. Number Theory **11** (1979), 69–75.

16. Paul Erdős, *Some of my favorite problems and results*, pp. 47–67 in: Ronald L. Graham, Jaroslav Nešetřil (eds.), *The Mathematics of Paul Erdős I*, Berlin: Springer, 1997.

17. Paul Erdős, *Some remarks on Euler's φ function*, Acta Arith. **4** (1958), 10–19.

18. F. Gassmann, *Bemerkungen zu der vorstehenden Arbeit von Hurwitz ('Über Beziehungen zwischen den Primidealen eines algebraischen Körpers und den Substitutionen seiner Gruppe')*, Math. Z. **25** (1926), 124–143.

19. Andrew Granville, K. Soundararajan, *A binary additive problem of Erdős and the order of 2 mod p^2*, Ramanujan J. **2** (1998), 283–298.

20. Richard K. Guy, *Unsolved problems in number theory*, Unsolved problems in intuitive mathematics **I**, New York: Springer 1994 (2nd edition).

21. Kenneth Ireland and Michael Rosen, *A classical introduction to modern number theory*, Graduate texts in mathematics **84**, New York: Springer, 1982.

22. G. Jaeschke, *On the smallest k such that all $k \cdot 2^N + 1$ are composite*, Math. Comp. **40** (1983), 381–384; Errata: Math. Comp. **45** (1985), 637.

23. Wilfrid Keller, *Factors of Fermat numbers and large primes of the form $k \cdot 2^n + 1$*, Math. Comp. **41** (1983), 661–673.

24. Wilfrid Keller, *The least prime of the form $k \cdot 2^n + 1$*, Abstracts Amer. Math. Soc. **9** (1988), 417–418.

25. V.L. Klee, Jr., *On a conjecture of Carmichael*, Bull. Amer. Math. Soc. **53** (1947), 1183–1186.

26. N. Klingen, *Arithmetical similarities*, Oxford: Oxford University Press, 1998.

27. Samuel E. LaMacchia, *Polynomials with Galois group PSL(2,7)*, Comm. Algebra **8** (1980), 983–992.

28. Douglas B. Meade, Charles A. Nicol, *Maple tools for use in conjecture testing and iteration mappings in number theory*, IMI Research Report **1993:06** (Department of Mathematics, University of South Carolina), 1993.

29. Ryozo Morikawa, *Some examples of covering sets*, Bull. Fac. Liberal Arts, Nagasaki Univ. **21** (1981), 1–4.

30. Oystein Ore, J. L. Selfridge, P. T. Bateman, *Euler's function*: Problem 4995 and its solution, Amer. Math. Monthly **70** (1963), 101–102.

31. R. Perlis, *On the equation* $\zeta_K(s) = \zeta_{K'}(s)$ J. Number Theory **9** (1977), 342–360.

32. R. Perlis, *On the class numbers of arithmetically equivalent fields*, J. Number Theory **10** (1978), 458–509.

33. A. de Polignac, *Recherches nouvelles sur les nombres premiers*, C. R. Acad. Sci. Paris Math. **29** (1849), 397–401; 738–739.

34. Carl Pomerance, *On Carmichael's conjecture*, Proc. Amer. Math. Soc. **43** (1974), 297–298.

35. P. Poulet, *Nouvelles suites arithmétiques*, Sphinx **2** (1932), 53–54.

36. H. Riesel, *Några stora primtal*, Elementa **39** (1956), 258–260.

37. Aaron Schlafly, Stan Wagon, *Carmichael's Conjecture on the Euler function is valid below* $10^{10000000}$, Math. Comp. **63** (1994), 415–419.

38. W. Sierpiński, *Sur un problème concernant les nombres* $k \times 2^n + 1$, Elemente der Mathematik **15** (1960), 63–74.

39. Bart de Smit, Robert Perlis, *Zeta functions do not determine class numbers*, Bull. Amer. Math. Soc. **31** (1994), 213–215.

40. R. G. Stanton, *Further results on covering integers of the form* $1 + k \cdot 2^n$ *by primes*, pp. 107–114 in: Kevin L. McAvaney (ed.), *Combinatorial Mathematics VIII*, Lecture Notes in Mathematics **884**, Berlin: Springer, 1981.

41. R. G. Stanton, H. C. Williams, *Computation of some number-theoretic coverings* pp. 8–13 in: Kevin L. McAvaney (ed.), *Combinatorial Mathematics VIII*, Lecture Notes in Mathematics **884**, Berlin: Springer, 1981.

42. http://www.prothsearch.net/sierp.html

43. http://www.prothsearch.net/rieselprob.html

Applications of the class field theory of global fields

*Claus Fieker**

School of Mathematics and Statistics
University of Sydney
Sydney, Australia
claus@maths.usyd.edu.au

Summary. Class field theory of global fields provides a description of finite abelian extensions of number fields and of function fields of transcendence degree 1 over finite fields. After a brief review of the handling of both function and number fields in Magma, we give an introduction to computational class field theory focusing on applications: We show how to construct tables of small degree extensions and how to utilize the class field theory to find curves with many rational points.

1 Introduction

Class field theory is concerned with abelian extensions of (number) fields, that is, finite Galois extensions K of $\mathbb{Q}$ for which the group of automorphisms is abelian. After establishing the theory for number fields, it was realized that almost all of it could be extended to certain infinite extensions of finite fields. Thus the concept of global fields was born as the class of fields for which "the theory works". In this chapter we will develop the class field theory for both function fields and number fields in Magma [3], [4, Chapters 52, 55, 56, 58, 59] and give some applications to number theory and coding theory.

We begin by briefly describing Magma's interface to number fields, starting with their representation. Next we explain how class groups are represented and computed, with a particular focus on the practical aspects of the computation. This is followed by a discussion of ray class groups, which parametrize class fields. The fundamental statements of class field theory that are needed for the applications presented here are recalled, again, with a strong emphasis on Magma's view of the material.

---

*This research was supported in part by a grant from the Australian Research Council (A00104694)

The program is then repeated for global function fields: we recall the definitions and show how they are implemented and used in Magma. The presentation for function fields focuses mainly on the differences between number fields and function fields as they essentially behave in the same way. We introduce divisors, divisor class groups and divisor ray class groups.

As applications of the class field theory developed, we show how to compile certain tables of global fields indexed by degree, discriminant and Galois group. For function fields, we show how the theory may be used to find fields with "many rational places" which are of interest in certain applications such as in the construction of "good codes".

2 Number fields

We briefly recall the definition of number fields and summarize their properties with a strong emphasis on Magma's view of them.

2.1 Basic properties

A number field K is a finite extension of $\mathbb{Q}$. By the primitive element theorem, there is some $\alpha \in K$ such that $K = \mathbb{Q}(\alpha)$ holds. Since K is a finite extension, α is algebraic over $\mathbb{Q}$, so α is a root of an irreducible polynomial $f(t) \in \mathbb{Q}[t]$ and as a field, we have an isomorphism

$$K \to \mathbb{Q}[t]/\langle f \rangle : \alpha \mapsto t.$$

To create a number field in Magma* we reverse this process: we first define an irreducible polynomial $f(t)$ and then use it to create K and α:

```
>    Q := Rationals();
>    Qt<t> := PolynomialRing(Q);
>    f := t³−25;
>    K<a> := NumberField(f);
>    a³;
```

     25

Now we have a number field at our disposal. Note that K is *not* considered to be a subfield of the field of complex numbers $\mathbb{C}$! This is due to the fact that a number field K of degree $n := [K : \mathbb{Q}]$ has n different embeddings into $\mathbb{C}$ and there is no canonical embedding.

The most important subring of a number field is its ring of integers $\mathbb{Z}_K$, consisting of the elements $\beta \in K$ that are roots of monic polynomials $g \in \mathbb{Z}[t]$. It is well known that $\mathbb{Z}_K$ is a free $\mathbb{Z}$-module of rank $n := [K : \mathbb{Q}]$ so we can

---

*See the Preface to this volume for style conventions regarding Magma code; code appearing in this book is available at http://magma.maths.usyd.edu.au/magma/.

represent elements with respect to a fixed basis of $\mathbb{Z}_K$. A so-called *integral basis* can be computed, for example, using either the Round2 or the Round4 method [6, 20, 9, 19].

By allowing an integral denominator, we extend the basis representation of $\mathbb{Z}_K$ to all elements of K:

```
>    Z_K := RingOfIntegers(K);
>    Basis(Z_K, K);
         [
             1,
             a,
             1/5*a^2
         ]
```

The ring $\mathbb{Z}_K$ is an invariant of the field so, in particular, it does not depend on the choice of the representation of the field or the primitive element. However, for computational purposes the representation is very important. A "bad" choice of primitive element or defining polynomial can potentially slow down even simple operations, such as multiplication of two field elements, to the point where computation becomes impractical.

Elements of $\mathbb{Z}_K$ are printed as sequences of coefficients; elements of the field of fractions of $\mathbb{Z}_K$ are displayed as linear combination of the (named) basis elements; elements of K are printed as polynomials in a:

```
>    Z_K ! [1,2,3];
         [1, 2, 3]

>    $1/1;
         Z_K.1 + 2/1*Z_K.2 + 3/1*Z_K.3

>    K ! $1;
         1/5*(3*a^2 + 10*a + 5)

>    Z_K ! $1;
         [1, 2, 3]
```

Note that $1 denotes the last printed result and ! coerces some object of one structure into another. It is used here to change the representation, which depends on the parent object. Finally, Z_K.1, Z_K.2, Z_K.3 denote the first, second and third basis elements respectively in the fixed basis of $\mathbb{Z}_K$.

It is possible to build finite extensions of K and to consider *relative extensions*. As an example, we will construct the field $E := K[\zeta_3]$, where ζ_3 is a third root of unity:

```
>    E<z> := NumberField(Polynomial(K, CyclotomicPolynomial(3)));
>    E;
         Number Field with defining polynomial $.1^2 + $.1 + 1
         over K
```

```
>    E: Maximal;
```

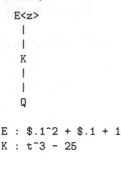

```
      E : $.1^2 + $.1 + 1
      K : t^3 - 25
```

```
>    z³;
```

```
      1
```

The $.1 is used to indicate the transcendental generator of the polynomial ring over K. As E is an extension of K, all invariants of E are automatically relative to K, so, for example, the norm of $z + a$ will be in E. In order to get the absolute norm over $\mathbb{Q}$ we have to indicate the 'target' ring as a second argument or use **AbsoluteNorm**:

```
>    Norm(z+a);
```

```
      a^2 - a + 1
```

```
>    Norm(z+a, Q);
```

```
      676
```

```
>    Norm(z+a, E);
```

```
      z + a
```

A major difference between absolute and relative extensions becomes visible when we look at the ring of integers: $\mathbb{Z}_E$ is a $\mathbb{Z}_K$-module, but since $\mathbb{Z}_K$, in general, is not a principal ideal domain (PID), it is no longer a free module. Instead, since $\mathbb{Z}_K$ is a Dedekind ring, the best we can achieve is an *almost-free* representation. There are fractional ideals $\mathfrak{a}_i$ of $\mathbb{Z}_K$ and elements $\omega_i \in E$ such that

$$\mathbb{Z}_E = \sum_{i=1}^{m} \mathfrak{a}_i \omega_i.$$

Recall that a fractional ideal of $\mathbb{Z}_K$ is a $\mathbb{Z}_K$-module $\mathfrak{a}$ for which there is some $d \in K$ such that $d\mathfrak{a}$ is an ideal in $\mathbb{Z}_K$. In particular, fractional ideals are, in general, neither ideals nor rings.

By replacing $\mathfrak{a}_i \omega_i$ by $(1/\lambda_i \mathfrak{a}_i)(\lambda_i \omega_i)$ for $0 \neq \lambda_i \in K$ (which does not change the sum), we see that in general

1. $\omega_i \notin \mathbb{Z}_E$
2. $\mathfrak{a}_i \not\subset \mathbb{Z}_K$.

Therefore, in general, integral elements will appear to have denominators (their coefficients are elements of some fractional ideal) and elements that appear to be integral may not be in $\mathbb{Z}_E$, if their integral coefficients are not contained in the proper coefficient ideals. The system $(\mathfrak{a}_i, \omega_i)_i$ $(1 \leq i \leq n)$ is called a *pseudo-basis* of $\mathbb{Z}_E$ [7, 12]. Strictly, an almost-free representation of the module should have $\mathfrak{a}_1 = \ldots = \mathfrak{a}_{n-1} = \mathbb{Z}_K$, but for our purposes the weaker version is sufficient.

It is possible to convert E and $\mathbb{Z}_E$ to absolute extensions. The resulting absolute field allows coercion between the different representations. To illustrate this we create an extension of $\mathbb{Q}$ isomorphic to E:

```
>    E_Q := AbsoluteField(E);
>    E_Q: Maximal;

            E_Q
             |
             |
             Q

    E_Q : t^6 + 3*t^5 + 6*t^4 - 43*t^3 - 69*t^2 + 78*t + 676

>    Z_E_Q := MaximalOrder(E_Q);
>    Z_E := MaximalOrder(E);
>    Z_E ! Z_E_Q.3;

        [[0, 0, 1], [0, 0, 0]]

>    E_Q ! $1;

        1/4680*(-25*E_Q.1^5 - 62*E_Q.1^4 - 124*E_Q.1^3 +
        1439*E_Q.1^2 + 1426*E_Q.1 - 1612)
```

Ideals and fractional ideals of orders are modules over the same coefficient ring as the order. As with the representation of the maximal order, ideals in absolute extensions are free $\mathbb{Z}$-modules whereas ideals in relative extensions are, in general, not free. Ideals can be represented in three different ways:

- By means of a (pseudo) basis.
- By means of two generators. The ring of integers of a number field is a Dedekind domain, and a straightforward application of the weak approximation theorem shows that every ideal can be generated using only two elements. Furthermore, since one of the elements can be chosen arbitrarily from the ideal, any ideal can be generated by an integer and some algebraic number.
- By means of a single generator when the ideal is principal.

The printed display of an ideal will change depending upon the information known about the ideal.

2.2 Class groups

The classification of class fields is achieved through the use of generalized class groups which are extensions of the standard class group. The "standard" class group $\mathrm{Cl} := \mathrm{Cl}_K$ of a number field K is defined as the quotient of the group of all ideals by the subgroup of principal ideals. An important theorem states that the class group is a finite group. Efficient algorithms are known for computing the class group, i.e., to compute the structure of the group and for mapping ideals into the abstract abelian group [6, 14].

Since the class group algorithm is also used to compute units, S-units, discrete logarithms and to decide whether an ideal is principal, it is important to understand roughly how the method works.

The class group being a finite group, is finitely generated and in fact, it is generated by the set $F := \{\mathfrak{p} \mid N(\mathfrak{p}) \leq B_1\}$ of all prime ideals having norm bounded by some constant B_1. The set F is called the *factor basis*. Frequently, in practice, since the true bound B_1 is either not known or is too large to be useful, we choose some convenient bound B_1 and work with the corresponding set F. Assuming that the class group is generated by F, its computation is now equivalent to finding a sufficiently large set of relations involving the factor basis elements. By a relation we mean an element $\gamma \in K$ such that

$$\gamma \mathbb{Z}_K = \prod_{\mathfrak{p} \in F} \mathfrak{p}^{v_\mathfrak{p}(\gamma)}.$$

The set of such relations, or more precisely the exponent vectors $(v_\mathfrak{p}(\gamma))_{\mathfrak{p} \in F}$, form a lattice Λ. Determining the class group is equivalent to being able to compute a basis for Λ. Note that finding a $\mathbb{Z}$-linear dependency $a = (a_i)_i$ between relations implies the existence of a unit:

$$\sum_i a_i (v_\mathfrak{p}(\gamma_i))_{\mathfrak{p} \in F} = 0 \iff \prod_i \gamma_i^{a_i} \in \mathbb{Z}_K^*.$$

The algorithm proceeds by constructing "small" elements in $\mathbb{Z}_K$ and testing if they factor over F. This procedure is repeated until sufficient relations are found. As a stopping condition, we compute an approximation to the Euler product

$$E_{B_2} := \prod (1 - \frac{1}{N(\mathfrak{p})}) \prod (1 - \frac{1}{p})$$

where the first product is over all prime ideals dividing prime numbers not exceeding B_2 and the second is over the prime numbers $\leq B_2$. The bound B_2 is chosen independently of B_1, where usually $B_2 > B_1$. The analytic class number formula gives a precise relation between the determinant of the lattice Λ (i.e. the class number $h_K := \#\mathrm{Cl}_K$), the regulator R_K of the unit group (generated by dependencies between the relations) and E:

$$E_\infty := \lim_{B_2 \to \infty} E_{B_2} = c_K h_K R_K$$

where c_K is an explicit constant depending on the signature, the discriminant and the number of torsion units in K. The search stops when this equality is approximately satisfied. To be more precise, we stop when

$$\frac{1}{\log 2} E_{B_2} \leq c_K h_K R_K \leq \log 2 E_{B_2}$$

holds. At this point, the class group is known under the assumption that B_1 is sufficiently large and that E_{B_2} is a good approximation to the true Euler product. Therefore, it remains to prove this.

Using methods from the geometry of numbers, it can be shown that every element in Cl can be represented by an ideal $\mathfrak{a}$ of norm $N(\mathfrak{a}) \leq M_K$, where the *Minkowski bound* is defined as

$$M_K := \frac{n!}{n^n} \left(\frac{4}{\pi}\right)^{r_2} \sqrt{|\operatorname{Disc}(K)|}.$$

In particular, this implies that setting $B_1 := M_K$ would guarantee that Cl_K is generated by F. Therefore, to verify the class group, it remains to show that for all prime ideals $\mathfrak{p}$ such that $B_1 \leq N(\mathfrak{p}) \leq M_K$, there is a relation $\gamma_{\mathfrak{p}}$ such that $\mathfrak{p}^{-1}\gamma_{\mathfrak{p}}$ can be factored over F. Since $\operatorname{Disc}(K)$ and therefore M_K grows rapidly with the degree of K, this process can be very slow due to the large number of primes that need to be considered.

On the other hand, assuming the GRH (generalized Riemann hypotheses), Bach [2] derived a much smaller bound B_K,

$$B_K := 12 \log^2(|\operatorname{Disc}(K)|)$$

which can be taken as the value of B_1. So again, for every prime $\mathfrak{p}$ between B_1 and B_K we have to find a relation between $\mathfrak{p}$ and F. An additional advantage of using bounds based on the GRH is that we can also bound the error term of the Euler product, that is we get bounds for the error $|E_\infty - E_{B_2}|$. Using this, we can compute a precise approximation for E_{B_2} and use this to verify both the class group and the unit group.

In general, the size of B_K makes it impossible to compute class groups that are guaranteed to be correct even assuming GRH. On the other hand, in practical applications where the class group is just an intermediate step, it is frequently possible to independently verify the final result so a provably correct class group is not required. In this case a much smaller bound, $B_1 \leq 1000$, is usually sufficient.

Note, however, by default Magma will always use M_K so that class group computations always return guaranteed results at the possible cost of extremely long execution times.

Magma represents class groups as finite abelian groups together with a map between the (abstract) group and the set of ideals of the number field:

```
>    K := NumberField(x^2 - 10);
```

```
>   Z_K := RingOfIntegers(K);
>   Cl, mCl := ClassGroup(Z_K);
>   Cl;
```
```
        Abelian Group isomorphic to Z/2
        Defined on 1 generator
        Relations:
            2*Cl.1 = 0
```
```
>   mCl;
```
```
        Mapping from: GrpAb: Cl to Set of ideals of Z_K
```

We can use this map to create ideals in any class, while *mCl* allows the calculation of inverse images, where *I* @@ *mCl* represents an arbitrary ideal in the class group:

```
>   I := mCl(Cl.1); I;
```
```
        Ideal of Z_K
        Two element generators:
            [2, 0]
            [2, 1]
```

```
>   I * K.1;
```
```
        Ideal of Z_K
        Two element generators:
            [0, 2]
            [10, 2]
```

```
>   I @@ mCl;
```
```
        Cl.1
```

```
>   (I²) @@ mCl;
```
```
        0
```

```
>   IsPrincipal(mCl(Cl.1));
```
```
        false
```

```
>   IsPrincipal(mCl(Cl.1)²);
```
```
        true
```

The unit group, which is computed at the same time as the class group, is also returned as an abstract abelian group together with a map, formally between the group and the order, but actually between the group and the units of $\mathbb{Z}_K$.

2.3 Ray class groups

The next step towards class field theory is the introduction of "generalized class groups", the so-called *ray class groups*. In order to define them, we need to introduce more terminology.

A number field $K = \mathbb{Q}(\alpha)$ of degree n has exactly n different embeddings into the complex numbers, each mapping α to a complex root of its minimal polynomial. An embedding is said to be *real* if the corresponding root is real, in which case the image of K will be a subfield of the real numbers.

Let $\mathfrak{m}_0$ be a fixed integral ideal and let $\mathfrak{m}_\infty$ a subset of the real embeddings of K into $\mathbb{R}$. A (congruence) *modulus* $\mathfrak{m}$ is the pair $(\mathfrak{m}_0, \mathfrak{m}_\infty)$. For $x \in K$ we write

$$x \equiv 1 \bmod^* \mathfrak{m} \iff \begin{cases} v_\mathfrak{p}(x-1) \geq v_\mathfrak{p}(\mathfrak{m}_0) & \text{for all } \mathfrak{p} \mid \mathfrak{m}_0 \\ s(x) > 0 & \text{for all } s \in \mathfrak{m}_\infty \end{cases}$$

Now let $K_\mathfrak{m} := \{x \in K \mid x \equiv 1 \bmod^* \mathfrak{m}\}$ and let $I^\mathfrak{m}$ be the set of ideals coprime to $\mathfrak{m}_0$. Then the ray class group modulo $\mathfrak{m}$ is defined as

$$\mathrm{Cl}_\mathfrak{m} := I^\mathfrak{m}/K_\mathfrak{m}.$$

Using standard techniques [17] it is easy to see that $\mathrm{Cl}_\mathfrak{m}$ fits into the exact sequence:

$$1 \to U_\mathfrak{m} \to (\mathbb{Z}_K/\mathfrak{m}_0)^* \times \{\pm 1\}^{\#\mathfrak{m}_\infty} \to \mathrm{Cl}_\mathfrak{m} \to \mathrm{Cl}_K \to 1$$

with $U_\mathfrak{m} := \mathbb{Z}_K^* \cap K_\mathfrak{m}$. This exact sequence allows us to compute $\mathrm{Cl}_\mathfrak{m}$ ([7, 16]) and also shows that this is a finite abelian group.

To compute the units of $\mathbb{Z}_K/\mathfrak{m}_0$ we apply the Chinese remainder theorem:

$$(\mathbb{Z}_K/\mathfrak{m}_0)^* = (\mathbb{Z}_K/\prod_\mathfrak{p} \mathfrak{p}^{v_\mathfrak{p}(\mathfrak{m}_0)})^* = \prod_\mathfrak{p} (\mathbb{Z}_K/\mathfrak{p}^{v_\mathfrak{p}(\mathfrak{m}_0)})^*.$$

Using Hasse's 1-units we have

$$(\mathbb{Z}_K/\mathfrak{p}^s)^* = (\mathbb{Z}_K/\mathfrak{p})^* \times (1+\mathfrak{p})/(1+\mathfrak{p}^s)$$

We can compute $(1+\mathfrak{p})/(1+\mathfrak{p}^s)$ by repeatedly using the exact sequence

$$1 \to (1+\mathfrak{p}^l)/(1+\mathfrak{p}^m) \to (1+\mathfrak{p}^k)/(1+\mathfrak{p}^m) \to (1+\mathfrak{p}^k)/(1+\mathfrak{p}^l) \to 1$$

(for $l \leq 2k$ and $m \leq 2l$) as well as

$$(1+\mathfrak{p}^l)/(1+\mathfrak{p}^m) \equiv (\mathfrak{p}^l/\mathfrak{p}^k)^+.$$

In particular, we obtain

$$\phi(\mathfrak{m}_0) := \#(\mathbb{Z}_K/\mathfrak{m}_0)^* = \prod_{\mathfrak{p} \mid \mathfrak{m}_0} (N(\mathfrak{p})-1)N(\mathfrak{p})^{v_\mathfrak{p}(\mathfrak{m}_0)-1}$$

which will be important subsequently.

All of the above sequences are hidden in two intrinsics: RayClassGroup computes Cl_m and RayResidueRing computes $(\mathbb{Z}_K/m_0)^* \times \{\pm 1\}^{\# m_\infty}$. As with the class group computation, the result will be an abstract abelian group and a map between this group and the (ideals of the) number field. In order to specify the defining modulus m we have to supply an integral ideal m_0 and a sequence $[s_1, \ldots, s_t]$ of the indices giving the real places in m_∞:

```
>    m_0 := 7*5*9*Z_K;
>    m_inf := [2];
>    R, mR := RayClassGroup(m_0, m_inf);
>    R;
```

```
        Abelian Group isomorphic to Z/2 + Z/6 + Z/24
        Defined on 3 generators
        Relations:
            6*R.1 = 0
            24*R.2 = 0
            2*R.3 = 0
```

```
>    mR;
```

```
        Mapping from: GrpAb: R to Set of ideals of Z_K
```

For two moduli $m = (m_0, m_\infty)$ and $n = (n_0, n_\infty)$ we write $m \leq n$ or $n \mid m$ if $n_0 \mid m_0$ and $m_\infty \supseteq n_\infty$. In this situation we have a canonical surjection

$$\phi_m^n : Cl_m \to Cl_n : xK_m \mapsto xK_n.$$

Thus it makes sense to define the *conductor* $\mathfrak{f}$ of a given ray class group Cl_m as the smallest n such that $Cl_m \to Cl_n$ is injective.

For moduli m and n, let U be a subgroup of Cl_m and let V be a subgroup of Cl_n and let $mn = (m_0 n_0, m_\infty \cup n_\infty)$. Then both projections $\phi_{mn}^m : Cl_{mn} \to Cl_m$ and $\phi_{mn}^n : Cl_{mn} \to Cl_n$ are surjective. We write $(Cl_m, U) \equiv (Cl_n, V)$ if U and V agree in Cl_{mn}, that is if $(\phi_{mn}^m)^{-1}(U) = (\phi_{mn}^n)^{-1}(V)$. Using this, the conductor of (Cl_m, U) is the smallest modulus n for which there is a subgroup $V \leq Cl_n$ such that $(Cl_m, U) \equiv (Cl_n, V)$.

It is important to remember that ray class groups depend upon both the class group and the unit group. If either is missing, it will be computed using Magma's default settings for unconditionally proven results. Thus for large fields it might be necessary to compute both the class group and the unit group conditionally in advance, since Magma will use them if they are already known — regardless of whether they have been proven or not.

2.4 Class field theory

Now that we have defined (almost) all objects needed to compute class fields, we are ready to start. As mentioned before, class fields will be parametrized by ray class groups and their quotients. Before stating the theorems, we will present an example. Starting with $k := \mathbb{Q}(\sqrt{10})$ (of class number 2), we will define the full ray class field corresponding to modulus $\mathfrak{m} = (\mathfrak{m}_o, \mathfrak{m}_\infty)$, where $\mathfrak{m}_0 = 5\mathbb{Z}_k$ and $\mathfrak{m}_\infty$ contains both real embeddings:

```
>    k := NumberField(x^2-10);
>    Z_k := RingOfIntegers(k);
>    R, mR := RayClassGroup(5*Z_k, [1,2]);
>    A := AbelianExtension(mR);
>    A;

         FldAb, defined by (<[5, 0]>, [1 2])
         of structure: Z/4
```

In order to see what A really looks like, we have to convert it to a number field:

```
>    K := NumberField(A);
>    K;

         Number Field with defining polynomial $.1^4 +
            (36*k.1 + 280) *$.1^2 + 4200*k.1 + 13820 over k

>    G, _, mG := AutomorphismGroup(A);
>    G;


         Abelian Group isomorphic to Z/4
         Defined on 1 generator
         Relations:
            4*G.1 = 0
```

Note that the group $\mathrm{Aut}(K/k)$ of k-automorphisms of K is isomorphic to $\mathrm{Cl}_\mathfrak{m}$, thus illustrating the first part of the correspondence between abelian fields and ray class groups. Furthermore, this isomorphism is available explicitly, being the extension of the map that sends unramified prime ideals to their Frobenius automorphism. The Frobenius automorphism $(\mathfrak{p}, K/k) \in \mathrm{Aut}(K/k)$ of a prime ideal $\mathfrak{p}$ is the unique automorphism such that

$$(\mathfrak{p}, K/k)(x) \equiv x^{N(\mathfrak{p})} \bmod \mathfrak{p}\mathbb{Z}_K$$

for all $x \in K$. We will demonstrate this for a prime ideal over 13:

```
>    lp := Decomposition(Z_k, 13);
>    p := lp[1][1];
>    Norm(p);

         13
```

We compare each automorphism with $x \mapsto x^{13}$ modulo $\mathfrak{p}\mathbb{Z}_K$, by looking for $\sigma \in \mathrm{Aut}(K/k)$ such that for $x := K.1$, (the primitive element of K) we have $\sigma(x) - x^{13} \in \mathfrak{p}\mathbb{Z}_K$:

```
>    Z_K := MaximalOrder(A);
```

```
>    P := Z_K !! p; // to define pZ_K
>    [ <s, (s @ mG)(K.1) − K.1¹³ in P> : s in G];
        [
            <0, false>,
            <G.1, false>,
            <2*G.1, false>,
            <3*G.1, true>
        ]
```

```
>    s₁ := FrobeniusAutomorphism(A, p);
>    s₂ := (3*G.1) @ mG;
>    s₁ eq s₂;
        true;
```

```
>    p @@ mR;
        3*R.1
```

So indeed, if we map $R.1$ to $G.1$ we get an isomorphism of finite groups. The image of a prime ideal in R gives (via this map) the corresponding Frobenius automorphism. The multiplicative extension of the Frobenius map is called the *Artin map* and is denoted by $(\mathfrak{a}, K/k)$. Thus, for $\mathfrak{a} = \prod \mathfrak{p}_i^{l_i}$ where all the primes occurring in the factorization are unramified, we have

$$(\mathfrak{a}, K/k) := \prod (\mathfrak{p}_i, K/k)^{l_i}.$$

In general we need to work with those subfields of the full ray class field which correspond to quotients of the ray class groups. This is necessary since the groups in question (and therefore the fields) become very large:

```
>    R, mR := RayClassGroup(8*3*5*7*Z_k);
>    #R;
        1536
```

```
>    A := AbelianExtension(mR);
>    A;
        FldAb, defined by (<[840, 0]>, [])
        of structure: Z/2 + Z/2 + Z/2 + Z/8 + Z/24
```

```
>    K := NumberField(A);
>    K;
        Number Field with defining polynomial [ $.1^2 - 3*k.1 -
            10, $.1^2 - 3*k.1 - 24, $.1^2 - k.1 - 8, $.1^8 -
            420*$.1^6 + 57860*$.1^4 - 2788320*$.1^2 + 21715280,
            $.1^8 - 41160*$.1^6 + (-14776440*k.1 + 408615900)*
            $.1^4 + (60243779120*k.1 - 1063468399280)*$.1^2 -
            28477255077040*k.1 + 246599356889480, $.1^3
            - 147*$.1 - 637] over k
```

Although it is possible to create the field K, its large degree makes it time consuming to construct and expensive to work with.

We are going to use the existence theorem of class field theory in the following form: There is a bijection between abelian extensions K/k and pairs $(\mathrm{Cl_m}, U)$ of subgroups of ray class groups (modulo $\equiv$) mapping unramified prime ideals to their Frobenius automorphism.

Suppose we only wish to work with a subfield L of degree 4. By the above theorem, L corresponds to a subgroup U of R that has a quotient of order 4. The correspondence yields $\mathrm{Aut}(L/k) \cong R/U$. In order to define L in Magma, we have to define U and R/U and provide a map $\phi : R/U \to I$ that behaves identically to the map returned by RayClassGroup but with R/U as the group.

```
>    U := sub<R | R.1, R.2, R.3, 4*R.4, R.5>;
>    q, mq := quo<R | U>; q;

         Abelian Group isomorphic to Z/4
         Defined on 1 generator
         Relations:
             4*q.1 = 0

>    φ := Inverse(mq)*mR; φ;

         Mapping from: GrpAb: q to Set of ideals of Z_k
         Composition of Mapping from: GrpAb: q to GrpAb: R and
         Mapping from: GrpAb: R to Set of ideals of Z_k

>    A₁ := AbelianExtension(φ); A₁;

         FldAb, defined by (<[840, 0]>, [])
         of structure: Z/4

>    NumberField(A₁);

         Number Field with defining polynomial $.1^4 + (-36*k.1 -
             280)*$.1^2 + 840*k.1 + 9020 over k
```

Using this technique we can investigate all cyclic degree 4 subfields of K to find, for example, the subfields of maximal and minimal discriminant:

```
>    L := Subgroups(R : Quot := [4]); #L;

         48

>    AL := [ AbelianExtension(Inverse(mq)*mR)
>              where _, mq := quo< R | i`subgroup > : i in L ];
>    Minimum([AbsoluteDiscriminant(i) : i in AL]);

         1024000000 3

>    Maximum([AbsoluteDiscriminant(i) : i in AL]);

         2498119335936000000 47
```

The function Subgroups(R: Quot := [4]) computes a list of all subgroups U of R such that the quotient R/U is isomorphic to $\mathbb{Z}/4\mathbb{Z}$.

Suppose now that $G \leq \text{Aut}(k/\mathbb{Q})$ is a subgroup of the group of $\mathbb{Q}$-automorphisms of k such that the modulus $\mathfrak{m}$ is invariant under the action of G so that G acts on $\text{Cl}_\mathfrak{m}$. Finally, let $k_0 := \text{Fix}(G) := \{x \in k \mid \sigma(x) = x \ \forall \sigma \in G\}$. Then L/k corresponding to $(\text{Cl}_\mathfrak{m}, U)$ is normal over k_0 if and only if G acts on U (and therefore on $\text{Cl}_\mathfrak{m}/U$). In this situation the elements of G can be extended to k_0-automorphisms of L/k_0. The groups are connected by the exact sequence:

$$1 \to \text{Cl}_\mathfrak{m}/U \to \text{Aut}(L/k_0) \to G \to 1.$$

We will use this result in two different ways. Firstly, it allows us to decide normality of extensions simply by looking at the ideal groups. Secondly, by using results from extension theory of groups, we can determine invariants of the resulting Galois group $\text{Aut}(K/k_0)$ by analyzing the action of G on $\text{Cl}_\mathfrak{m}/U$.

We apply this to search for normal (over $\mathbb{Q}$) degree 8 subfields of K occurring in the list *AL* and then further restrict the selection to non-central extensions:

```
>    N := [ x : x in AL | IsNormal(x : All)];
>    nC := [ x : x in N | not IsCentral(x : All)];
>    LG := [ AutomorphismGroup(x : All) : x in nC];
>    LP := [ CosetImage(x, sub<x | >) : x in LG];
>    [ TransitiveGroupIdentification(x) : x in LP ];

        [ 4, 4, 4, 4, 5, 5, 5, 5 ]

>    TransitiveGroupDescription(8, 4);
        D_8(8)=[4]2

>    TransitiveGroupDescription(8, 5);
        Q_8(8)
```

The parameter *All* indicates that all $\mathbb{Q}$-automorphisms of the base field k will be considered. Otherwise, IsNormal would be trivially true for all class fields. It is possible to restrict to a subgroup of the full automorphism group here.

For later applications we note some formal properties of the class field K associated to the pair $(\text{Cl}_\mathfrak{m}, U)$:

1. *The primes of k that are ramified in K/k are exactly those dividing the conductor $\mathfrak{f}_{K/k}$ of $(\text{Cl}_\mathfrak{m}, U)$.*

2. *A prime $\mathfrak{p}$ is wildly ramified if and only if $\mathfrak{p}^2$ divides $\mathfrak{f}_{K/k}$.*

3. *For cyclic extensions of prime degree l we have the following formula for the relative discriminant:*

$$\mathfrak{d}_{K/k} = \mathfrak{f}_{K/k}^{l-1}$$

   In general, the discriminant can be computed from $(\text{Cl}_\mathfrak{m}, U)$ but the formula is more involved [7].

4. *An unramified prime $\mathfrak{p}$ has degree $f = f(K/k) = \text{ord}_{\text{Cl}_\mathfrak{m}/U}(\mathfrak{p}U)$. In particular $\mathfrak{p}$ has degree 1, and is therefore totally split, if and only if $\mathfrak{p} \in U$.*

3 Global function fields

Global function fields are similar to number fields in a very precise way. In practical terms this means that class field theory as outlined above applies in the same way for function fields as it does for number fields. For Magma this implies that the user interface and the presentation of fields is closely related to those used in the case of number fields. Internally, a great deal of the code is shared between the two modules. In addition to the number theoretic view, global function fields also admit a geometric point of view as plane curves. Although it is possible in Magma to take this viewpoint, our presentation emphasizes the number theoretic view. However, the geometric point is sufficiently strong to enforce some difference in the language.

3.1 Basic properties

Global function fields are finite separable extensions of $\mathbb{F}_q(x)$, where $\mathbb{F}_q$ is a finite field with $q = l^r$ elements, l a prime. A good way of viewing global function fields is to think of $\mathbb{F}_q(x)$ as the equivalent of $\mathbb{Q}$ and of $\mathbb{F}_q[x]$ as the equivalent of $\mathbb{Z}$. Setting up a function field is more complicated than setting up a number field, but once created, it behaves very much like a number field. We will define an extension of $k := \mathbb{F}_4$. The multiplicative group of k is cyclic of order 3, and we take $\omega := k.1$ as a generator for it. The function field will be $\mathbb{F}_q(x)[y]$ where $y^4 - (\omega x^2)y^2 + y - (\omega^3 x^3) = 0$:

```
>    k := GF(4);
>    kx<x> := FunctionField(k);
>    kxy<y> := PolynomialRing(kx);
>    K := FunctionField(y^4−k.1*x^2*y^2 + y − x^3*k.1^3);
```

As in the number field case, elements in K are printed as polynomials in the primitive element. Since we did not assign a name, the primitive element will be printed as $K.1$.

Although it is possible to formulate class field theory in the same way as for number fields, namely using ideals and embeddings, it is more convenient to use the geometric language of places and divisors. A *place* of a function field K is an equivalence class of valuations defining a common topology. It is possible to identify places with prime ideals of suitable subrings of K. Almost all of the places can be represented in terms of prime ideals of the integral closure of $k[x]$ in K, the *finite maximal order*. Exceptions are the places corresponding to the real and complex embeddings in the number field case: they have to be represented using prime ideals of the integral closure of $k[1/x]$ (the valuation ring of the degree valuation). The primes of the finite maximal order are called *finite places*, while the non-finite places are called *infinite places*.

From a theoretical point of view the distinction is artificial, as there is no difference between the finite and the infinite places since, by making the change of variable, $x \mapsto 1/x$, all the infinite places become finite. Thus the classification of a place as infinite or finite depends upon the representation rather than being an invariant of the field. It is important to remember that once K is created, the classification of finite and infinite places is fixed.

In Magma this also leads to the fact that there are two different rings of integers attached to each function field. Thus, **MaximalOrderFinite** represents the integral closure of $\mathbb{F}_q[x]$ in K and **MaximalOrderInfinite** represents the integral closure of the valuation ring of the degree valuation. The latter having only a finite number of prime ideals, is a semi-local Dedekind ring and therefore a principal ideal domain. The PID property of the "infinite maximal order" is not used in Magma. Since orders in function fields are subrings of their respective maximal orders, we also distinguish between finite and infinite orders.

Elements of orders are represented with respect to a fixed basis. Since $\mathbb{F}_q[x]$ considered as a polynomial ring over a field is Euclidean, finite orders always have a free basis over $\mathbb{F}_q[x]$. Similarly, since valuation rings are Euclidean, the infinite orders also have a free representation.

```
>    MF := MaximalOrderFinite(K);
>    MF;

          Maximal Equation Order of K over Univariate Polynomial
             Ring in x over GF(2^2)

>    MI := MaximalOrderInfinite(K);
>    MI;

          Maximal Equation Order of K over Valuation ring of
             Univariate rational function field over GF(2^2)
             with generator 1/x

>    Basis(MF);
          [
                1,
                K.1,
                K.1^2,
                K.1^3
          ]

>    Basis(MI);
          [
                1,
                1/x*K.1,
                1/x^2*K.1^2,
                1/x^3*K.1^3
          ]
```

As in the case of number fields, a function field can be repeatedly extended:

```
>    Kt<t> := PolynomialRing(K);
>    E := FunctionField(t^2+t+K.1);
>    E: Maximal;
```

```
E
|
| t^2 + t + K.1
|
K
|
| y^4 + k.1*x^2*y^2 + y + x^3
|
Univariate rational function field over GF(2^2)
Variables: x
```

Also, as for number fields, invariants of the field and the orders depend upon the representation, so that, for example, discriminants are ideals while norms and traces are relative to the coefficient field.

3.2 Places and divisors

As mentioned above, one of the important differences between number fields and function fields, is that since all places give rise to discrete valuations and are, from a geometric point of view, of equal importance, there are no truly distinguished infinite places. Rather than using prime ideals with the additional problem of the correct choice of the ring, we instead use places. The same idea is extended to ideals: since ideals are (in the number field case) uniquely representable as a product of prime ideals, we will replace them by divisors and use additive notation instead of multiplicative: A divisor D is a formal linear combination of places P_i

$$D = \sum n_i P_i$$

where $n_i \in \mathbb{Z}$ and all but finitely many n_i are 0.

One way of viewing divisors is as factored ideals. In fact, all divisors can be represented as a pair of ideals, one from the finite part (an ideal in the finite maximal order) and the other combining the places and prime ideals of the infinite maximal order. Formally, the correspondence is obtained by replacing each place by its prime ideal and the linear combination by a power product:

$$\sum n_i P_i \leftrightarrow (\prod \mathfrak{p}_i^{n_i}, \prod (\mathfrak{p}_j^\infty)^{n_j})$$

where the $\mathfrak{p}_i$ are the finite prime ideals, and $\mathfrak{p}_j^\infty$ denotes an infinite prime.

To create divisors we combine places of low degree or decompose functions (elements of the function field):

```
>   lp1 := Places(K, 1);
>   lp1;
        [(1/x, 1/x*K.1 + 1/x), (1/x, 1/x*K.1 + (k.1^2*x + 1)/x),
        (x, K.1), (x, K.1 + 1), (x, K.1 + k.1),
        (x, K.1 + k.1^2), (x + k.1, K.1 + x + k.1^2),
        (x + k.1^2, K.1 + x), (x + 1, K.1 + x + k.1^2)]
```

```
>    Ip₂ := Places(K, 2);
>    D := 10 * Ip₂[2] − Ip₁[1];
         10*(x^2 + x + k.1^2, K.1 + x^2 + k.1*x + 1) −
         (1/x, 1/x*K.1 + 1/x)

>    Ideal(Ip₁[3]);
         Ideal of MF
         Generators:
         x
         K.1

>    Ideal(Ip₁[1]);
         Ideal of MI
         Generators:
         1/x
         1/x*K.1 + 1/x

>    Ideals(D);
         Ideal of MF
         Basis:
         [x^20 + x^18 + k.1*x^16 + x^12 + x^10 + k.1*x^8 +
           k.1*x^4 + k.1*x^2 + k.1^2   0      0    0]
         [k.1*x^18 + k.1*x^16 + x^14 + x^12 + x^8 + k.1*x^6 +
           x^3 + 1  1    0    0]
         [x^18 + x^16 + k.1^2*x^12 + x^10 + x^6 + x^4 +
           k.1*x^2   0   1    0]
         [k.1*x^18 + k.1*x^17 + k.1*x^16 + k.1^2*x^15 +
           k.1^2*x^14 + x^11 + x^10 + k.1^2*x^9 + x^7 +
           k.1*x^6 + x^4 + k.1*x^3 + k.1*x^2 + k.1^2*x + 1
           0   0       1]
         Fractional ideal of MI
         Basis:
         [1/x   0    0    0]
         [ 0  1/x    0    0]
         [ 0    0  1/x    0]
         [ 0  k.1    0    1]
         Denominator: 1/x
```

In addition to invariants arising from the number theoretic point of view, function fields also admit geometric invariants such as the genus and invariants based on the Riemann–Roch theorem. Although geometric invariants are representation independent, the time taken to compute them depends heavily on the representation chosen. In general, the representation best suited for invariant computation and general arithmetical operations is the representation as a simple extension of a rational function field. In Magma, this representation can be computed using **RationalExtensionRepresentation**. As in the case of **AbsoluteField** on the number field side, the newly created field allows coercion to and from the original field:

```
>    Ea := RationalExtensionRepresentation(E);
>    Ea: Maximal;
        Ea
        |
        | y^8 + (k.1*x^2 + 1)*y^4 + (k.1*x^2 + 1)*y^2 +
        |     + y + x^3
        |
        Univariate rational function field over GF(2^2)
        Variables: x

>    E ! Ea.1;
        E.1

>    Ea ! (E.1+K.1);
        Ea.1^2
```

As with the number fields, the creation of the rational extension representation and the conversion between it and the relative extension can be very time consuming. In addition, the representation as a tower of fields is often more compact.

3.3 Ray divisor class groups

The map
$$K^* \to D : x \mapsto (x) := \sum v_P(x)P$$

allows us to map functions into divisors to get the subgroup of *principal divisors* of the full divisor group. The quotient group of divisors modulo principal divisors is defined as the divisor class group of K, and it may be regarded as the analogue of the ideal class group of a number field. Their computation follows the same method as outlined for the number fields [15]. A notable difference is that the GRH for function fields is known to be true, so we can use the improved bounds that it gives. In general, divisor class groups and ray divisor class groups of function fields behave similarly to their counterparts in number fields. However, there is one important difference:

```
>    C, mC := ClassGroup(K);
>    C;
        Abelian Group isomorphic to Z/185 + Z
        Defined on 2 generators
        Relations:
            185*$.1 = 0
```

The class group has a free part and is thus an infinite group. This reflects the fact that function fields have divisors of arbitrarily large degree and that principal divisors are always of degree 0. The *degree* of a divisor $D = \sum v_P(D)P$

is defined as $\deg(D) = \sum v_P(D) \deg(P)$, where the degree of a place is the degree of its residue class field O_P/P over the constant field k.

In order to introduce the ray divisor class groups, let us fix an effective divisor M, that is $M = \sum n_i P_i$ with $n_i \geq 0$ for all i. As for the number fields we define

$$x \equiv 1 \bmod^* M \iff v_P(x-1) \geq v_P(M)$$

for all $P|M$. The ray class group Cl_M is then defined as the quotient of divisors coprime to M by the principal divisors (x) generated by $x \equiv 1 \bmod^* M$. It fits into a exact sequence:

$$1 \to \mathbb{F}_q^* \to \prod (O_P/P^{v_P(M)})^* \to \mathrm{Cl}_M \to \mathrm{Cl} \to 1$$

The product $\prod(O_P/P^{v_P(M)})^*$ can be evaluated using 1-units. The function field situation is easier than the number field case, as one is able to write down an explicit basis for the local factors. A short analysis of the groups shows that

$$(O_P/P^r)^* \cong (O_P/P)^* \oplus (\mathbb{F}_q)^{n_1} \oplus (\mathbb{F}_{q^2})^{n_2} \oplus \dots$$

where the sum is finite and depends on r only [1].

In Magma, ray divisor class groups are represented in the same way as for number fields: as an abelian group and a map between the group and the group of divisors coprime to M:

```
>    M := 2*lp₁[2]+3*lp₂[3];
>    R, mR := RayClassGroup(M);
>    R;
        Abelian Group isomorphic to Z/2 + Z/2 + Z/4 + Z/4 +
            Z/11100 + Z
        Defined on 6 generators
        Relations:
            2*R.1 = 0
            2*R.2 = 0
            4*R.3 = 0
            4*R.4 = 0
            11100*R.5 = 0
```

Again there is a free part inherited from the divisor class group that we have to factor out to get finite quotients.

3.4 Class field theory

Class fields are abelian extensions of function fields. In exactly the same way as for number fields, these extensions are parametrized by the finite quotients of ray class groups and the splitting behaviour of the places is governed by their order in the class group. Since the fields are uniquely determined by the class groups, we are able to compute invariants by just analyzing the groups. In particular, we are able to compute the degree, genus, discriminant divisor, the exact constant field, and the number of places of any degree *without* computing defining equations. This makes class field techniques very powerful tools for constructing fields with many rational places. Also, since this only requires computations in the base field, we are able to compute invariants for much larger fields:

```
>    U := sub< R | lp₁[1] @@ mR >;
>    degree := #quo<R | U>; degree;
         710400
```

```
>    Genus(M, U);
         4043361
```

```
>    DiscriminantDivisor(M, U);
           1184000*(1/x, 1/x*K.1 + (k.1^2*x + 1)/x) +
             2030560*(x^2 + k.1^2*x + k.1^2, K.1 +
             x^2 + x + k.1^2)
```

```
>    DegreeOfExactConstantField(M, U);
         1
```

```
>    NumberOfPlacesOfDegreeOne(M, U);
         710400
```

Even if the field is small enough for direct computation, it is much faster to use the indirect methods:

```
>    U := sub<R | U, lp₂[4] @@ mR, lp₁[4] @@ mR, lp₂[1] @@ mR,
>                 lp₂[2] @@ mR >;
>    degree := #quo<R | U>; degree;
         4
```

```
>    time Genus(M, U);
         17
         Time: 0.290
```

```
>    NumberOfPlacesOfDegreeOne(M, U);
>    E := AbelianExtension(M, U);
>    time Genus(E);
         17
         Time: 8980.890
```

4 Applications

The applications of class field theory (CFT) presented here are based on two properties: as CFT can be used to construct large families of non-isomorphic fields it can be used for the compilation of extensive tables of number (or function) fields with a given Galois group or a given degree [8, 5]. Other approaches based on the geometry of numbers have the disadvantage of only finding a large number of candidates for defining polynomials that have to be further processed [10, 11, 18].

The other important property of CFT is that some invariants that are hard to compute from the polynomials can be easily read off the class group representation. For example, while it is difficult to construct a field with a large number of rational places, it is rather easy to compute the precise number once the parameterizing class group is found.

In what follows, we give two examples for the construction of families of fields and one of a probabilistic approach to good codes.

4.1 C_l-fields

We will demonstrate one approach to solve the following problem:

> *Given a field k, find 'all' cyclic extensions of degree l, where l is a prime.*

Formulated in this way, the problem obviously has no finite solution, so we restrict further:

> *Given a number field k, find all cyclic degree l extensions K of absolute discriminant $|\operatorname{Disc}(K/\mathbb{Q})| \leq B$ for some bound B.*

In the function field case this becomes

> *Given a function field k, find all cyclic degree l extensions K of genus $g \leq B$ for some bound B having the same exact constant field as k.*

Both problems are very similar - they differ only in minor details. So let us start with the number field case.

Class field theory gives us the following information about the relative discriminant of K/k:

$$\operatorname{Disc}(K/k) = \mathfrak{f}_{K/k}^{l-1}$$

Basic number theory tells us:

$$|\operatorname{Disc}(K/\mathbb{Q})| = |\operatorname{Disc}(k/\mathbb{Q})|^l N(\operatorname{Disc}(K/k)).$$

Putting this together, we get a bound for the conductor $\mathfrak{f}_{K/k}$:

$$N(\mathfrak{f}_{K/k}) \leq \frac{B}{|\operatorname{Disc}(k/\mathbb{Q})|^l} =: B'$$

so the task is reduced to enumerating all integral ideals of norm $\leq B'$ — which is a finite problem. For each ideal $\mathfrak{f}$ we then have to enumerate all subgroups U of $\mathrm{Cl}_{\mathfrak{f}}$ that give rise to a cyclic quotient of size l having conductor $\mathfrak{f}$ and then compute the fields. To avoid repetition in our list, for each subgroup U we will compute the conductor of $(\mathrm{Cl}_{\mathfrak{f}}, U)$ and only proceed if the conductor is $\mathfrak{f}$.

The method thus outlined will indeed find all fields K/k subject to our conditions only once, so it provides us with an algorithmic solution. However, we can easily make further improvements to the method: We need to find ideals such that the ray class group has a cyclic factor of degree l, so the ray class number has to be divisible by l. Referring back to the exact sequence used to compute $\mathrm{Cl}_{\mathfrak{m}}$, this means that we have only to consider prime ideals $\mathfrak{p}$ such that $N(\mathfrak{p}) - 1$ is divisible by l (which immediately implies $l|p-1$) or such that $l \in \mathfrak{p}$ - which will greatly reduce the number of useless candidates.

A second optimization concerns the exponents of the prime in $\mathfrak{f}$. We already know that a prime $\mathfrak{p}$ is wildly ramified if and only if $\mathfrak{p}^2|\mathfrak{f}$. In addition it is known that if $\mathfrak{p}$ is wildly ramified then $\min \mathfrak{p} \cap \mathbb{Z}_{>0} \leq l$ since only primes smaller than the degree can be wildly ramified.

So we essentially have to enumerate only *square-free* ideals $\mathfrak{f}$! One can further analyze the contribution of the wildly ramified primes and give tight bounds for the exponents. Let $\mathfrak{p}$ be a prime in k that is wildly ramified in K/k, $\mathfrak{p}\mathbb{Z}_K = \prod \mathfrak{P}_i^{e_i}$. Serre ([21]) asserts that $v_{\mathfrak{p}}(\mathfrak{d}_{K/k}) \leq l - 1 + l v_{\mathfrak{p}}(l)$ so that we get $v_{\mathfrak{p}}(\mathfrak{f}_{K/k}) \leq 1 + \frac{l v_{\mathfrak{p}}(l)}{l-1}$. This implies that the potentially wildly ramified primes are those that divide l only.

To enumerate all (square-free) ideals of norm $\leq B$ we generate all prime ideals of norm not exceeding B by decomposing all small prime numbers.

```
>    k := NumberField(x^2 - 10);
>    B := 5*10^7; l := 3;
>    Z_k := MaximalOrder(k);
>    Bs := Floor(B / AbsoluteDiscriminant(Z_k)^l); Bs;

         781

>    lp := [ x : x in [2..Bs] | IsPrime(x) and x mod l eq 1 ];
>    LP := &cat [ [ x[1] : x in Decomposition(Z_k, y)|
>                          Norm(x[1]) le Bs ] : y in lp];
```

Next we have to enumerate the square-free ideals of norm not exceeding Bs:

```
>    L := [1*Z_k];
>    for i in LP do
>        b := Floor(Bs / Norm(i));
>        L cat:= [ x*i : x in L | Norm(x) le b];
>    end for;
>    #L;

         76
```

Finally, we have to supplement those ideals by including the "exceptional" primes $\mathfrak{p}_{3,1}$ and $\mathfrak{p}_{3,2}$ with larger exponents for the wild ramification.

```
>    p31 := Decomposition(Z_k, 3)[1][1];
>    p32 := Decomposition(Z_k, 3)[2][1];
>    b3 := Floor(1 + 3*Valuation(Z_k ! 3, p31));
>    b3;
         4
>    L31 := [p31^i : i in [1..4]];
>    L32 := [p32^i : i in [1..4]];
>    L cat:= &cat [ [ x*i : x in L | Norm(x) le b ]
>              where b := Floor(Bs/Norm(i)) : i in L31 ];
>    L cat:= &cat [ [ x*i : x in L | Norm(x) le b ]
>              where b := Floor(Bs/Norm(i)) : i in L32 ];
>    #L;
         192
```

Now we can set up the class fields:

```
>    LA := [ ];
>    for i in L do
>      "Now doing: ", i;
>      R, mR := RayClassGroup(i);
>      I := Subgroups(R: Quot := [3]);
>      "with ", #I, " possible subgroups";
>      for j in I do
>        _, mq := quo<R | j`subgroup>;
>        A := AbelianExtension(Inverse(mq)* mR);
>        if Conductor(A) ne i then continue; end if;
>        Append(~LA, A);
>      end for;
>    end for;
```

```
         Now doing:  Principal Ideal of Z_k
         Generator:
             [1, 0]
         with  0  possible subgroups
         Now doing:  Principal Ideal of Z_k
         Generator:
             [7, 0]
         with  1  possible subgroups
         ...
         Now doing:  Ideal of Z_k
         Basis:
         [ 9 72]
         [ 0 81]
         with  1  possible subgroups
```

> #*LA*;

    44

So there are far fewer fields than there are candidate conductors.

If B is much larger, the simple approach will not work as the list of ideals will become prohibitively large. Also, using just a simple loop makes it impossible to restart the job or distribute the computations. It is an interesting combinatorial problem to do this efficiently.

The corresponding computation for function fields is not difficult. Here we are interested in the following question:

Given a global function field k, find all cyclic extensions K of degree l such that the genus of K is bounded by B.

In order to get similar bounds for the degree of the possible conductors, we use the Hurwitz-Genus-Formula [22]

$$2g_K - 2 = \frac{[K : k]}{[\mathrm{ECF}(K) : \mathrm{ECF}(k)]}(2g_k - 2) + \deg(\mathrm{Diff}(K/k))$$

where $\mathrm{ECF}(K)$ denotes the algebraic closure of $\mathbb{F}_q$ in K, that is the (exact) constant field of K and $\mathrm{Diff}(K/k)$ denotes the different divisor. From the general theory we know that $N(\mathrm{Diff}(K/k)) = \mathrm{Disc}(K/k) = \mathfrak{f}_{K/k}^{l-1}$ and $\deg(\mathrm{Diff}(K/k)) = \deg N(\mathrm{Diff}(K/k))$. Combining this we get

$$2g_K - 2 = l(2g_k - 2) + (l - 1)\deg f_{K/k}$$

if we require $\mathrm{ECF}(K) = \mathrm{ECF}(k) = \mathbb{F}_q$. This now implies

$$\deg f_{K/k} \leq \frac{2B - 2 - l(2g_k - 2)}{l - 1}$$

The remainder is completely analogous to the number field case. We start by collecting all places of "small" degree of $\mathbb{F}_q(x)$, split them in k to get all "small" places in k, then compile lists of divisors of small degree in k followed by the conductor computation and the definition of the abelian extensions.

4.2 S_3-fields

Solvable extensions can, in principle, be obtained by computing towers of abelian extensions. Although tempting from a theoretical point of view, this is in general not very efficient as we will see in the sequel. However, for small solvable groups this is feasible.

We will investigate the following problem:

Given a field k find all cubic extensions K/k such that $\mathrm{Gal}(K/k) \cong S_3$, in particular, we want to construct non-normal fields.

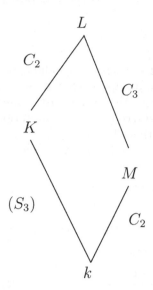

Fig. 1. S_3 structure

We summarize properties of S_3 fields K/k, with the assistance of figure 1. The normal closure of the field K/k will be denoted by L, so L/k is a sextic field with a Galois group isomorphic to S_3 and L/K is a quadratic extension. There is a quadratic field M/k such that L/M is a cyclic cubic field. So one way of constructing K is to find "all" quadratic extensions M of k (which we can do easily using the approach of the previous section), then find all C_3 extensions L of those (again, using the previous section) and finally for all L that are constructed this way and that are normal and non-abelian, find K as a subfield. Although this would work, it is not very efficient. The great majority of sextic fields constructed this way will not be normal.

To get a more efficient version we analyze the various relative discriminants more carefully and thus get restrictions on the possible conductors. The starting point for this situation is the theorem of Klüners ([8]):

$$\mathrm{Disc}(K/k) = \mathrm{Disc}(M/k) N_{M/k}(\mathfrak{f}_{L/M}).$$

In fact, one can do even better ([7]) and show that $\mathfrak{f}_{L/M} = \mathfrak{b}\mathbb{Z}_M$ for some ideal $\mathfrak{b}$ of $\mathbb{Z}_k$. Thus we get $\mathrm{Disc}(K/k) = \mathfrak{a}\mathfrak{b}^2$, where $\mathfrak{a}$ is a conductor of a quadratic extension M/k and $\mathfrak{b}$ is a cubic conductor.

Combining the above result with the algorithm from the previous section gives us strong restrictions on the ideals: both have to be square-free (outside $2, 3$), their norm is bounded, and, since the ray class group of $\mathfrak{b}$ has to admit a C_3 quotient, only primes $\mathfrak{p}$ with $N_{L/\mathbb{Q}}(\mathfrak{p}\mathbb{Z}_L) \equiv 1 \bmod 3$ need to be considered.

A second criterion used to reduce the search comes from the Galois group. Suppose $\mathfrak{m}$ is a module in an normal extension L/k, $U < \mathrm{Cl}_\mathfrak{m}$ is a subgroup and M/k corresponds to $(\mathrm{Cl}_\mathfrak{m}, U)$. Then

- M/k is normal if and only if $\sigma(U) = U$ for all $\sigma \in \mathrm{Aut}(L/k)$.
- M/k is central if and only if $\sigma(\mathfrak{a})U = \mathfrak{a}U$ for all $\sigma \in \mathrm{Aut}(L/k)$.

So an abelian extension of a normal field is normal if the corresponding group is fixed as a subset and central if it is point-wise fixed. Since in the case of extensions M/k of cyclic fields, central extensions and abelian extension coincide, we have a fast check for L/k being normal and non-abelian which implies $\mathrm{Gal}(L/k) = S_3$.

We use a nested loop: an outer loop to find all quadratic extensions M of k followed by an inner loop to find the cyclic cubic extensions L/M such that L/k is normal but non-abelian.

Assuming that the procedure of the last section is implemented as the function CyclicExtensions with arguments k, l and B we can summarize the method in the following program:

```
B := 10^6;
S_3 := [ ];
LC := CyclicExtensions(k, 2, B);
for i in LC do
  L := NumberField(i);
  σ := Automorphisms(L)[2];
  La := AbsoluteField(L);
  MC := CyclicExtensions(La, 3, B);
  for j in MC do
    if Conductor(j) ne σ(Conductor(j)) then
      continue;
    end if;
    c := j`DefiningGroup;
    a := c`m_0;
    U := Kernel(c`GrpMap);
    S := InducedAutomorphism(c`RcgMap, σ, Minimum(a));
    if U ne S(U) then continue; end if;
    q, mq := quo< Domain(c`RcgMap) | U >;
    if q.1 ne S(q.1) then continue; end if;
    Append(∼S_3, j);
  end for;
end for;
```

Again, since the basis of the above procedure is purely class field theoretic, it applies equally to function fields if we use the Hurwitz-Genus-Formula to relate the conductors and discriminants to the genera of the fields in question.

4.3 Codes

A very important application of function fields is the construction of error-correcting codes. Error-correcting codes are used to transmit data over unreliable communication lines. They are used to encode the data in such a way that a certain number of transmission errors can be detected and corrected on the receiving end. Obviously this means that there is some redundancy in the data transmitted. A widely used application for such a code is the storage of information on CDs where in spite of scratches or even holes, the player can correct the data read from the surface.

Mathematically, codes are frequently represented as subspaces of finite dimensional vector spaces over a finite field. The error-correction capabilities depend on "how close" different elements of the subspace can be, so a good code distributes the elements of the subspace ("code words") so they are maximally separated in the ambient space. The quality of a code is thus a function of the dimension of the ambient space, the minimal distance between code words, and the size of the finite field in question.

One of the most fruitful ways to define codes is through the use of function fields (or equivalently plane curves) over finite fields. Goppa [13] suggested the following idea:

Fix a global function field k, a set S of rational places and a divisor D with support coprime to S. By evaluating the elements of the Riemann–Roch space $\mathcal{L}(D)$ of D at the elements of S, we define a subspace of $\mathbb{F}_q^n$ and thus a linear code.

Goppa was able to relate the parameters of the code to the geometric invariants of the field and the dimension of D. So to find a good code, i.e., a code that can correct many errors in comparison to its length, we have to find a function field

- with many rational places (as the length of the code depends on that)
- of small genus, as the genus links the dimensions of D, the length of the code and the minimal distance.

Class field theory, as outlined above, allows us to construct a function field as an abelian extension of a fixed base field in terms of the base field where we can predict the invariants. Not surprisingly, several of the best known codes are constructed using class field theoretic methods, with the asymptotically best codes being obtained as towers of class fields.

Here we will demonstrate how to get reasonably good "small" codes using a probabilistic method. The idea is straightforward. We choose a small base field k with a reasonable number of rational places. Next we choose some small degree non-rational places to define a ray class group (and thus a large extension) and consider subgroups generated by some (or all) of the rational places thus guaranteeing that those places will split completely in the extension. Then we compute the exact number of rational places and the genus and see if the field is suitable.

```
>   F<w> := GF(4);
>   Ft<t> := FunctionField(F);
>   Fty<y> := PolynomialRing(Ft);
>   k := FunctionField(y²+y+t³);
```

The field $k = \mathbb{F}_4(x)[t]/(y^2 + ty + t^3)$ will act as our base field. The divisor D will be constructed randomly out of places of degree 1, 3 and 4.

```
>   lp₁ := Places(k, 1);
>   lp₃ := Places(k, 3) cat Places(k, 4);
>   opt := 0;
>   while opt lt 3 do
>       l₁ := SequenceToSet(lp₁);
>       D := 0*lp₁[1];
>       for j in [1..Random(#lp₁ div 2)] do
>           l := Random(l₁);
>           D +:=l;
>           Exclude(~l₁, l);
>       end for;
>       D +:= &+ [ Random(lp₃) : j in [1..9−#lp₁+#l₁] ];
```

Now that we have a "random" divisor, we need to define the corresponding ray class group. As the group is likely to be too large, we also need to find a "random" subgroup. Here we search for a subgroup U such that the quotient of the ray class group by it has at most 10 elements. If the group is too large, we simply ignore it.

```
>       R, mR := RayClassGroup(D);
>       U := sub<R | [ x@@mR : x in l₁ ]>;
>       q := quo<R | U>;
>       if #q eq 1 then continue; end if;
>       while #q gt 10 do
>           U := sub< R | U, Random(R) >;
>           q := quo< R | U >;
>       end while;
>       if #q eq 1 then continue; end if;
```

Next we compute the invariants of the class field corresponding to R/U. To find our target field, we compute the genus g and the number of rational places n. We accept a field if the ratio n/g improves the previously found largest ratio. We terminate the search if the ratio is at least 3. The last number printed in each line is the degree of the extension, that is the size of R/U.

```
>       g := Genus(D, U);
>       n := NumberOfPlacesOfDegreeOne(D, U);
>       g, n, n/g*1.0, #q;
>       if n/g gt opt then
>               opt := n/g; D_opt := D; U_opt := U;
>       end if;
```

```
>   end while;
        24  27  1.1250000000000000000000000000000  3
        33  25  0.7575757575757575757575757575756  3
        31  27  0.8709677419354838709677419354  3
        67  51  0.7611940298507462686567164179  9
        39  48  1.2307692307692307692307692307538  6
        67  54  0.8059701492537313432835820895  9
        21  27  1.2857142857142857142857142857108  3
        5  16  3.1999999999999999999999999999959  2
```

Now we have the parameters for our best field in D_{opt} and U_{opt}. Next we create the field as an actual function field and find all rational places:

```
>   K := AbelianExtension(D_opt, U_opt);
>   Kr := RationalExtensionRepresentation(K);
>   L := Places(Kr, 1);
>   #L;
        16

>   Genus(Kr);
        5
```

To define a code from this function field, we also need a divisor that is coprime to the rational places. Here again, we do a random search and record the minimal distances. All codes computed here will be of length 16. For each dimension we record the largest minimal distance found.

```
>   L_3 := Places(Kr, 2) cat Places(Kr, 3);
>   D := &+L;
>   d := [ 0 : i in [1..#L] ];
>   for i in [1..100] do
>     G := &+[ Parent(D) | Random(-1,2)*Random(L_3) :
>         j in [1..Random(7)] ];
>     dd := Dimension(G) - Dimension(G-D);
>     if dd eq 0 then continue; end if;
>     C := AGCode(L, G);
>     m := MinimumDistance(C);
>     if d[17-dd] lt m then
>       d[17-dd] := m;
>     end if;
>   end for;
>   d;
        [ 16, 0, 11, 9, 0, 8, 6, 5, 5, 0, 4, 3, 0, 2, 2, 2 ]
```

Comparing this with the codes stored in the database of linear codes [4, 115.13], we see that we "rediscovered" some of the best known codes!

```
>   be := [ ];
```

```
>    for d in [0..15] do
>       c, f := BestKnownLinearCode(F, 16, d);
>       be[d+1] := MinimumDistance(c);
>    end for;
>    be; d;

        [ 16, 16, 12, 12, 11, 9, 8, 8, 7, 6, 5, 4, 4, 3, 2, 2 ]
        [ 16,  0, 11,  9,  0, 8, 6, 5, 5, 0, 4, 3, 0, 2, 2, 2 ]
```

References

1. Roland Auer, *Ray class fields of global function fields with many rational places*, PhD thesis, Oldenburg: Carl von Ossietzky Universität Oldenburg, Fachbereich Mathematik, 1999.
2. Eric Bach, *Explicit bounds for primality testing and related problems*, Math. Comput. **55**-191 (1990), 355–380.
3. Wieb Bosma, John Cannon, Catherine Playoust, *The Magma algebra system I: The user language*, J. Symbolic Comput. **24** (1997), 235–265.
   See also the Magma home page at http://magma.maths.usyd.edu.au/magma/.
4. John Cannon, Wieb Bosma (eds), *Handbook of Magma Functions, V2.11*, Sydney, 2004. The Magma Group http://magma.maths.usyd.edu.au/.
5. Henri Cohen, Francisco Diaz y Diaz, Michel Olivier, *Construction of tables of quartic number fields*, pp. 257–268 in: Wieb Bosma (ed.), *Algorithmic Number Theory*, Proceedings 4th international symposium (ANTS-IV), Leiden, the Netherlands, July 2-7, 2000, Lect. Notes Comput. Sci. **1838**, Berlin: Springer, 2000.
6. Henri Cohen, *A Course in Computational Algebraic Number Theory*, Graduate Texts in Mathematics **138**, Berlin-Heidelberg-New York: Springer, 1993.
7. Henri Cohen, *Advanced Topics in Computational Number Theory*, Graduate Texts in Mathematics **193**, Berlin-Heidelberg-New York: Springer, 2000.
8. Claus Fieker, Jürgen Klüners, *Minimal discriminants for fields with small Frobenius groups as Galois groups*, J. Number Theory, **99**-2 (2003), 318–337.
9. David Ford, Pascal Letard, *Implementing the Round Four maximal order algorithm*, J. Theo. Nombres Bordeaux, **6**-1 (1994), 39–80.
10. David Ford, Michael E. Pohst, *The totally real A₆ extension of degree 6 with minimum discriminant*, Exp. Math. **2**-3 (1993), 231–232.
11. David Ford, Michael E. Pohst, Mario Daberkow, Nasser Haddad, *The S₅ extensions of degree 6 with minimum discriminant*, Exp. Math. **7**-2 (1998), 121–124.
12. Carsten Friedrichs, *Berechnung von Maximalordnungen über Dedekindringen*, Dissertation, Technische Universität Berlin, 2000.
    http://www.math.tu-berlin.de/~kant/publications/diss/diss_cf.ps.gz.
13. V. D. Goppa, *Codes on algebraic curves*, Dokl. Akad. Nauk SSSR **259**(6) 1981, 1289–1290.
14. Florian Heß, *Zur Klassengruppenberechnung in algebraischen Zahlkörpern*, Diplomarbeit, Technische Universität Berlin, 1996.
    http://www.math.tu-berlin.de/~kant/publications/diplom/hess.ps.gz
15. Florian Heß, *Zur Divisorenklassengruppenberechnung in globalen Funktionenkörpern*, Dissertation, Technische Universität Berlin, 1999.
    http://www.math.tu-berlin.de/~kant/publications/diss/diss_FH.ps.gz

16. Florian Heß, Sebastian Pauli, Michael E. Pohst, *Computing the multiplicative group of residue class rings*, Math. Comp. **72**-243 (2003), 1531–1548 (electronic).
17. Serge Lang, *Algebraic Number Theory*, Graduate Texts in Mathematics **110** (2nd edition), New York: Springer, 1994.
18. Michel Olivier, *Corps sextiques primitifs. (Primitive sextic fields)*. Ann. Inst. Fourier, **40**-4 (1990), 757–767.
19. Sebastian Pauli, *Efficient Enumeration of Extensions of Local Fields with Bounded Discriminant*, PhD thesis, Concordia University, 2001.
    http://www.math.tu-berlin.de/~pauli/papers/phd.ps.gz
20. Michael E. Pohst, Hans Zassenhaus, *Algorithmic Algebraic Number Theory*, Encyclopaedia of Mathematics and its Applications, Cambridge: Cambridge University Press, 1989.
21. J.-P. Serre, *Local Fields*, New York: Springer, 1995.
22. Henning Stichtenoth, *Algebraic Function Fields and Codes*, Berlin: Springer-Verlag, 1993.

Some ternary Diophantine equations of signature $(n, n, 2)$

Nils Bruin

Department of Mathematics
Simon Fraser University
Burnaby BC, Canada
bruin@member.ams.org

Summary. In this article, we will determine the primitive integral solutions x, y, z to equations of the form

$$x^n + y^n = Dz^2$$

with $n = 4, 5, 6, 7, 9, 11, 13, 17$ and $D \in \{2, 3, 5, 6, 10, 11, 13, 17\}$.

These equations form the small exponent cases of the equations considered by Bennett and Skinner in [1], where their modular techniques do not apply.

The computations necessary form a nice showcase of the arithmetic geometric functionality in the Magma computer algebra system. We will show how to construct curves, how to test curves for local solvability, how to analyse elliptic curves over number fields and how to use Chabauty-techniques to determine the rational points on a curve.

1 Introduction

The following result is stated in the paper [1] by Bennett and Skinner.

Theorem 1.1. *If $n \geq 4$ is an integer and*

$$D \in \{2, 3, 5, 6, 10, 11, 13, 17\}$$

then the equation

$$x^n + y^n = Dz^2$$

has no solutions in nonzero coprime integers (x, y, z) with, say, $x > y$, unless $(n, D) = (4, 17)$ or $(n, D, x, y, z) \in \{(5, 2, 3, -1, \pm 11), (5, 11, 3, 2, \pm 5)\}$.

In that paper the authors use techniques based on Galois representations on torsion subgroups of elliptic curves and modular forms to prove a large part

---
[1] The research described in this paper is partly funded by NSERC and the University of Sydney.

of this theorem, but these methods do not apply to all combinations of n, D that occur in the statement and for each $n \leq 17$, there are some values of D for which they refer to this paper rather than give a proof. Indeed, here we we will prove the following result.

Theorem 1.2. *If $n \in \{5, 6, \ldots, 17\}$ and $D \in \{2, 3, 5, 6, 10, 11, 13, 17\}$ then the equation*

$$x^n + y^n = Dz^2$$

has no solutions in coprime nonzero integers except those arising from the identities $1^n + 1^n = 2 \cdot 1^2$, $3^5 - 1^5 = 2 \cdot 11^2$ and $3^5 + 2^5 = 11 \cdot 5^2$.

We will also prove

Proposition 1.3. *The equation $x^4 + y^4 = Dz^2$ has no integral solutions for $D \in \{3, 5, 6, 10, 11, 13\}$. For $D = 2$, the only integral solutions with $\gcd(x, y, z) = 1$ are $(x, y, z) = (\pm 1, \pm 1, \pm 1)$. For $D = 17$, it has infinitely many integral solutions with distinct values of x/y.*

We will use the proofs to introduce the reader to some of the very powerful tools that Magma [2] offers for solving arithmetic geometric questions. The article is laid out in the following way.

As an introduction we give an easy proof to Proposition 1.3. It shows the basic mechanisms that are available in Magma to define arithmetic geometric objects and answer questions about them. We try to point out that many questions can be formulated and answered using Magma in a language that is very close to the one that mathematicians are used to.

Next, we review some mathematical concepts and constructions that will prove indispensable in the rest of the paper. We recall a theorem from [7] that translates questions like the one in Theorem 1.2 to questions about rational points on some algebraic curves.

In Section 4, we apply those results to $x^5 + y^5 = Dz^2$ and, using Magma, obtain some curves that parametrise the primitive solutions to the equation under consideration. We then construct elliptic subcovers of those curves such that an application to them of the methods from [7] yields the rational points on the original curves. We defer the actual application to Section 7.

We trust that after this demonstration of the problem solving capability of Magma, the reader will be interested in knowing some of the algorithms employed. In Section 5 we give a full account of the algorithms the author has implemented in Magma to test schemes for local solvability. A highlight is an algorithm that decides local solvability of hyperelliptic curves in time that is essentially independent of the size of the residue class field in the odd residue characteristic case.

In Section 6, we explain how 2-Selmer groups and 2-isogeny Selmer groups of elliptic curves over number fields can be computed, how they can be used to bound the free ranks of Mordell–Weil groups and how they can be used to find generators for Mordell–Weil groups. We also explain how one can do this in Magma using the implementation by the author, based on [4].

In Section 7, we explain how the techniques first introduced in [3] can be applied to use elliptic curves over number fields to find the rational points on curves. We outline how one can use the implementation by the author to prove the result in Theorem 1.2 for $n = 5$.

In the last section, we give an outline of successful strategies to solve the remaining cases from Theorem 1.2. For full details and a transcript of the Magma session that obtains all computational results in this paper, we refer the reader to the electronic resource [8].

2 Proof of Proposition 1.3

To get a taste for things to come, we first prove Proposition 1.3. It is very straightforward. First note that any solution to $x^4 + y^4 = Dz^2$ corresponds to a rational point $(u, v) = (x/y, Dz/y)$ on the curve

$$D(u^4 + 1) = v^2.$$

We can simply ask* Magma to compute for each desired value of D, whether this curve has any points over, say, $\mathbb{Q}_2$ (see Section 5.4).

```
>    _<x> := PolynomialRing(Rationals());
>    Dset := {2, 3, 5, 6, 10, 11, 13, 17};
>    {D : D in Dset | IsLocallySolvable(HyperellipticCurve(D*(x^4+1)), 2)};
        { 2, 17 }
```

So just by testing local solvability at 2, we have already proved the lemma for all values of D except 2 and 17. Let's first consider $D = 2$. Clearly, the curve $2u^4 + 2 = v^2$ has a rational point $(u, v) = (1, 2)$, so it is isomorphic to an elliptic curve E. The rational points of an elliptic curve form a finitely generated group. Magma can compute an upper bound on the free rank of that group (see Section 6) and as it turns out, it is 0.

```
>    C2 := HyperellipticCurve(2*(x^4+1));
>    p0 := C2 ! [1,2];
>    E, C2toE := EllipticCurve(C2, p0);
>    RankBound(E);
        0

>    #TorsionSubgroup(E);
        4
```

---

*See the Preface to this volume for style conventions regarding Magma code; code appearing in this book is available at http://magma.maths.usyd.edu.au/magma/.

We find an upper bound of 0 on the free rank, so $E(\mathbb{Q})$ consists entirely of torsion points, of which there are 4. Indeed, there are 4 obvious points:

$$(u, v) = (1, 2), (-1, 2), (1, -2), (-1, -2).$$

and these all correspond to solutions with $x = \pm y$.

For $D = 17$ we proceed similarly. We find the point $(u, v) = (2, 17)$ on the curve $17u^4 + 17 = v^2$. This time we find an upper bound of 2 on the free rank.

```
>    C17 := HyperellipticCurve(17*(x^4+1));
>    p0 := C17 ! [2, 17];
>    E, C17toE := EllipticCurve(C17, p0);
>    RankBound(E);
        2
```

In fact, Magma can find two independent points on E (note that the command MordellWeilGroup in principle could return a group of smaller rank, so one should always check that the rank of the returned group corresponds to the expected rank).

```
>    G, GtoE := MordellWeilGroup(E);
>    G;
        Abelian Group isomorphic to Z/2 + Z/2 + Z + Z
        Defined on 4 generators
        Relations:
            2*G.1 = 0
            2*G.2 = 0

>    [ Inverse(C17toE)(GtoE(g)) : g in OrderedGenerators(G) ];
        [ (-1 : 17 : 2), (-2 : -17 : 1), (13 : 697 : 2),
        (314 : 3097553 : 863) ]
```

The last solution corresponds to the primitive solution:

$$(2 \cdot 157)^4 + 863^4 = 17 \cdot (182209)^2$$

and, using the group law on E, arbitrarily many can be constructed.

3 Construction of parametrising curves

In this section, we recall a result from [15] in an explicit form, occurring in [3], which relates integer solutions (x, y, z) of equations like $x^n + y^n = Dz^m$ with $\gcd(x, y, z) = 1$ to rational points on some algebraic curves.

First we need some notation.

Let $f(x, y) \in \mathbb{Z}[x, y]$ be a square-free homogeneous form of degree n and assume for simplicity that $f(x, 1)$ is monic of degree n, i.e, assume that

$f(x, y) = x^n + y(\cdots)$. We construct the algebra $A = \mathbb{Q}[\theta] = \mathbb{Q}[x]/f(x, 1)$. This allows us to express f as a *norm form*

$$f(x, y) = \mathrm{N}_{A[x,y]/\mathbb{Q}[x,y]}(x - \theta y).$$

Let S be a finite set of rational primes and let K be a number field. For a prime $\mathfrak{p}$ of K we write $\mathfrak{p} \nmid S$ if $\mathfrak{p}$ does not extend any prime in S to K. Following [22], we define

$$K(p, S) := \{a \in K^* : v_{\mathfrak{p}}(a) \equiv 0 \pmod{p} \text{ for all primes } \mathfrak{p} \nmid S\}/K^{*p}.$$

Following Magma's terminology, we refer to this set as the (p, S)-*Selmer group* of K. It is a finite, effectively computable group. An algorithm for computing it is described in [20] and the implementation and optimizations in Magma are due to Fieker [16].

Since f is square-free, the algebra A is isomorphic to a direct product of number fields $K_1 \times \cdots \times K_r$. We generalise the notation above to

$$A(p, S) := K_1(p, S) \times \cdots \times K_r(p, S) \subset A^*/A^{*p}.$$

Furthermore, we will identify elements of $A(p, S)$ with some set of representatives in A^*.

Since $\{1, \theta, \ldots, \theta^{n-1}\}$ forms a $\mathbb{Q}[Z_0, \ldots, Z_{n-1}]$-basis of the vector space $A[Z_0, \ldots, Z_{n-1}]$, for any $\delta \in A^*$ there are unique homogeneous forms $Q_i = Q_{\delta,i} \in \mathbb{Q}[Z_0, \ldots, Z_{n-1}]$ of degree m such that

$$Q_0 + Q_1\theta + \cdots + Q_{n-1}\theta^{n-1} = \delta(Z_0 + Z_1\theta + \cdots + Z_{n-1}\theta^{n-1})^m.$$

We define the projective curve

$$C_\delta := \{Q_2 = Q_3 = \cdots = Q_{n-1} = 0\} \subset \mathbb{P}^{n-1}$$

and the map $\phi_\delta : C_\delta \to \mathbb{P}^1$ defined by

$$\phi_\delta : (Z_0 : \cdots : Z_{d-1}) \mapsto -\frac{Q_0(Z_0, \ldots, Z_{n-1})}{Q_1(Z_0, \ldots, Z_{n-1})}.$$

From [3, Theorem 3.1.1], it follows that C_δ is absolutely irreducible of genus $1 + m^{n-2}(\frac{1}{2}n(m - 1) - m)$ and that ϕ_δ is a Galois cover with Galois-group $(\mathbb{Z}/m\mathbb{Z})^{n-1}$, ramified exactly at $\{(x : y) \in \mathbb{P}^1(\overline{\mathbb{Q}}) : f(x, y) = 0\}$. For given nonzero integer D and $m \geq 1$, we consider the equation

$$f(x, y) = Dz^m.$$

Let S be a finite set of primes containing the prime divisors of $D \, \mathrm{Disc}(f)$.

$$\Delta := \{\delta \in A(m, S) : \mathrm{N}_{A/\mathbb{Q}}(\delta)/D \in \mathbb{Q}^{*m}\}.$$

A consequence of [3, Lemma 3.1.2] is

Theorem 3.1. *Let $f, D, m, \theta, A, \Delta$ be defined as above. Then*

$$\left\{(x : y) : x, y, z \in \mathbb{Z}, f(x, y) = Dz^m, \gcd(x, y, z) = 1\right\} \subset \bigcup_{\delta \in \Delta} \phi_\delta(C_\delta(\mathbb{Q}))\}.$$

One can easily recover (x, y, z) from $(x : y)$ in the following way.

Let $(x_0 : y_0) \in \mathbb{P}^1(\mathbb{Q})$. If (x, y, z) is a solution with $(x : y) = (x_0 : y_0)$, then there is a $\lambda \in \mathbb{Q}^*$ such that

$$x = \lambda x_0,$$

$$y = \lambda y_0,$$

$$z = \sqrt[m]{\frac{\lambda^n f(x_0, y_0)}{D}}.$$

Given x_0, y_0, it is straightforward to determine for which values of λ the above system has solutions with $\gcd(x, y, z) = 1$ and what those solutions are.

The strategy for proving Theorem 1.2 is to determine the relevant curve C_δ for each of the combinations (n, D) and to find the rational points on C_δ. In some cases we get away with only constructing a subcover of $C_\delta/\mathbb{P}^1$.

In the particular, when m divides n, then the weighted projective equivalence classes of solutions to $f(x, y) = Dz^m$ are in bijection with the rational points of the curve given by the weighted projective model $C' : f(x, y) = Dz^m$, where (x, y, z) have weights $(1, 1, n/m)$. In the special case that $m = 2$, we see that C' is a double cover of a projective line. The curves C_δ are twists of the unramified cover of C' obtained by embedding C' in its Jacobian $\mathrm{Jac}(C')$ and taking the pullback along the multiplication-by-2 map on $\mathrm{Jac}(C')$. These properties are explained and exploited in [9] and in a trivial way in Section 8.

In the particular case that $m = 2$ and $n = 4$, we recover the multiplication-by-2 covers of curves of genus 1 that play a role in 2-descents and 4-descents. If f has a rational root, then C' is isomorphic to its Jacobian and the C_δ are the homogeneous spaces that play a role in 2-descents as described in Section 6.

If f does not have a rational root, then C' can still be expressed as a $\mathbb{Z}/2\mathbb{Z} \times \mathbb{Z}/2\mathbb{Z}$ cover of its Jacobian. The C_δ are then homogeneous spaces associated to a 4-descent on the Jacobian. See [19] and [26].

4 The equation $x^5 + y^5 = Dz^2$

We begin putting the construction from Section 3 into Magma. In our case $f = x^5 + y^5$. We model this by defining a univariate polynomial in Magma. This allows us to construct the algebra straight away.

```
>    _<x> := PolynomialRing(Rationals());
>    f := x^5+1;
>    A<θ> := quo< Parent(x) | f >;
```

Next we construct the rings $A[Z_0, \ldots, Z_4]$, $\mathbb{Q}[Z_0, \ldots, Z_4]$ and the corresponding projective space. Note that, while mathematically

$$\mathbb{Q}[\theta][Z_0, \ldots, Z_4] \simeq \mathbb{Q}[Z_0, \ldots, Z_4][\theta]$$

in a canonical way, this is not the case in a computer algebra system. We first construct the left hand side (*PA*) and then obtain the right hand side (*AP*) using SwapExtension, available from [8]. We also get the appropriate isomorphism *swap*. We then extract $\mathbb{Q}[Z_0, \ldots, Z_4]$ as the base ring of *AP*.

```
>    PA<Z0A,Z1A,Z2A,Z3A,Z4A> := PolynomialRing(A, 5);
>    AP<thetaP>, swap := SwapExtension(PA);
>    _<Z0,Z1,Z2,Z3,Z4> := BaseRing(AP);
>    P4 := Proj(BaseRing(AP));
>    P1 := ProjectiveSpace(Rationals(), 1);
```

Given $\delta \in A$, it is now straightforward to construct C_δ and ϕ_δ. We present a Magma routine that takes δ as input and returns both C_δ and ϕ_δ. Note that Coefficients on an element of *AP* returns a sequence of coefficients with respect to the power basis $[1, \theta, \ldots, \theta^4]$, i.e., the Q_i for us.

```
    function Cdelta(δ)
        g := δ*(&+[PA.i*θ^(i-1): i in [1..5]])^2;
        Q := Coefficients(swap(g));
        Crv := Scheme(P4, [Q[3], Q[4], Q[5]]);
        φ := map< Crv → P1 | [Q[1], -Q[2]] >;
        return Crv, φ;
    end function;
```

Given D, we can compute the set Δ as well. For that, we need to represent $\mathbb{Q}$ as a number field Q and A as an algebra over Q. We also compute the decomposition of $A = \mathbb{Q} \times K$, where $K = \mathbb{Q}(\zeta)$, the field generated by a primitive fifth root of unity. By giving an explicit representation to AbsoluteAlgebra, we make sure the system uses that representation.

```
>    Q := NumberField(x-1: DoLinearExtension);
>    OQ := IntegerRing(Q);
>    Qx<xQ> := PolynomialRing(Q);
>    AQ := quo< Qx | Polynomial(Q, f) >;
>    AQtoA := hom< AQ → A | [θ] >;
>    K<ζ> := NumberField(x^4 - x^3 + x^2 - x + 1);
>    OK := IntegerRing(K);
>    Aa, toAa := AbsoluteAlgebra(AQ: Fields := {Q,K});
```

We can now compute Δ in the following way. We take S to be the set of primes that divide $D \operatorname{Disc}(f)$. We compute the subgroup of $A(2, S)$ that has square norm and we translate it over the class of D in $A(2, S)$.

```
    function DeltaForD(D)
        S := Support(D*Discriminant(f)*OQ);
```

```
slmA, slmAmap := pSelmerGroup(AQ, 2, S);
slmQ, slmQmap := pSelmerGroup(2, S);
slmNorm := map< slmA → slmQ | a ↦
                slmQmap(Norm(a@@slmAmap))>;
slmSquareNorm := Kernel( hom< slmA → slmQ |
                [ slmNorm(a): a in OrderedGenerators(slmA) ] >);
classD := slmAmap(D);
return {AQtoA( (d−classD)@@slmAmap):d in slmSquareNorm};
end function;
```

We gather all δs that are relevant together (remember we can always recover the corresponding D from $N_{A/\mathbb{Q}}(\delta)$) and throw out any for which C_δ is not locally solvable at 2, 5 or 11 (other primes turn out to make no further contributions).

```
>    Dset := {2, 3, 5, 6, 10, 11, 13, 17};
>    BigDelta := &join{ DeltaForD(D) : D in Dset };
>    Δ := BigDelta;
>    Δ := { δ : δ in Δ | IsLocallySolvable( Cdelta(δ), 2:
>                        AssumeIrreducible, AssumeNonsingular) };
>    Δ := { δ : δ in Δ | IsLocallySolvable( Cdelta(δ), 5:
>                        AssumeIrreducible, AssumeNonsingular) };
>    Δ := { δ : δ in Δ | IsLocallySolvable( Cdelta(δ), 11:
>                        AssumeIrreducible, AssumeNonsingular) };
```

This eliminates already 4/5th of the parametrising curves. Now we construct a subcover, derived from the ring homomorphism $m_1 : A \to \mathbb{Q}$ given by $\theta \mapsto -1$. From $x - \theta y = \delta c_0^2$, it follows that $f(x, y) = N_{A/\mathbb{Q}}(\delta) N_{A/\mathbb{Q}}(c_0)^2$. Hence,

$$x^4 - x^3 y + x^2 y^2 - xy^3 + y^4 = f(x, y)/m_1(x - \theta y) = N(\delta)/m_1(\delta) N(c_0)^2.$$

Putting $d = N(\delta)/m_1(\delta)$, it follows that C_δ covers

$$E_d : u^4 - u^3 + u^2 - u + 1 = dv^2.$$

We compute which curves occur.

```
>    m₁ := hom< A → Rationals() | −1 >;
>    { PowerFreePart( Norm(δ)/m₁(δ), 2 ) : δ in Δ };
        { 1, 5, 55 }
```

Unfortunately, E_d has infinitely many rational points for $d = 1, 55$. For $d = 5$ we do get some useful information.

```
>    E₅ := HyperellipticCurve(5*(x⁴−x³+x²−x+1));
>    p₀ := E₅ ! [−1, 5];
>    ell := EllipticCurve(E₅, p₀);
>    RankBound(ell);
        0
```

```
>    #TorsionSubgroup(ell);

     2
```

We see that E_5 only has two rational points $(-1, \pm 5)$. It is straightforward to check that these correspond to the obvious solutions $(x, y, z) = (1, -1, 0), (-1, 1, 0)$ for $x^5 + y^5 = Dz^2$. If we take heed of these solutions, we can discard any δ for which C_δ covers E_5.

```
>    Δ := { δ: δ in Δ | PowerFreePart( Norm(δ)/m₁(δ), 2) ne 5 };
>    #Δ;

     16
```

For these remaining values, we use the same idea as above, but now we use the map $m_2 : A \to K$ given by $\theta \mapsto \zeta$. We define $d = \mathrm{N}(\delta)/m_2(\delta)$ and we obtain the subcover of $C_\delta/\mathbb{P}^1$ defined by

$$E_d : u^4 + \zeta u^3 + \zeta^2 u^2 + \zeta^3 u + \zeta^4 = dv^2$$

where the cover $\phi_\delta : C_\delta \to \mathbb{P}^1$ induces the cover $u : F_d \to \mathbb{P}^1$. Since the value of d only matters up to squares, we take unique representatives by mapping through $K(2, S')$ for an appropriate S'.

```
>    m₂ := hom< A → K | ζ >;
>    slmK, slmKmap := pSelmerGroup(2, Support(2*3*5*11*13*17*OK));
>    dset := { K | (slmKmap(Norm(δ)/m₂(δ)))@@slmKmap : δ in Δ };
>    dset;
          {
              2*zeta^3 - 2*zeta^2 - 2,
              1,
              15*zeta^3 - 5*zeta^2 + 8*zeta - 17,
              -3*zeta^3 - 7*zeta^2 - 8*zeta - 9,
              -2*zeta^2 - 2
          }
```

For our subsequent operations, it is beneficial to compute a Weierstrass model of E_d using the point $(u, v) = (-1, 0)$. We express the function u in the coordinates of that model. We obtain

$$E_d : Y^2 = X^3 - d(3\zeta^3 + \zeta - 1)X^2 - d^2(\zeta^2 + \zeta + 1)X$$

and

$$u = \frac{-X + d(\zeta^3 - 1)}{X - d(\zeta^3 + \zeta)}.$$

Using Magma, one can verify this using a few lines of code. Notice that the elliptic curve is represented as a *projective* curve.

```
>    Kd<d> := RationalFunctionField(K);
>    KdX<X> := PolynomialRing(Kd);
>    FEd₁ := (X⁴+ζ*X³+ζ²*X²+ζ³*X+ζ⁴)/d;
```

```
>    Ed₁ := HyperellipticCurve(FEd₁);
>    Ed₂, toEd₂ := EllipticCurve(Ed₁, Ed₁ ! [−1, 0]);
>    umap := map<Ed₁ → P₁| [Ed₁.1, Ed₁.3] >;
>    FEd := X³+(−3∗ζ³−ζ+1)∗d∗X²+(−ζ²−ζ−1)∗d²∗X;
>    Ed<xE,yE,zE> := EllipticCurve(FEd);
>    bl, toEd := IsIsomorphic(Ed₂, Ed);
>    u := Expand(Inverse(toEd₂∗toEd)∗umap);
>    u: Minimal;
```

$$(xE : yE : zE) \to ((-2*zeta\char`\^3 + 3*zeta\char`\^2 - 3*zeta$$
$$+ 2)/d\char`\^2*xE + (-zeta\char`\^3 - 2*zeta + 2)/d*zE :$$
$$(2*zeta\char`\^3 - 3*zeta\char`\^2 + 3*zeta - 2)/d\char`\^2*xE +$$
$$(2*zeta\char`\^2 - zeta + 2)/d*zE)$$

We have the following diagram of covers.

Clearly,

$$\phi_\delta(C_\delta(\mathbb{Q})) \subset u(E_d(K)) \cap \mathbb{P}^1(\mathbb{Q}).$$

We can now complete the proof of Theorem 1.2 for $n = 5$ by computing the right hand side of the above inclusion. By techniques we will explain in Section 7, we find the following table.

d	$u(E_d(K)) \cap \mathbb{P}^1(\mathbb{Q})$
1	$\{-1, 0, \infty\}$
$-2\zeta^2 - 2$	$\{-1, 1\}$
$2\zeta^3 - 2\zeta^2 - 2$	$\{-3, -1, -1/3\}$
$-3\zeta^3 - 7\zeta^2 - 8\zeta - 9$	$\{-1, 3/2\}$
$15\zeta^3 - 5\zeta^2 + 8\zeta - 17$	$\{-1, 2/3\}$

It is straightforward to check that all of these values for x/y lead to solutions with $xyz = 0$ or solutions that are mentioned in Theorem 1.2.

5 Deciding local solvability

As we have seen in Sections 2 and 4, the first step in solving arithmetic geometric questions often involves deciding if, for a projective variety X over a number field K, the set $X(K_\mathfrak{p})$ is empty for some prime $\mathfrak{p}$. In this section, we outline several algorithms that have been implemented in Magma by the author to test local solvability. They include tools for determining the $K_\mathfrak{p}$-points of separated 0-dimensional schemes, $K_\mathfrak{p}$-solvability of complete intersections,

$K_{\mathfrak{p}}$-solvability of smooth projective curves, given by a possibly singular planar model and $K_{\mathfrak{p}}$-solvability of hyperelliptic curves.

For the rest of this section, $\mathcal{O}$ will be a complete local ring of characteristic 0 with maximal ideal $\mathfrak{p}$ and finite residue field $\mathcal{O}/\mathfrak{p}$. We write π for a generator of $\mathfrak{p}$ and L for the field of fractions of $\mathcal{O}$. We use $\nu : L^* \to \mathbb{Z}$ to denote the normalised valuation, i.e., $\nu(\pi^e) = e$ and use the customary extension $\nu(0) = \infty$.

For any object f (vector, matrix, polynomial) defined over $\mathcal{O}$, we write $\bar{f}$ for the corresponding reduced object over $\mathcal{O}/\pi$. We also write $\nu(f)$ for the minimum of $\nu(c)$, where c runs through the coefficients of f.

5.1 Determining $X(L)$ for a reduced 0-dimensional projective scheme

Let $\mathbb{P}^n$ be n-dimensional projective space over K with variables $(X_0 : \ldots : X_n)$ and let X be a reduced 0-dimensional projective scheme, defined by

$$f_1 = \cdots = f_m = 0,$$

where $f_i \in K[X_0, \ldots, X_m]$. Without loss of generality, we can assume $f_i \in \mathcal{O}[X_0, \ldots, X_m]$.

Note that any point in $\mathbb{P}^n(L)$ has a representative $(x_0 : \cdots : x_n)$ such that for some $0 \le N \le n$, we have

$$(x_1, \ldots, x_{N-1}, x_N, x_{N+1}, \ldots, x_n) \in \mathcal{O} \times \cdots \times \mathcal{O} \times \{1\} \times \mathfrak{p} \times \cdots \times \mathfrak{p}$$

Hence, it is sufficient to solve the problem of finding $\mathcal{O}$-*integral* points on an *affine* separated 0-dimensional scheme Y, given by equations $f_i \in \mathcal{O}[\mathbf{y}] = \mathcal{O}[y_1, \ldots, y_n]$.

In principle, one could solve the problem in the same way as one does for exactly representable fields like $\mathbb{Q}$, $\mathbb{F}_q$ and number fields, by using resultants and univariate factorisation. In practice, however, the objects considered are not exactly represented and it is almost impossible to make such algorithms numerically stable. Therefore, we will present an algorithm here that simply builds solutions one π-adic digit at the time, until the solution is verifiably separated and Hensel liftable. We simply reduce the system of equations to $\mathcal{O}/\mathfrak{p}$, determine the solutions over that finite field and interpret what these solutions mean over $\mathcal{O}$.

The first step is to pick $g_i \in \mathcal{O}[\mathbf{y}]$ such that

$$(g_1, \ldots, g_m)\mathcal{O}[\mathbf{y}]$$

is as close as possible to

$$I = (f_1, \ldots, f_m)K[\mathbf{y}] \cap \mathcal{O}[\mathbf{y}].$$

Let

$$M_f = \left(\frac{\partial}{\partial y_j} f_i \right)_{i,j} \in \mathcal{O}^{m \times n}$$

Algorithm Saturate($f_1, \ldots, f_m$):

1. **repeat**
2.   Let $T \in \mathrm{GL}_m(\mathcal{O})$ such that $\overline{T(M_f(0, \ldots, 0))}$ is in row echelon form.
3. $(f_1, \ldots, f_m)^t \leftarrow T(f_1, \ldots, f_m)^t$.
4.   **for** $i \in \{1, \ldots, m\}$:
5. $f_i \leftarrow f_i / \pi^{\nu(f_i)}$
6. **until** in step 5 no f_i was changed.
7. **return**($f_1, \ldots, f_m$)

This algorithm does not always find generators of I. However, if $(0, \ldots, 0)$ is sufficiently close to a non-singular point of Y, then for all $f \in I$, the minimal valuation of the coefficients of f is attained by the coefficient of a linear term. It is clear that in this situation, the algorithm will find g_i that generate I.

In practice, the coefficients of f_i are only given up to finite precision. Hence, in step 5, it might happen that a coefficient has no precision left. In that case, an error should be generated.

It is now straightforward to determine the integer points of Y up to some precision bound r:

Algorithm IntegerPoints($r, f_1, \ldots, f_m$):

1. $(f_1, \ldots, f_m) \leftarrow$ Saturate($f_1, \ldots, f_m$).
2. Let $V \subset \mathcal{O}^n$ be a set of representatives of the solutions of $\overline{f_1} = \cdots = \overline{f_m} = 0$ in $\mathcal{O}/\mathfrak{p}$.
3. Let $V_0 \subset V$ represent the points over $\mathcal{O}/\mathfrak{p}$ with 0-dimensional tangent space.
4. $W \leftarrow \{\}$; $V_1 = V \setminus V_0$
5. **for** $v \in V_0$:
6. $v \leftarrow$ Hensel lift of v to precision r using a suitable subset $\{f_{i_1}, \ldots, f_{i_n}\}$
7.   **if** for all i we have $\nu(f_i(v)) \geq r$ **then**
8.   Add v to W
9.   **else**
10.   Discard v
11. **for** $v \in V_1$:
12. $g_i \leftarrow f_i(v_1 + \pi y_1, \ldots, v_n + \pi y_n)$ for $i = 1, \ldots, m$
13.   **for** $w \in$ IntegerPoints($r - 1, g_1, \ldots, g_m$):
14.   Add $(v_1 + \pi w_1, \ldots, v_n + \pi w_n)$ to W
15. **return** W

Obviously, if Y has some higher multiplicity $\mathcal{O}$-point, then successive approximations to it will be in V_1 and never in V_0. The algorithm recurses infinitely. Therefore, if IntegerPoints gets called with $r \leq 0$, an error should be generated, indicating that the scheme Y has points that do not separate below the requested precision level. An alternative is to return such points as *non-separating* approximations to possible solutions and return them. These give the user neighbourhoods that could not be resolved at the requested precision.

It should also be clear, and this is an essential problem, that step 7 only tests that v is *approximately* on Y. While v can be uniquely lifted to arbitrary precision r' using $\{f_{i_1}, \ldots, f_{i_n}\}$ (provided the f_i themselves are given to sufficient precision), it may be that this lift does not satisfy the other equations to precision r', but that it did to precision r. Obviously, if Y is presented as a complete intersection and $n = m$, then this problem will not arise. Otherwise, the best one can do is to assume that the user supplies a sufficiently high r to begin with.

5.2 Determining solvability of complete intersections

Let $X \subset \mathbb{P}^n$ be a complete intersection defined over a number field K of dimension d. We assume that X is equidimensional, which means that its maximal components are all of dimension d. This condition is certainly met if X is irreducible. Let X' denote the reduced singular subscheme.

Let L be the completion of K at some finite prime. First we assume $X'(L)$ is empty. In this case, our task is to decide if $X(L)$ contains any non-singular points. We follow the same approach as in Section 5.1 and note that, again, it is sufficient to solve the problem for integral points on affine complete intersections Y:

Algorithm HasNSIntegralPoints($f_1, \ldots, f_{n-d}$):

1. $(f_1, \ldots, f_{n-d}) \leftarrow$ Saturate($f_1, \ldots, f_{n-d}$).
2. Let $V \subset \mathcal{O}^n$ be a set of representatives of the solutions of $\overline{f_1} = \cdots = \overline{f_{n-d}} = 0$ in $\mathcal{O}/\mathfrak{p}$.
3. **if** any of the points in V represent a point over $\mathcal{O}/\mathfrak{p}$ with a d-dimensional tangent space:
4.   **return** *true*
5. **for** $v \in V$:
6. $g_i \leftarrow f_i(v_1 + \pi y_1, \ldots, v_n + \pi y_n)$ for $i = 1, \ldots, n - d$
7.   **if** HasNSIntegralPoints($g_1, \ldots, g_{n-d}$):
8.   **return** *true*
9. **return** *false*

Since we assume that Y is a complete intersection, the problem of step 7 in Section 5.1 does not arise.

The strategy to determine if $X(L)$ is empty is now straightforward:

Algorithm ClHasPoints(X,L):

1. if $X'(L)$ is nonempty, then $X(L)$ is nonempty
2. otherwise, use HasNSIntegralPoints on the affine patches of X to decide if $X(L)$ is nonempty.

Obviously, step 1 can only be decided if X' is of one of the types we have considered before, i.e., X' is empty, $\dim X' = 0$ or X' is a complete intersection. One may be able to show that $X'(L)$ is empty by showing that some complete intersection containing X' has no points over L, but the converse does not hold.

In order to compute X', it is essential that X is represented exactly over some field allowing exact arithmetic, because only then do Gröbner basis algorithms allow for the computation of the radical of an ideal.

5.3 Solvability of smooth curves

In this section we consider a reduced scheme $X \subset \mathbb{P}^2$ given by a single equation $f(x, y, z) = 0$. We present an algorithm to determine the local solvability of the desingularisation $\tilde{X}$ of X. As in Section 5.2, we note that it is sufficient to solve the problem for integral points on affine curves $Y : f(x, y) = 0$ and to a large extent, the algorithm is the same. The only difference occurs in how singular points in $Y(\mathcal{O})$ are treated. Instead of considering them as rational points, we blow up Y at the singularity and remove the exceptional component. We then look for L-valued points on the resulting scheme.

First, let us study how to blow up an affine curve in $(0,0)$. Therefore, let $f \in \mathcal{O}[x, y]$ describe a curve in $\mathbb{A}^2$ with a singularity at $(0,0)$ and no other $\mathcal{O}$-valued singularities. We resolve the singularity by blowing up $\mathbb{A}^2$ at $(0,0)$. This means we take the inverse image under

$$\beta : \{xv = yu\} \subset \mathbb{P}^1 \times \mathbb{A}^2 \to \mathbb{A}^2$$
$$(u : v; x, y) \qquad \mapsto (x, y)$$

Note that any point $(u : v; x, y)$ that has an image $(x, y) \in \mathbb{A}^2(\mathcal{O})$ under β has a representative of one of the forms $(u : 1; x, y)$ or $(1 : \pi v; x, y)$ with $u, v \in \mathcal{O}$. Hence, any integral point on $\beta^{-1}Y$ is covered by an integral point on one of $\beta_1^{-1}Y$ or $\beta_2^{-1}Y$, where

$$\beta_1 : \quad \mathbb{A}^2 \quad \to \quad \mathbb{A}^2, \qquad \beta_2 : \quad \mathbb{A}^2 \quad \to \quad \mathbb{A}^2,$$
$$(u, y) \to (uy, y) \qquad\qquad (v, x) \to (x, \pi x v)$$

To remove the exceptional component from $\beta_1^{-1}Y : f(uy, y) = 0$, compute

$$Y_1 : f_1(u, y) = f(uy, y)/u^{\text{(highest possible power)}}.$$

The curve Y_1 may have new singularities, but since Y_1 is isomorphic to Y outside $u = 0$, any integral-valued singularities will have $u = 0$. The singular points of Y_1 can be easily described as

$$Y_1'(\mathcal{O}) = \left\{ (u, 0) : u \in \mathcal{O} \text{ and } f_1(u, 0) = \frac{\partial f_1}{\partial u}(u, 0) = \frac{\partial f_1}{\partial y}(u, 0) = 0 \right\}$$

and can be computed using univariate root finding for polynomials over $\mathcal{O}$. Of course, in practice, an expression like "$= 0$" should be interpreted as "is indistinguishable from 0 at the given precision". For $\beta_2^{-1}Y$ we can proceed similarly.

Given a list S of integral-valued singularities, one can check the desingularisation of $Y : f(x,y) = 0$ for integral points:

Algorithm HasSmoothIntegralPoints(f, S):

1. $f \leftarrow f/\pi^{\nu(f)}$
2. **if** $S = \{(x_0, y_0)\}$:
3. $f \leftarrow f(x + x_0, y + y_0)$
4. $f_1 \leftarrow f(uy, y)/u$ (highest possible power)
5. $S_1 \leftarrow \left\{ (u,0) : u \in \mathcal{O} \text{ and } f_1(u,0) = \frac{\partial f_1}{\partial u}(u,0) = \frac{\partial f_1}{\partial y}(u,0) = 0 \right\}$
6.   **if** HasSmoothIntegralPoints(f_1, S_1): **return** *true*
7. $f_2 \leftarrow f(x, \pi x v)/v$ (highest possible power)
8. $S_2 \leftarrow \left\{ (v,0) : v \in \mathcal{O} \text{ and } f_2(v,0) = \frac{\partial f_2}{\partial v}(v,0) = \frac{\partial f_2}{\partial x}(v,0) = 0 \right\}$
9.   **if** HasSmoothIntegralPoints(f_2, S_2): **return** *true*
10. **else**
11.   Let $V \subset \mathcal{O}^2$ be a set of representatives of the solutions of $\overline{f} = 0$ in $\mathcal{O}/\mathfrak{p}$.
12.   **if** a point in V represents a nonsingular point over $\mathcal{O}/\mathfrak{p}$: **return** *true*
13.   **for** $(x_0, y_0) \in V$:
14. $g \leftarrow f(x_0 + \pi x, y_0 + \pi y)$
15. $S' \leftarrow \{((x_1 - x_0)/\pi, (y_1 - y0)/\pi) : (x_1, y_1) \in S\}$
16.   Remove any non-integral entries from S'
17.   **if** HasSmoothIntegralPoints(g, S'): **return** *true*
18. **return** *false*

To determine local solvability of the desingularisation of a reduced projective plane curve $X \subset \mathbb{P}^2$, one can determine the reduced singular subscheme X' of X, find $X'(L)$ using Section 5.1 and apply HasSmoothIntegralPoints to each affine patch of X, using $X'(L)$ to initialise S.

5.4 Solvability of hyperelliptic curves

In this section, we adopt Magma's terminology and understand *hyperelliptic curve* to mean *nonsingular double cover of* $\mathbb{P}^1$. Some geometric hyperelliptic curves can be represented in this category (but not the ones that have a twisted $\mathbb{P}^1$ as a canonical model). Conics and some curves of genus 1 also fit in this category.

We represent such a curve as a nonsingular curve in weighted projective space $\mathbb{P}_{(1,d,1)}$ with coordinates (x, y, z) and a model of the form

$$C : y^2 + h(x,z)y = f(x,z).$$

Over fields of odd characteristic we can complete the square and without loss of generality, we can assume $h = 0$. In this case, the non-singularity of C means that $f(x,z)$ is a square-free form of degree $2d$ and a simple application of Riemann-Hurwitz shows that C is of genus $d - 1$.

Of course, to decide if a hyperelliptic curve has points over L, one could cover it with two non-singular affine patches and use Section 5.3. One can also use [3, Appendix A.2], which is slightly more efficient. Both these algorithms are essentially polynomial in $\#\mathcal{O}/\mathfrak{p}$, though. We can do better if $\mathcal{O}/\mathfrak{p}$ is of odd characteristic and satisfies

$$(\#\mathcal{O}/\mathfrak{p}) - 2(d-1)\sqrt{\#\mathcal{O}/\mathfrak{p}} > 0.$$

We generalise an algorithm that is presented for $d = 2$ in [21] and [19]. It is based on the fact that a curve defined over a finite field of large cardinality compared to the genera of the components, must be *very* singular not to have any nonsingular rational points.

Since multiplying f with an even power of π does not change the local solvability of $y^2 = f(x, z)$, we can assume $f \in \mathcal{O}[x, z]$ with $0 \le \nu(f) \le 1$.

Note that if $\nu(f) = 0$, any point $(\overline{x} : \overline{y} : \overline{z}) \in \mathbb{P}_{(1,d,1)}(\mathcal{O}/\mathfrak{p})$ satisfying $\overline{y}^2 = \overline{f}(\overline{x}, \overline{z})$ with $\overline{y} \ne 0$ or $(\overline{x} : \overline{z})$ a zero of $\overline{f}$ of multiplicity 1 is a nonsingular point on $\overline{C}$ and hence Hensel-lifts to a point in $C(L)$. If we can show such a point exists, then $C(L)$ is not empty. We distinguish the following cases.

1. $\nu(f) = 1$. If $x, z \in \mathcal{O}$ such that $f(x, z)$ is a square, then in particular, $\nu(f(x, z)) \equiv 0 \bmod 2$. Hence, $\overline{(x : z)}$ must be a root of $\overline{f/\pi}$ in $\mathcal{O}/\mathfrak{p}$. For any such root we take a representative $(x_0, z_0) \in \mathcal{O}^2$ and we test $y^2 = f(x_0 + \pi x, z_0 + \pi z)/\pi^2$ for local solvability. If any of those cases is solvable, then so is the original equation. If none is, or if no roots are available, then $y^2 = f(x, z)$ has no solutions.

2. $\nu(f) = 0$ and $\overline{f} = \alpha(g(\overline{x}, \overline{z}))^2$ with α a non-square in $\mathcal{O}/\mathfrak{p}$. If $x, z \in \mathcal{O}$ such that $f(x, z)$ is a square, then $g(\overline{x}, \overline{z}) = 0$. Hence we take representatives $(x_0, z_0) \in \mathcal{O}^2$ for the roots of g and test $y^2 = f(x_0 + \pi x, z_0 + \pi z)$ for local solvability. If any of those cases is solvable, then so is the original equation. If none is, or if no roots are available, then $y^2 = f(x, z)$ has no solutions.

3. $\nu(f) = 0$ and $\overline{f} = \alpha(g(\overline{x}, \overline{z}))^2$ with α a nonzero square in $\mathcal{O}/\mathfrak{p}$. We take $(x_0, z_0) \in \mathcal{O}$ to represent a non-root of g in $\mathbb{P}^1(\mathcal{O}/\mathfrak{p})$. Note that g has at most d roots, while $\#\mathbb{P}^1(\mathcal{O}/\mathfrak{p}) = \#\mathcal{O}/\mathfrak{p} + 1 > 2(d-1)$ points, so this is easy. Then $f(x_0, z_0)$ is a square, because it represents a non-zero square in $\mathcal{O}/\mathfrak{p}$. Therefore, the original equation is solvable.

4. In all other cases, $\overline{f} = g_1(\overline{x}, \overline{z})(g_2(\overline{x}, \overline{z}))^2$, where g_1 is square-free and $\deg(g_1) + 2\deg(g_2) = 2d$. The curve $D : y_1^2 = g_1(\overline{x}, \overline{z})$ is a hyperelliptic curve over $\mathcal{O}/\mathfrak{p}$ of genus $(\deg(g_1) - 2)/2$ and hence, by the Hasse–Weil bounds, has at least

$$(\#\mathcal{O}/\mathfrak{p}) - 2(d-1)\sqrt{\#\mathcal{O}/\mathfrak{p}} + (2\deg g_2 + 2)\sqrt{\#\mathcal{O}/\mathfrak{p}}$$

points. It follows that D must have points $(\overline{x}, y_1, \overline{z})$ with $g_2(\overline{x}, \overline{z}) \ne 0$. Since the higher multiplicity roots of $\overline{f}$ are exactly the roots of g_2, it follows that

$\bar{y}^2 = f(\bar{x}, \bar{z})$ has a non-singular point, which is Hensel-liftable. It follows that $C(L)$ is non-empty.

These cases lead directly to a recursive algorithm, where the most difficult operation is factorisation of univariate polynomials of degree at most $2d$ over a finite field. The branching degree of the algorithm is bounded by $2d$ as well and not (as is the algorithm in Section 5.3), essentially by $\#\mathcal{O}/\mathfrak{p}$.

6 Mordell–Weil groups of elliptic curves

The *Mordell–Weil group* of an abelian variety A over a number field K is the set of K-rational points $A(K)$. The abelian variety structure of A induces a group structure on $A(K)$. A celebrated theorem of Weil, which for elliptic curves over $\mathbb{Q}$ was already proved by Mordell, states that $A(K)$ is a finitely generated commutative group. Actually determining $A(K)$, even in the case where A is an elliptic curve and $K = \mathbb{Q}$, is still more an art than a science. However, even an artist works better if he has proper tools available. In this chapter, we introduce the tools that Magma offers to determine Mordell–Weil groups of elliptic curves over number fields. The Magma implementation is based on [4].

First, we review some of the fundamental definitions connected to the subject. We do not give much detail, since many other excellent descriptions already exist (see for instance [22]). Computational concerns that arise specifically when applying the methods outlined here to elliptic curves over number fields are addressed in [23].

Let E be an elliptic curve over a number field K. In order to bound the free rank of $E(K)$, we bound the size of $E(K)/2E(K)$. For this, we use the *2-Selmer group* of E over K. From the exact Galois-cohomology sequence

$$0 \to E(K)/2E(K) \to \mathrm{H}^1(K, E[2]) \to \mathrm{H}^1(K, E)$$

we derive a set that approximates the image of $E(K)/2E(K)$ in $\mathrm{H}^1(K, E[2])$ *everywhere locally*. We define $\mathrm{S}^{(2)}(E/K)$ to be the intersection of the kernels of $\mathrm{H}^1(K, E[2]) \to \mathrm{H}^1(K_p, E)$ for *all* primes p of K:

$$0 \to \mathrm{S}^{(2)}(E/K) \to \mathrm{H}^1(K, E[2]) \to \prod_p \mathrm{H}^1(K_p, E).$$

Clearly, $\mathrm{S}^{(2)}(E/K)$ provides a sharp bound, unless $\mathrm{H}^1(K, E[2])$ maps to any cocycle in $\mathrm{H}^1(K, E)$ that trivialises under all restrictions $\mathrm{Gal}(K_p) \subset \mathrm{Gal}(K)$. The group consisting of such cocycles is called the *Shafarevich–Tate group* $\mathrm{III}(E/K)$ and we have the exact sequence

$$0 \to E(K)/2E(K) \to \mathrm{S}^{(2)}(E/K) \to \mathrm{III}(E/K)[2] \to 0.$$

The group $S^{(2)}(E/K)$, as a Galois-module, can be represented in the following way (see [10]). For an elliptic curve

$$E : y^2 + a_1 xy + a_3 y = x^3 + a_2 x^2 + a_4 x + a_6,$$

we define the following algebra:

$$A[\theta] = K[X]/(X^3 + a_2 X^2 + a_4 X + a_6 + (a_1 X + a_3)^2/4).$$

The Galois-module $H^1(K, E[2])$ can be identified with the subgroup of A^*/A^{*2} consisting of the elements of square norm and for some suitable, effectively computable, set S of primes of K, we have $S^{(2)}(E/K) \subset A(2, S)$. The map $\mu : E(K) \to S^{(2)}(E/K)$ is induced by $(x, y) \mapsto x - \theta$ where $x - \theta \in A^*$.

Magma computes $S^{(2)}(E/K)$ by computing the local images of

$$E(K_p) \to A^*/A^{*2} \otimes K_p$$

and computing the elements from $A(2, S)$ of square norm that land in these local images. Wherever possible, elements of $A(2, S)$ are left in product representation, to avoid coefficient blowup.

As an example, we compute $S^{(2)}(E_d/K)$ for $d = 2\zeta^3 - 2\zeta^2 - 2$, as defined in Section 4.

```
>   _<x> := PolynomialRing(Rationals());
>   K<ζ> := NumberField(x⁴−x³+x²−x+1);
>   OK := IntegerRing(K);
>   d := 2*ζ³−2*ζ²−2;
>   E<X,Y,Z> := EllipticCurve(
>               [ 0, (−3*ζ³−ζ+1)*d, 0, (−ζ²−ζ−1)*d², 0 ]);
>   two := MultiplicationByMMap(E, 2);
>   μ, tor := IsogenyMu(two);
>   S2E, toS2E := SelmerGroup(two); S2E;
            Abelian Group isomorphic to Z/2 + Z/2 + Z/2 + Z/2
            Defined on 4 generators in supergroup:
                S2E.1 = $.1 + $.2 + $.6 + $.7 + $.8 + $.9
                S2E.2 = $.2 + $.4 + $.7 + $.8
                S2E.3 = $.1 + $.2 + $.5 + $.7
                S2E.4 = $.3 + $.9
            Relations:
                2*S2E.1 = 0
                2*S2E.2 = 0
                2*S2E.3 = 0
                2*S2E.4 = 0
```

So we see that $E(K)/2E(K) \subset (\mathbb{Z}/2\mathbb{Z})^4$. Part of this corresponds to the image of the torsion subgroup of E.

```
>   Etors, EtorsMap := TorsionSubgroup(E);
>   sub<S2E | [toS2E(μ(EtorsMap(g))): g in OrderedGenerators(Etors)]>;
```

```
Abelian Group isomorphic to Z/2 + Z/2
Defined on 2 generators in supergroup S2E:
    $.1 = S2E.3 + S2E.4
    $.2 = S2E.1 + S2E.4
Relations:
    2*$.1 = 0
    2*$.2 = 0
Mapping from: Abelian Group isomorphic to Z/2 + Z/2
Defined on 2 generators in supergroup S2E:
    $.1 = S2E.3 + S2E.4
    $.2 = S2E.1 + S2E.4
Relations:
    2*$.1 = 0
    2*$.2 = 0 to GrpAb: S2E
```

We conclude that the free rank of $E(K)$ is at most 2. We look for rational points on E, up to some tiny bound and we see that the found points already generate $E(K)/2E(K)$.

```
>    V := RationalPoints(E: Bound := 5);
>    sub<S2E | [toS2E(μ(P)) : P in V]> eq S2E;
        true
```

We then select some minimal subset of V that generates $E(K)/2E(K)$ and construct a group homomorphism from an abstract abelian group into G.

```
>    gs := [ E ! [0, 0], E ! [−2*ζ³ − 2*ζ + 2, 0],
>        E ! [−2*ζ³, −4*ζ²], E ! [−2*ζ³ − 4*ζ + 4, −4*ζ³ + 4*ζ] ];
>    assert S2E eq sub<S2E | [toS2E(μ(g)) : g in gs]>;
>    G := AbelianGroup([2, 2, 0, 0]);
>    mwmap := map< G → E | g:->&+[ c[i]*gs[i]: i in [1..#gs] ]
>                 where c := Eltseq(g) >;
```

In fact, we could have left this all to the system and just executed:

```
>    success, G, mwmap := PseudoMordellWeilGroup(E);
>    assert success;
```

Here, it is of the utmost importance to check that *success* is true. Only then is there a guarantee that the returned group is of finite (odd) index in $E(K)$. If the value false is returned, then only a subgroup is returned that will itself be 2-saturated in $E(K)$ (meaning that, if $2P \in G$ and $P \in E(K)$ then $P \in G$ as well), but need not be of finite index.

In fact, the computation done by **PseudoMordellWeilGroup** is not completely equivalent to the computation we did above. By default, if possible, **PseudoMordellWeilGroup** uses a *2-isogeny descent* (see [22]). For any non-trivial element of $E[2](K)$, there is an associated *2-isogeny*

$$\phi : E \to E',$$

together with a dual isogeny $\hat{\phi} : E' \to E$, such that $\hat{\phi} \circ \phi = 2|_E$. In complete analogy to the 2-Selmer group, we define the ϕ-Selmer group by considering the exact sequence

$$0 \to E'(K)/\phi E(K) \to \mathrm{H}^1(K, E[\phi]) \to \mathrm{H}^1(K, E)$$

and we define $\mathrm{S}^{(\phi)}(E/K)$ by insisting on exactness of

$$0 \to \mathrm{S}^{(\phi)}(E/K) \to \mathrm{H}^1(K, E[\phi]) \to \prod_p \mathrm{H}^1(K_p, E).$$

From the exact sequence

$$\begin{aligned}0 \to E[\phi](K) \to E[2](K) &\to E'[\hat{\phi}](K) \to \\ E'(K)/\phi E(K) &\to E(K)/2E(K) \to E(K)/\hat{\phi}E'(K) \to 0\end{aligned}$$

it follows that

$$4\#E(K)/2E(K) = \#E'(K)/\phi E(K) \cdot \#E(K)/\hat{\phi}E'(K) \cdot \#E[2](K).$$

Therefore, we can use ϕ-Selmer groups to bound the free rank of $E(K)$ as well. One can compute ϕ-Selmer groups in the same way as 2-Selmer groups.

```
>   φ := TwoIsogeny(E ! [0,0]);
>   Sphi, toSphi := SelmerGroup(φ);
>   phihat := DualIsogeny(φ);
>   Sphihat, toSphihat := SelmerGroup(phihat);
>   4*#S2E, #Sphi, #Sphihat, #TwoTorsionSubgroup(E);
        64 2 8 4
```

Apart from providing an upper bound on the rank of $E(K)$, Selmer groups also contain information about possible generators of $E(K)$. To access this information, it is useful to interpret $\mathrm{S}^{(2)}(E/K) \subset \mathrm{H}^1(K, E[2])$ as a set of twists of the cover $E \xrightarrow{2} E$. The second return value of IsogenyMu gives a map that computes such a cover from an element of $\mathrm{H}^1(K, E[2])$. The covering space is represented as an intersection X of two quadrics in $\mathbb{P}^3$, with a map $\phi : X \to E$. If the cover represents an element from $\mathrm{S}^{(2)}(E/K)$, however, one can construct a model of X of the form $C : v^2 = f_0 u^4 + \cdots + f_4$. A call to Quartic realises this. One can then search for points on C, which can be mapped back to E.

```
>   for δ in [s:s in S2E| Order(s) ne 1] do
>     print "Looking at delta =",δ@@toS2E;
>     ψ := tor(δ@@@toS2E);
>     XX := Domain(ψ);
>     C, CtoXX := Quartic(XX);
>     V := RationalPoints(C:Bound:=20);
>     print "Points found on quartic:",V;
```

```
>     W := {ψ(CtoXX(v)):v in V};
>     print "points on the curve:", W;
>     assert forall{P: P in W | toS2E(μ(P)) eq δ};
>     end for;
```

Here is part of the output:

```
Looking at delta = 1/4*(-3*zeta^3 - 16*zeta^2 + 30*zeta - 20)*
     theta^2 + 1/2*(-8*zeta^3 + 24*zeta^2 - 23*zeta + 10)*
     theta + 2*zeta^3 - 2*zeta^2
Points found on quartic:
     {@ (1 : -24*zeta^3 + 34*zeta^2 - 17*zeta - 5 : 0) @}
points on the curve:
     { (1/5*(-22*zeta^3 - 14*zeta + 14) : 1/5*(-20*zeta^3 +
          12*zeta^2 - 4*zeta + 24) : 1) }
```

Note, however, that it is a rarity for it to make sense to search for points on C as computed. The model computed for C generally does not have particularly small coefficients and there is no reason to expect that the point we are looking for will be easier to find on C than on E. Over $\mathbb{Q}$, a rather satisfactory solution to this problem has been found in the form of a proper minimization and reduction theory [25], [13]. For other number fields, a satisfactory theory is woefully lacking and Magma leaves it to the art and ingenuity of the user to find a suitable model from the returned one.

The same functionality is available for 2-isogenies as well. Here, the cover corresponding to an element in the Selmer group naturally has a model of the form $C : v^2 = f_0 u^4 + f_2 u^2 + f_4$, and therefore it does make sense to look for rational points on the covering curve. Therefore, the routine PseudoMordellWeilGroup uses the following default strategy:

1. If a 2-isogeny is available, this is chosen as isogeny ϕ, Otherwise full multiplication-by-2 is used as ϕ.
2. The ϕ-Selmer group is computed and, if $\psi \neq 2$, then also the $\hat{\phi}$-Selmer group is computed.
3. The image of the torsion subgroup is determined in the computed Selmer groups.
4. The elliptic curve is searched for rational points up to a preset bound and, if relevant, also the 2-isogenous curve is searched. If the found points already generate the Selmer group(s), we are done.
5. Otherwise, if ϕ is a 2-isogeny or if the elliptic curve is defined over $\mathbb{Q}$, the covers corresponding to elements of the Selmer group that are not represented by rational points are constructed (and, if reduction is available, reduced) and searched for points.
6. If this still leaves some elements of the Selmer group(s) not corresponding to found rational points, then false is returned, together with the group generated by the found points. Otherwise, true is returned.

One can override the default choice of isogeny and whether or not homogeneous spaces should be used for searching for rational points.

If $\text{III}(E/K)[2]$ is nontrivial, then obviously neither a 2-descent nor a 2-isogeny descent will provide a sharp bound on $E(K)/2E(K)$. In this situation, a 4-descent may give more information ([19] and [26]). For $K = \mathbb{Q}$, Tom Womack has implemented routines to perform such a computation in Magma. Another option consists of using the Cassels–Tate pairing to obtain more information (see [11]).

Alternatively, one may use *visualisation* (see [14]) to obtain more information. See [6] for an explicit approach using Magma.

7 Chabauty methods using elliptic curves

In this section, we show how, given an elliptic curve E over a number field K and a map $u : E \to \mathbb{P}^1$, one can try to determine $\{p \in E(K) : u(p) \in \mathbb{P}^1(\mathbb{Q})\}$. The method is an adaptation of Chabauty's partial proof of Mordell's conjecture [12] and is described in [7] and [3]. A similar method applied to bi-elliptic genus 2 curves is described in [17]. We quickly review the theory here.

We write $\mathcal{O}$ for the ring of integers of K. and we fix models of E and $\mathbb{P}^1$ over $\mathcal{O}$. We choose a prime p such that the primes $\mathfrak{p}_1, \ldots, \mathfrak{p}_t$ of K over p are unramified and such that the cover $u : E \to \mathbb{P}^1$ has good reduction at each $\mathfrak{p}_i$, as a scheme morphism over $\mathcal{O}$.

Let p be a rational prime which is unramified in K. Let $\mathfrak{p}_1, \ldots, \mathfrak{p}_t$ be the primes of K over p. Suppose that $u : E \to \mathbb{P}^1$ has good reduction at all $\mathfrak{p}_i$. We write $\mathcal{O}$ for the ring of integers of K, $E(\mathcal{O}/\mathfrak{p}_i)$ for the points in the special fibre of E, considered as a scheme over $\mathcal{O}_{\mathfrak{p}_i}$ and $E^{(1)}(K_{\mathfrak{p}_i})$ for the kernel of reduction:

$$0 \to E^{(1)}(K_{\mathfrak{p}_i}) \to E(K_{\mathfrak{p}_i}) \xrightarrow{\rho_i} E(\mathcal{O}/\mathfrak{p}_i) \to 0$$

Let $g_1, \ldots, g_r \in E(K)$ be generators of the free part of $E(K)$. Then if $P_0 = T + n_1 g_1 + \cdots + n_r g_r \in E(K)$ has $u(P_0) \in \mathbb{P}^1(\mathbb{Q})$, then certainly (abusing notation), $u(\rho_i(P_0)) \in \mathbb{P}^1(\mathbb{F}_p)$ and in fact $u(\rho_i(P_0)) = u(\rho_j(P_0))$. The points $P_0 \in E(K)$ define a collection of cosets of

$$\Lambda_p = \bigcap_{i=1}^t \left(E(K) \cap E^{(1)}(K_{\mathfrak{p}_i}) \right)$$

Let V_p be this coset collection and let $b_1, \ldots, b_r$ be generators of Λ_p. In Magma, both V_p and Λ_p are easily computed.. We take the the same elliptic curve as in the previous chapter, together with its (finite index subgroup of the) Mordell–Weil group and the cover suggested in Section 4.

```
>    P₁ := ProjectiveSpace(Rationals(),1);
>    u := map< E → P₁ | [−X + (ζ³ − 1)*d*Z, X+(−ζ³−ζ)*d*Z] >;
>    V₃ := RelevantCosets(mwmap, u, Support(3*OK));
>    Λ₃ := Kernel(V₃[1]);
```

```
>    GmodLambda₃ := Codomain(V₃[1]);
>    V₃;
```

```
<Mapping from: GrpAb: G to GrpAb: GmodLambda3, {
    0,
    11*GmodLambda3.2,
    GmodLambda3.2,
    GmodLambda3.1 + 10*GmodLambda3.2,
    5*GmodLambda3.2,
    7*GmodLambda3.2,
    GmodLambda3.1 + 2*GmodLambda3.2
}>
```

As is clear, the coset data is returned as a tuple consisting of the map $G \rightarrow G/\Lambda_p$, together with the collection of cosets, represented as elements of G/Λ_p. We can compute a similar coset collection V_q and intersect it with V_p. This gives a new coset collection mod $\Lambda_p + \Lambda_q$. Alternatively, one could project $V_p \cap V_q$ down to get again a coset collection modulo Λ_p. This is what in Magma is called a Weak coset intersection.

```
>    V₁₁ := RelevantCosets(mwmap, u, Support(11*OK));
>    V₃₁₁ := CosetIntersection(V₃, V₁₁: Weak);
>    V₃₁₁;
```

```
<Mapping from: GrpAb: G to GrpAb: GmodLambda3, {
    0,
    11*GmodLambda3.2,
    GmodLambda3.2,
    5*GmodLambda3.2,
    7*GmodLambda3.2
}>
```

In order to bound the number of points $P \in E(K)$ with $u(P) \in \mathbb{P}^1(\mathbb{Q})$, we make use of the formal group description of the group structure on E. Let $b_1, \ldots, b_r$ be generators of $\Lambda_p \subset E(K)$. In terms of formal power series, there are isomorphisms

$$\text{Exp}_E : K[[z]] \rightarrow E(K[[z]]), \ \text{Log}_E : E(K[[z]]) \rightarrow K[[z]],$$

where z is a local coordinate on E around the origin. These power series converge on $E^{(1)}(K_{\mathfrak{p}})$ for unramified primes of odd residue characteristic and establish an isomorphism $E^{(1)}(K_{\mathfrak{p}}) \simeq \mathfrak{p}\mathcal{O}_{\mathfrak{p}}$. Therefore, for each prime $\mathfrak{p}_i$ we obtain a power series

$$\theta_{P_0,i}(n_1, \ldots, n_r) = u\left(P_0 + \text{Exp}_E\left(\sum_{j=1}^r n_j \text{Log}_E(b_j)\right)\right) \in \mathcal{O}_{\mathfrak{p}_i}[[n_1, \ldots, n_r]].$$

If $u(P_0 + n_1 b_1 + \cdots + n_r b_r)$, then $\theta_{P_0,i}(n_1, \ldots, n_r) \in \mathbb{Q}_p$ and $\theta_{P_0,i}(n_1, \ldots, n_r) = \theta_{P_0,j}(n_1, \ldots, n_r)$. Using that $\mathcal{O}_{\mathfrak{p}_i}$ is a finite $\mathbb{Z}_p$-module, we can decompose

$$\mathcal{O}_{\mathfrak{p}_i}[[n_1, \ldots, n_r]] = \oplus \mathbb{Z}_p[[n_1, \ldots, n_r]]$$

and express the above equations as $[K : \mathbb{Q}] - 1$ equations in $\mathbb{Z}_p[[n_1, \ldots, n_r]]$. We can do this in Magma:

```
>    P₀ := mwmap(G.3+G.4);
>    u(P₀);
         (-3 : 1)

>    θ := ChabautyEquations(P₀, u, mwmap, Support(3*OK));
>    PrintToPrecision(θ[1], 1); "\n";
>    PrintToPrecision(θ[2], 1); "\n";
>    PrintToPrecision(θ[3], 1);
         O(3^5) - (3 + O(3^5))*$.1 + (3^2*10 + O(3^5))*$.2
         O(3^5) - (3*29 + O(3^5))*$.1 - (3*32 + O(3^5))*$.2
         O(3^5) - (3^2*5 + O(3^5))*$.1 - (3^4 + O(3^5))*$.2
```

A consequence of the shape of $\mathrm{Exp}_E(z)$ is that the power series returned by **ChabautyEquations** have the property that the coefficient c of a monomial of total degree d satisfies $\mathrm{ord}_p(c) \geq d - \lfloor \mathrm{ord}_p(d!) \rfloor$. In particular, from the power series above, one can see that any integral solution (n_1, n_2) must satisfy $n_1 \equiv n_2 \equiv 0 \bmod 3$ and, since

$$\det \begin{pmatrix} -1 & 0 \\ -29 & -32 \end{pmatrix} \not\equiv 0 \bmod 3,$$

by Hensel's lemma such an integral solution lifts uniquely to $(0, 0)$. In other words, there is at most one point in the coset $G_3 + G_4 + \Lambda_3$ that has a rational image under u. One can do similar arguments for the other fibres of reduction:

```
>    N, V, R, C := Chabauty(mwmap, u, 3: Aux := {7});
>    assert N eq #V;
>    assert #C[2] eq 0;
>    R;
         4

>    V;
         {
             0,
             G.3 - G.4,
             -G.3 + G.4,
             G.3 + G.4,
             -G.3 - G.4
         }

>    { EvaluateByPowerSeries(u, mwmap(P)) : P in V };
         { (-1 : 1), (-1/3 : 1), (-3 : 1) }
```

To interpret the above results, consider that in the previous computations, we have not really used that we have generators of $E(K)$. In fact, for this particular example, we don't know we have. We only know we have generators of some finite odd index subgroup G. For the finite field arguments, we only need that the $[E(K) : G]$ is prime to $[E(\mathcal{O}/\mathfrak{p}_i) : \rho_i(G)]$ for each of the i. Since the power series argument works for $n_1, n_2 \in \mathbb{Z}_p$, we only need that that $[E(K) : G]$ is prime to p as well. However, when computing $\mathrm{Log}_E(b_j)$, we can often already deduce that $p \nmid [E(K) : G]$.

We only need G to be q-saturated in $E(K)$ for finitely many l. The l that are encountered during the computations, are collected as prime divisors of R. In our case, this is only 2 and since we already know G to be 2-saturated in $E(K)$, any conclusions we draw from G will also be valid for $E(K)$. The interpretation of the other return values can be stated as follows.

$$\#\{P \in E(K) : u(P) \in \mathbb{P}^1(\mathbb{Q})\} \leq \mathsf{N}$$

$$\mathsf{V} \subset \{P \in E(K) : u(P) \in \mathbb{P}^1(\mathbb{Q})\} \subset \mathsf{V} \cup \mathsf{C}$$

Here, C is a coset collection of the type we described before. Note that if #V=N then all inequalities above are identities.

The routine **Chabauty** only tries a limited number of techniques to determine p-adic solution and only with finite precision. It uses an adaptation of the algorithm in Section 5.1 to find solutions of multiplicity 1 and it uses a generalisation of [3, Lemma 4.5.1] to test if the solution $(0,\ldots,0)$ is the only integral solution of possibly higher multiplicity. It may therefore fail to produce a finite bound at all. In that case, $\mathsf{N} = \infty$ is returned.

As an advanced example, we also give the computation for $d = -3\zeta^3 - 7\zeta^2 - 8\zeta - 9$.

```
>   d := -3*ζ^3-7*ζ^2-8*ζ-9;
>   E<X,Y,Z> := EllipticCurve(
>        [ 0, (-3*ζ^3-ζ+1)*d, 0, (-ζ^2-ζ-1)*d^2, 0 ]);
>   P1 := ProjectiveSpace(Rationals(), 1);
>   u := map<E → P1 | [-X + (ζ^3 - 1)*d*Z, X+(-ζ^3-ζ)*d*Z]>;
>   success, G, mwmap := PseudoMordellWeilGroup(E);
>   assert success;
>   [ mwmap(P) : P in OrderedGenerators(G) ];
          [ (-41*zeta^3 + 18*zeta^2 - 14*zeta + 42 : 0 : 1),
            (0 : 0 : 1), (-5*zeta^3 + 6*zeta^2 + 9 :
             -69*zeta^3 + 17*zeta^2 - 34*zeta + 60 : 1),
            (-6*zeta^3 + 3*zeta^2 - zeta + 6 : 57*zeta^3 -
             16*zeta^2 + 27*zeta - 51 : 1), (-36*zeta^3 +
             8*zeta^2 - 20*zeta + 32 : 10*zeta^3 + 72*zeta^2 +
             60*zeta + 62 : 1) ]

>   N, V, R, C := Chabauty(mwmap, u, 3);
>   C31, R31 := RelevantCosets(mwmap, u, Support(31*OK));
```

```
>    R := LCM(R, R31);
>    Cnew := CosetIntersection(C, C31: Weak);
>    assert #Cnew[2] eq 0;
>    R;

          2

>    V;

          {
               0,
               G.4 - G.5,
               -G.4 + G.5
          }

>    { EvaluateByPowerSeries(u, mwmap(P)) : P in V };
          { (3/2 : 1), (-1 : 1) }
```

An interesting feature of this example is, that the 3-adic argument by itself is not sufficient. We see that there are two 3-adic "ghost" solutions. The 3-adic computation did come up with a rather precise 3-adic approximation of these putative solutions. The cosets are disjoint from V_{31}, so we proved that they indeed only correspond to $\mathbb{Z}_3$-solutions and not rational ones.

Incidentally, specifying 31 as an "auxiliary" prime, such that V_3 and V_{31} get intersected before the 3-adic argument, would have solved this particular equation as well, as would 191 by itself.

The other 3 values of d mentioned in Section 4 can be solved in a similar way, either with $p = 31$ or $p = 191$.

8 The equations $x^n + y^n = Dz^2$ for $n = 6, 7, 9, 11, 13, 17$

The proof of Theorem 1.2 for the remaining cases is straightforward and, in many cases, easier than for $n = 5$, because there are no non-trivial solutions. For each n, we outline a successful strategy. For full details, we refer the reader to the accompanying electronic resource [8].

$x^6 + y^6 = Dz^2$: Since 6 is even, we can reduce the genus (and the number) of the curves to consider tremendously. Note that a solution with $y \neq 0$ corresponds to a rational point on the genus 2 curve $Y^2 = DX^6 + D$. For $D \in \{2, 3, 5, 6, 10, 11, 13, 17\}$, we conclude that only for $D = 2$ does this curve have points over $\mathbb{Q}_2$ and $\mathbb{Q}_7$. Following the same approach as in [5], we write $2X^6 + 2 = (2X^2 + 2)(X^4 - X^2 + 1)$ and we conclude that any point (X, Y) corresponds to a solution (X, Y_1, Y_2) of

$$dY_1^2 = 2X^2 + 2, \ dY_2^2 = X^4 - X^2 + 1$$

for $d \in \mathbb{Q}(2, \{2, 3\})$. Only for $d = 1$ does this system of equations have solutions over $\mathbb{Q}_2$. The curve $Y_2^2 = X^4 - X^2 + 1$ only has rational points with $X \in \{-1, 0, 1, \infty\}$.

$x^7 + y^7 = Dz^2$: We note that any solution corresponds to a solution to

$$C_d : Y^2 = d(X^6 - X^5 + X^4 - X^3 + X^2 - X + 1)$$

for some $d \in \mathbb{Q}(2, S)$, where S contains the prime divisors of $7D$. For the relevant values of D, only $d = 1, 7$ yield curves with points over $\mathbb{R}, \mathbb{Q}_2, \mathbb{Q}_7, \mathbb{Q}_{11}$.

With [24] it is straightforward to check that $\text{Jac}(C_7)(\mathbb{Q})$ is of free rank 1 and using Stoll's implementation of [18], (3-adically), one finds that all rational points have $X = -1$.

For C_1, one uses [7] and the techniques outlined in Section 7 to show that all rational points have $X \in \{-1, 0, 1, \infty\}$

$x^9 + y^9 = Dz^2$: We factor:

$$\begin{aligned} y_1^2 &= d_1(x^6 - x^3 z^3 + z^6) \\ y_2^2 &= d_2(x^2 - xz + z^2) \\ y_3^2 &= D d_1 d_2 (x + 1) \end{aligned}$$

and note that any primitive solution (x, y, z) gives rise to a solution of the system above for $d_1, d_2 \in \mathbb{Q}(2, S)$, where S contains the prime divisors of $3D$. Furthermore, because $\gcd(x, z) = 1$, we have $\gcd(d_1, d_2) \mid 3$.

We can dehomogenize the first two equations and if we test them for simultaneous solvability over $\mathbb{R}, \mathbb{Q}_2, \mathbb{Q}_3, \mathbb{Q}_5$, we are only left with $d_1 \in \{1, 3\}$. The first equation gives rise to a curve of genus 2 with a Mordell–Weil group of free rank 1. Again, [18] yields that all points have $X \in \{-1, 1, 0, \infty\}$.

$x^{11} + y^{11} = Dz^2$: Using the same argument as for $n = 7$, we find all that solutions correspond to rational points on

$$C_d : Y^2 = d(X^{10} - X^9 + \cdots - X + 1)$$

for $d \in \{1, 11\}$. Rather than applying [9] to C_d directly, we substitute $(U, V) = ((X^2 + 1)/X, Y + Y/X)$ to find the covered curve

$$D_d : V^2 = d(U^6 + U^5 - 6U^4 - 5U^3 + 9U^2 + 5U - 2).$$

For $d = 1$ the free rank of the Mordell–Weil group is bounded above by 2 and for $d = 11$ the free rank is bounded by 1, but we could not find a generator. Using the techniques from [7], we find that $U(D_1(\mathbb{Q})) \subset \{-2, -1, 2, \infty\}$ and that $U(D_{11})(\mathbb{Q})) \subset \{-2, -1, 1, 2, \infty\}$. From this, it follows easily that C_d only has rational points above $X \in \{-1, 0, 1, \infty\}$ for $d = 1, 11$.

$x^{13} + y^{13} = Dz^2$: Using the same argument as for $n = 7$, we find that all solutions correspond to rational points on

$$C_d : Y^2 = d(X^{12} - X^9 + \cdots - X + 1)$$

for $d \in \{1, 13\}$. For $d = 13$ we substitute $(U, V) = ((X^2 + 1)/X, Y/X^3)$ to obtain

$$D_d : V^2 = 13(U^6 - U^5 - 5U^4 + 4U^3 + 6U^2 - 3U - 1).$$

Following [18] yields that $U(D_{13}(\mathbb{Q})) \subset \{-2\}$, which corresponds to $X = -1$.

For $d = 1$ we get a bound on the Mordell–Weil rank of 2, so we use that over $\mathbb{Q}(\beta)$ with $\beta^3 - \beta^2 - 4\beta + 1 = 0$, we have a quartic factor

$$X^4 + \beta X^3 + (\beta^2 + \beta - 1)X^2 + \beta X + 1$$

of $(X^{13} + 1)/(X + 1)$. Using [9] and [7] we find $X(C_1(\mathbb{Q})) \subset \{0, 1, \infty\}$.

$x^{17} + y^{17} = Dz^2$: Using the same argument as for $n = 7$, we find that all solutions correspond to rational points on

$$C_d : Y^2 = d(X^{16} - X^9 + \cdots - X + 1)$$

for $d \in \{1, 17\}$. Over $K = \mathbb{Q}(\beta)$ with $\beta^4 + \beta^3 - 6\beta^2 - \beta + 1 = 0$ we have

$$R(X) := X^4 + \beta X^3 + 1/2(-\beta^3 + 6\beta + 1)X^2 + \beta X + 1$$

with $N_{K/\mathbb{Q}} R(X) = (X^{17} + 1)/(X + 1)$. Hence, any rational point on C_d has an X-coordinate corresponding to a rational point on

$$D_\delta : V^2 = \delta R(X).$$

for some $\delta \in K(2, S)$ with $d N_{K/\mathbb{Q}}(\delta)$ a square, where S contains the primes above $2 \cdot 17 \cdot \delta$.

Local arguments show that only $\delta = 1, \beta^3 + 2\beta^2 - 3\beta + 1$ need consideration. The techniques from Section 7 then show that $X(C_d(\mathbb{Q})) \subset \{-1, 0, 1, \infty\}$ for $d = 1, 17$.

References

1. Michael A. Bennett, Chris M. Skinner, *Ternary Diophantine equations via Galois representations and modular forms*, Canad. J. Math. **56**-1 (2004), 23–54.
2. Wieb Bosma, John Cannon, Catherine Playoust, *The Magma algebra system I: The user language*, J. Symbolic Comput. **24** (1997), 235–265.
   See also the Magma home page at http://magma.maths.usyd.edu.au/magma/.
3. N. R. Bruin, *Chabauty methods and covering techniques applied to generalized Fermat equations*, Dissertation, University of Leiden, Leiden, 1999, CWI Tract **133**, Amsterdam: Stichting Mathematisch Centrum Centrum voor Wiskunde en Informatica, 2002.

4. Nils Bruin, *Algae, a program for 2-Selmer groups of elliptic curves over number fields*, see `http://www.cecm.sfu.ca/~bruin/ell.shar`

5. Nils Bruin, *The Diophantine equations $x^2 \pm y^4 = \pm z^6$ and $x^2 + y^8 = z^3$*, Compositio Math. **118** (1999), 305–321.

6. Nils Bruin, *Visualising Sha[2] in abelian surfaces*, Math. Comp. **73** (2004), 1459–1476 (electronic).

7. Nils Bruin, *Chabauty methods using elliptic curves*, J. reine angew. Math. **562** (2003), 27–49.

8. Nils Bruin, *Transcript of computations*, available from `http://www.cecm.sfu.ca/~bruin/nn2`, 2003.

9. Nils Bruin, E. Victor Flynn, *Towers of 2-covers of hyperelliptic curves*, Technical Report PIMS-**01-12**, PIMS, 2001.
   `http://www.pims.math.ca/publications/#preprints`

10. J. W. S. Cassels, *Diophantine equations with special reference to elliptic curves*, J. London Math. Soc. **41** (1966), 193–291.

11. J. W. S. Cassels, *Second descents for elliptic curves*, J. reine angew. Math. **494** (1998), 101–127. Dedicated to Martin Kneser on the occasion of his 70th birthday.

12. Claude Chabauty, *Sur les points rationnels des variétés algébriques dont l'irrégularité est supérieure à la dimension*, C. R. Acad. Sci. Paris **212** (1941), 1022–1024.

13. J. E. Cremona, *Reduction of binary cubic and quartic forms*, LMS J. Comput. Math. **2** (1999), 64–94 (electronic).

14. John E. Cremona, Barry Mazur, *Visualizing elements in the Shafarevich-Tate group*, Experiment. Math. **9**-1 (2000), 13–28.

15. Henri Darmon, Andrew Granville, *On the equations $z^m = F(x, y)$ and $Ax^p + By^q = Cz^r$*, Bull. London Math. Soc. **27**-6 (1995) 513–543.

16. Claus Fieker, *p-Selmer groups of number fields*, Private communication.

17. E. Victor Flynn, Joseph L. Wetherell, *Finding rational points on bi-elliptic genus 2 curves*, Manuscripta Math. **100**-4 (1999), 519–533.

18. E.V. Flynn, *A flexible method for applying Chabauty's theorem*, Compositio Mathematica **105** (1997), 70–94.

19. J. R. Merriman, S. Siksek, N. P. Smart, *Explicit 4-descents on an elliptic curve*, Acta Arith. **77**-4 (1996), 385–404.

20. Bjorn Poonen, Edward F. Schaefer, *Explicit descent for Jacobians of cyclic covers of the projective line*, J. reine angew. Math. **488** (1997), 141–188.

21. Samir Siksek, *Descent on curves of genus 1*, PhD thesis, University of Exeter, 1995.

22. Joseph H. Silverman, *The Arithmetic of Elliptic Curves*, GTM **106**, Berlin: Springer-Verlag, 1986.

23. Denis Simon, *Computing the rank of elliptic curves over number fields*, LMS J. Comput. Math. **5** (2002), 7–17 (electronic).

24. Michael Stoll, *Implementing 2-descent for Jacobians of hyperelliptic curves*, Acta Arith. **98**-3 (2001), 245–277.

25. Michael Stoll, John E. Cremona, *Minimal models for 2-coverings of elliptic curves*, LMS J. Comput. Math. **5** (2002), 220–243 (electronic).

26. Tom Womack, *Four descent on elliptic curves over $\mathbb{Q}$*, PhD thesis, University of Nottingham, 2003.

Studying the Birch and Swinnerton-Dyer conjecture for modular abelian varieties using Magma

William Stein

Department of Mathematics
Harvard University
Cambridge MA, USA
was@math.harvard.edu

Summary. In this paper we describe the Birch and Swinnerton-Dyer conjecture in the case of modular abelian varieties and how to use Magma to do computations with some of the quantities that appear in the conjecture. We assume the reader has some experience with algebraic varieties and number theory, but do not assume the reader has proficiency working with elliptic curves, abelian varieties, modular forms, or modular symbols. The computations give evidence for the Birch and Swinnerton-Dyer conjecture and increase our explicit understanding of modular abelian varieties.

1 Introduction

In this paper we describe the Birch and Swinnerton-Dyer conjecture in the case of modular abelian varieties and how to use Magma [7] to do computations with some of the quantities that appear in the conjecture. We assume the reader has some experience with algebraic varieties and number theory, but do not assume the reader has proficiency working with elliptic curves, abelian varieties, modular forms, or modular symbols.

In Section 2 we quickly survey abelian varieties, modular forms, Hecke algebras, modular curves, and modular Jacobians, then discuss Shimura's construction of abelian varieties attached to modular forms. In Section 3 we survey many quantities associated to an abelian variety, including the Mordell–Weil group, torsion subgroup, regulator, Tamagawa numbers, real volume, and Shafarevich–Tate group, and use these to state the full Birch and Swinnerton-Dyer conjecture for modular abelian varieties. Section 4 contains some computational results from other papers about the Birch and Swinnerton-Dyer conjecture.

The rest of the paper is about how to use the package that I wrote for Magma [36] to carry out an explicit computational study of modular abelian varieties. Section 5 is about modular symbols and how to compute with them in Magma. In Section 6 we state a theorem that allows us to use Magma to compute subgroups of Shafarevich–Tate groups of abelian varieties. In Section 7 we discuss computation of special values of L-functions. Section 8 is about computing Tamagawa numbers, and in Section 9 we describe how to compute a divisor and multiple of the order of the torsion subgroup. All these computations are pulled together in Section 10 to obtain a conjectural divisor and multiple of the order of the Shafarevich–Tate group of a modular abelian variety of dimension 20. We finish with Section 11, which contains an example in which the level is composite and elements of the Shafarevich–Tate group only becomes "visible" at higher level.

Taken together, these computations give evidence for the Birch and Swinnerton-Dyer conjecture and increase our explicit understanding of modular abelian varieties.

2 Modular abelian varieties

An elliptic curve E over the rational numbers $\mathbb{Q}$ is a one-dimensional commutative compact algebraic group. Such a curve is usually given as the projective closure of an affine curve $y^2 = x^3 + ax + b$, with a and b in $\mathbb{Q}$. The points over the real numbers $\mathbb{R}$ of $y^2 = x^3 - x + 1$ are illustrated in Figure 1.

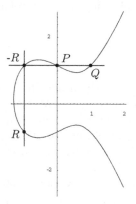

Fig. 1. Adding $P = (0,1)$ to $Q = (1,1)$ to get $R = (-1,-1)$ on $y^2 = x^3 - x + 1$

If P and Q are two distinct points on E, we find their sum as follows: draw the unique line through them and let (x, y) be the third point of intersection of this line with E. Then the sum of P and Q is $R = (x, -y)$, as illustrated in Figure 1. For more about elliptic curves, see [34, 35].

This paper is about abelian varieties, which are compact (commutative) algebraic groups of dimension possibly greater than 1. For example, the Cartesian product of two elliptic curves is an abelian variety of dimension 2.

Explicit equations for abelian varieties are vastly more complicated than for elliptic curves, so algorithms for computing with abelian varieties without recourse to explicit algebraic equations are of great value. In this paper we focus on such algorithms in the case when the abelian variety is endowed with extra structure coming from modular forms.

A cuspidal modular form of weight 2 for

$$\Gamma_0(N) = \left\{ \begin{pmatrix} a & b \\ c & d \end{pmatrix} \in \mathrm{SL}_2(\mathbb{Z}) : N \mid c \right\}$$

is a holomorphic function $f(z)$ on the upper half plane such that for all $\gamma = \begin{pmatrix} a & b \\ c & d \end{pmatrix} \in \Gamma_0(N)$ we have

$$f\left(\frac{az+b}{cz+d} \right) = (cz+d)^2 f(z),$$

and which satisfies certain vanishing conditions at the cusps (see [14, pg. 42] for a precise definition). We denote the finite dimensional complex vector space of all cuspidal modular forms of weight 2 for $\Gamma_0(N)$ by $S_2(\Gamma_0(N))$. Because $\begin{pmatrix} 1 & 1 \\ 0 & 1 \end{pmatrix} \in \Gamma_0(N)$, cuspidal modular forms have a Fourier series representation

$$f(z) = \sum_{n=1}^{\infty} a_n q^n = \sum_{n=1}^{\infty} a_n e^{2\pi i n z}.$$

The Hecke algebra

$$\mathbb{T} = \mathbb{Z}[T_1, T_2, T_3, \ldots] \subset \mathrm{End}(S_2(\Gamma_0(N)))$$

is a commutative ring that is free of rank equal to $\dim_{\mathbb{C}} S_2(\Gamma_0(N))$ (for the definition and basic properties of the Hecke operators T_n, see [14, §3] and the references therein).

A newform is a modular form

$$f = q + \sum_{n \geq 2} a_n q^n$$

that is a simultaneous eigenvector for every element of the Hecke algebra and such that the coefficients $\{a_p : p \nmid N\}$ are not the prime-index coefficients of another eigenform of some level that strictly divides N.

The group $\Gamma_0(N)$ acts as a discrete group of linear fractional transformations on the upper half plane; the quotient of the upper half plane by this action is a non-compact Riemann surface. Its compactification has the structure of algebraic curve over $\mathbb{Q}$, i.e., the compactification is the set of complex points of an algebraic curve $X_0(N)$ defined by polynomial equations with coefficients in $\mathbb{Q}$.

A divisor on an algebraic curve X is an element of the free abelian group generated by the points of X. For example, if f is a rational function on X then

$$(f) = \text{(formal sum of poles of } f) - \text{(formal sum of zeros of } f)$$

is a divisor on X, where the sums are with multiplicity. Two divisors D_1 and D_2 are linearly equivalent if there is a rational function f on X such that $D_1 - D_2 = (f)$. The Jacobian J of an algebraic curve X is an abelian variety of dimension equal to the genus (number of holes in the Riemann surface $X(\mathbb{C})$) of X such that the underlying group of J is naturally isomorphic to the group of divisor classes of degree 0 on X. Let $J_0(N)$ denote the Jacobian of $X_0(N)$.

Similarly, let

$$\Gamma_1(N) = \left\{ \begin{pmatrix} a & b \\ c & d \end{pmatrix} \in \mathrm{SL}_2(\mathbb{Z}) : N \mid c \text{ and } a \equiv 1 \pmod{N} \right\},$$

define $X_1(N)$ similarly, and let $J_1(N)$ be the Jacobian of $X_1(N)$.

A modular abelian variety is an abelian variety A for which there exists a surjective morphism $J_1(N) \to A$. Modular abelian varieties are appealing objects to study. For example, it is a deep theorem that every elliptic curve over $\mathbb{Q}$ is modular (see [8, 42, 43]), and this implies Fermat's Last Theorem (see [26, Cor. 1.2]). In [28], Ken Ribet conjectured that the simple abelian varieties over $\mathbb{Q}$ of "GL$_2$-type" are exactly the simple modular abelian varieties. A closely related conjecture of Serre (see [30, pg. 179] and [29]) asserts that every odd irreducible Galois representation

$$\rho : \mathrm{Gal}(\overline{\mathbb{Q}}/\mathbb{Q}) \to \mathrm{GL}_2(\overline{\mathbb{F}}_p)$$

is "modular"; this conjecture is equivalent to the assertion that ρ can be realized (up to twist) as the action of $\mathrm{Gal}(\overline{\mathbb{Q}}/\mathbb{Q})$ on a subgroup of the points on some $J_1(N)$ (see [29, §3.3.1] for a partial explanation). Though Serre's conjecture is still far from proved, it implies Ribet's conjecture (see [28, Thm. 4.4]).

We now return to considering $\Gamma_0(N)$, though we could consider $\Gamma_1(N)$ for everything in the rest of this section. The Hecke algebra $\mathbb{T}$, which we introduced above as a ring of linear transformations on $S_2(\Gamma_0(N))$, also acts via endomorphisms on $J_0(N)$.

In order to construct Galois representations attached to modular forms, Goro Shimura (see [32, §1] and [33, §7.14]) associated to each newform $f = \sum a_n q^n$ a simple abelian variety A_f defined over $\mathbb{Q}$. Let I_f be the ideal of elements of $\mathbb{T}$ that annihilate f. Then

$$A_f = J_0(N)/I_f J_0(N).$$

The dimension of A_f equals the degree of the field generated over $\mathbb{Q}$ by the coefficients a_n of f. Note that A_f need not be simple over $\overline{\mathbb{Q}}$.

We will frequently mention the dual $A_f^\vee$ below. The dual can be considered as an abelian subvariety of $J_0(N)$, by using that Jacobians are canonically self dual and the dual of the quotient map $J_0(N) \to A_f$ is an inclusion $A_f^\vee \hookrightarrow J_0(N)$. Note that $A_f^\vee$ is the connected component of the intersection of the kernels of all elements of I_f.

We say that a newform g is a Galois conjugate of f if there is σ in $\mathrm{Gal}(\overline{\mathbb{Q}}/\mathbb{Q})$ such that $g = \sum \sigma(a_n)q^n$. If g is a Galois conjugate of f, then $A_f = A_g$; if g is not a conjugate of f then the only homomorphism from A_f to A_g is the zero map. (A nonzero homomorphism $A_f \to A_g$ would induce an isogeny of Tate modules, from which one could deduce that f and g are Galois conjugate.)

We will concern ourselves almost entirely with these modular abelian varieties attached to newforms, because, as mentioned above, there are a number of algorithms for computing with them that do not require explicit algebraic equations (see [2, 3, 10, 11, 16, 20, 37, 39]). Also, it follows from standard results about constructing spaces of cusp forms from newforms, which can be found in [4, 23], that every modular abelian variety is isogenous to a product of abelian varieties of the form A_f. (An isogeny of abelian varieties is a surjective homomorphism with finite kernel.)

3 The Birch and Swinnerton-Dyer Conjecture

In the 1960s Bryan Birch and Peter Swinnerton-Dyer did computations with elliptic curves at Cambridge University on the EDSAC computer (see, e.g., [5]). These computations led to earth-shattering conjectures about the arithmetic of elliptic curves over $\mathbb{Q}$. Tate [41] formulated their conjectures in a more functorial way that generalized them to abelian varieties over global fields (such as the rational numbers). We now state their conjectures below for modular abelian varieties over $\mathbb{Q}$.

Let A_f be a modular abelian variety. Mordell and Weil proved that the abelian group $A_f(\mathbb{Q})$ of rational points on A_f is finitely generated, so it is isomorphic to $\mathbb{Z}^r \times T$ where T is the finite group $A_f(\mathbb{Q})_{\mathrm{tor}}$ of all elements of finite order in $A_f(\mathbb{Q})$. The exponent r is called the Mordell–Weil rank of A_f.

If f is a newform, the L-function of f is defined by the Dirichlet series $L(f,s) = \sum_{n \geq 1} a_n n^{-s}$. Hasse showed that $L(f,s)$ has an analytic continuation to a holomorphic function on the whole complex plane. The Hasse–Weil L-function of A_f is

$$L(A_f, s) = \prod L(g, s)$$

where the product is over the Galois conjugates g of f. The analytic rank of A_f is $\mathrm{ord}_{s=1} L(A_f, s)$.

We are now ready to state the first part of the conjecture.

Conjecture 3.1 (Birch and Swinnerton-Dyer) *The analytic rank of A_f is equal to the Mordell–Weil rank of A_f.*

Remark 3.2

1. It is an open problem to give, with proof, an example of an elliptic curve with analytic rank at least 4. No examples with analytic rank at least 3 were known until the deep theorem of [17, Prop. 7.4].
2. When A_f is an elliptic curve, Conjecture 3.1 is the Clay Mathematics Institute Millennium Prize Problem from arithmetic geometry [18], so it has received much publicity.

In order to explain the conjecture of Birch and Swinnerton-Dyer about the leading coefficient of $L(A_f, s)$ at $s = 1$, we introduce the regulator, real volume, Tamagawa numbers, and Shafarevich–Tate group of A_f. Most of what we say below is true for a general abelian variety over a global field; the notable exceptions are that we do not know that the L-function is defined on the whole complex plane, and there are hardly any cases in general when the Shafarevich–Tate group is known to be finite.

Let $A_f(\mathbb{Q})/\text{tor}$ denote the quotient of $A_f(\mathbb{Q})$ by its torsion subgroup, so $A_f(\mathbb{Q})/\text{tor}$ is isomorphic to $\mathbb{Z}^r$, where r is the Mordell–Weil rank of A_f. The height pairing is a nondegenerate bilinear pairing

$$h : A_f(\mathbb{Q})/\text{tor} \times A_f^\vee(\mathbb{Q})/\text{tor} \to \mathbb{R}.$$

The regulator Reg_{A_f} of A_f is the absolute value of the determinant of any matrix whose entries are $h(P_i, P_j')$, where $P_1, \ldots, P_r$ are a basis for $A_f(\mathbb{Q})/\text{tor}$ and the $P_1, \ldots, P_r$ are a basis for $A_f^\vee(\mathbb{Q})/\text{tor}$. When $A_f(\mathbb{Q})$ has rank zero, the regulator is 1.

We use a certain integral model of A_f to define the real volume and Tamagawa numbers of A_f. The Néron model $\mathcal{A}$ of A_f, whose existence was established by Néron in [25] (see also [6, Ch. 1]), is a canonical object associated to A_f that is defined over $\mathbb{Z}$. The Néron model can be reduced modulo p for every prime p, and when base extended to $\mathbb{Q}$, the Néron model is isomorphic to A_f. The Néron model $\mathcal{A}$ is determined, up to unique isomorphism, by the following properties, which the reader unfamiliar with schemes can safely ignore: $\mathcal{A}$ is a smooth commutative group scheme over $\mathbb{Z}$ such that whenever S is a smooth scheme over $\mathbb{Z}$ the restriction map

$$\text{Hom}(S, \mathcal{A}) \to \text{Hom}_{\mathbb{Q}}(S_{\mathbb{Q}}, A)$$

is a bijection.

The real volume Ω_{A_f} of A_f is the absolute value of the integral over $A_f(\mathbb{R})$ of $h_1 \wedge \cdots \wedge h_d$ where $h_1, \ldots, h_d$ are a basis for the holomorphic 1-forms on $\mathcal{A}$. Using various identifications as in [1, §2.2.2] one sees that the $\mathbb{Z}$-span M of $h_1, \ldots, h_d$ can be viewed as a submodule of

$$W = S_2(\Gamma_0(N), \mathbb{Z}) \cap (\mathbb{C}f_1 \oplus \cdots \oplus \mathbb{C}f_d)$$

where $f_1, \ldots, f_d$ are the $\text{Gal}(\overline{\mathbb{Q}}/\mathbb{Q})$ conjugates of f. We call the index of M in W the Manin constant of A_f, and conjecture (see [1]) that the Manin constant is 1. This conjecture would imply that a basis $h_1, \ldots, h_d$ can be computed, since W can be computed.

The reduction modulo p of $\mathcal{A}$ is an algebraic group $\mathcal{A}_{\mathbb{F}_p}$ over the finite field $\mathbb{F}_p$ with p elements. If p does not divide N, then this group is connected, but when p divides N, the reduction $\mathcal{A}_{\mathbb{F}_p}$ need not be connected. Let

$$\varPhi_{A,p} = \mathcal{A}_{\mathbb{F}_p}/\mathcal{A}^0_{\mathbb{F}_p}$$

be the finite group of components. The Tamagawa number of A_f at p, denoted c_p, is the number of $\mathbb{F}_p$-rational components of the reduction of $\mathcal{A}$ modulo p, so $c_p = \#\varPhi_{A,p}(\mathbb{F}_p)$.

The only object left to define before we state the second part of the Birch and Swinnerton-Dyer conjecture is the Shafarevich–Tate group of A_f. This is a group that measures the failure of a certain local-to-global principle for A_f. To give an exact description, we let $\mathrm{H}^1(\mathbb{Q}, A_f)$ be the first Galois cohomology group of A_f, which is a torsion group with infinitely many elements of any order bigger than 1 (see [31] for a proof in the case when A_f is an elliptic curve; the top of page 278 of [9] also purports to contain a proof). More precisely, $\mathrm{H}^1(\mathbb{Q}, A_f)$ is the set of equivalence classes of maps $c : \mathrm{Gal}(\overline{\mathbb{Q}}/\mathbb{Q}) \to A_f(\overline{\mathbb{Q}})$, with finite image, such that $c(\sigma\tau) = c(\sigma) + \sigma c(\tau)$, and two classes c_1 and c_2 are equivalent if there exists P in $A_f(\overline{\mathbb{Q}})$ such that $c_1(\sigma) - c_2(\sigma) = \sigma(P) - P$ for all $\sigma \in \mathrm{Gal}(\overline{\mathbb{Q}}/\mathbb{Q})$. For each prime p we define $\mathrm{H}^1(\mathbb{Q}_p, A_f)$ analogously, but with the rational numbers $\mathbb{Q}$ replaced by the p-adic numbers $\mathbb{Q}_p$. Also, we allow $p = \infty$, in which case $\mathbb{Q}_p = \mathbb{R}$. Then

$$\mathrm{III}(A_f) = \ker\left(\mathrm{H}^1(\mathbb{Q}, A_f) \longrightarrow \bigoplus_{\text{primes } p \leq \infty} \mathrm{H}^1(\mathbb{Q}_p, A_f) \right).$$

We are now ready to state the full Birch and Swinnerton-Dyer conjecture for modular abelian varieties A_f.

Conjecture 3.3 *Let $A = A_f$ be a modular abelian variety attached to a newform, and let $r = \mathrm{ord}_{s=1} L(A, s)$ be the analytic rank of A. Then*

$$\frac{L^{(r)}(A, 1)}{r!} = \frac{\prod c_p \cdot \Omega_A \cdot \mathrm{Reg}_A}{\#A(\mathbb{Q})_{\mathrm{tor}} \cdot \#A^\vee(\mathbb{Q})_{\mathrm{tor}}} \cdot \#\mathrm{III}(A).$$

Recall that $L^{(r)}(A_f, 1)$ makes sense at $s = 1$ because A_f is attached to a modular form. Also Kato established in [19, Cor. 14.3] that if $L(A_f, 1) \neq 0$ then $\mathrm{III}(A_f)$ is finite, and Kolyvagin–Logachev ([21, Thm. 0.3]) proved that if f is a modular form in $S_2(\Gamma_0(N))$ and $\mathrm{ord}_{s=1} L(f, s) \leq 1$, then $\mathrm{III}(A_f)$ is finite. When the theorems of Kato, Kolyvagin, and Logachev do not apply, we do not know even one example of a modular abelian variety A_f for which $\mathrm{III}(A_f)$ is provably finite. John Tate once remarked that Conjecture 3.3 (for arbitrary abelian varieties) relates the value of a function where it is not known to be defined to the order of a group that is not known to be finite.

The rest of this paper is about how to use Magma to gather computational evidence for Conjecture 3.3, a task well worth pursuing. Elliptic curves are naturally surrounded by modular abelian varieties, so we want to understand modular abelian varieties well in order to say something about Conjectures 3.1–3.3 for elliptic curves. Doing explicit computations about these conjectures results in stimulating tables of data about modular abelian varieties, which could never be obtained except by direct computation. Until [2, 16] there were very few nontrivial computational examples of Conjecture 3.3 for abelian varieties in the literature, so it is important to test the conjecture since we might find a counterexample. Trying to compute information about a conjecture stimulates development of algorithms and theorems about that conjecture. Finally, our computations may lead to refinements of Conjecture 3.3 in the special case of modular abelian varieties; for example, most objects in Conjecture 3.3 are modules over the Hecke algebra so there should be more precise module-theoretic versions of the conjecture.

4 Some computational results

In [2] we use Magma to compute some of the arithmetic invariants of the 19608 abelian variety quotients A_f of $J_0(N)$ with $N \leq 2333$. Over half of these A_f have analytic rank 0, and for these we compute a divisor and a multiple of the order of $\text{III}(A_f)$ predicted by Conjecture 3.3. We find that there are at least 168 abelian varieties A_f such that the Birch and Swinnerton-Dyer Conjecture implies that $\#\text{III}(A_f)$ is divisible by an odd prime, and we use Magma to show that for 37 of these the odd part of the conjectural order of $\text{III}(A_f)$ divides $\#\text{III}(A_f)$ by constructing nontrivial elements of $\text{III}(A_f)$ using visibility theory. The challenge remains to show that the remaining 131 abelian varieties A_f have odd part of $\text{III}(A_f)$ divisible by the odd part of the conjectural order of $\text{III}(A_f)$ (we successfully take up this challenge for one example of level 551 in Section 11 of the present paper).

In [10, §2 and §7] we investigate Conjecture 3.1–3.3 when A_f is a quotient of $J_1(p)$ with p prime. In particular, we compute some of the invariants of every A_f for $p \leq 71$.

It was once thought by some mathematicians that Shafarevich–Tate groups of abelian varieties would have order a perfect square (or at least twice a perfect square). This is false, as we showed in the paper [38], where we use Magma to prove that for every odd prime $p < 25000$ there is an abelian variety whose Shafarevich–Tate group has order pn^2 with n an integer.

Much of the data mentioned above is of interest even if the full Birch and Swinnerton-Dyer conjecture were known, since this data could probably never be discovered without considerable computation, even assuming the conjectures were true.

The rest of this paper is about how to use Magma to do computations with newform quotients A_f of $J_0(N)$ as in [2]. These computations involve modular symbols, which underly most algorithms for working with modular abelian varieties. (While this book was in press, I added an interface to Magma for computing directly with modular abelian varieties, so that no explicit mention of modular symbols is required. Nonetheless, the functions described in this chapter should still work as illustrated below.)

Remark 4.1 From a computational point of view, it is difficult to give evidence for Conjecture 3.1 when the dimension is greater than 1 in cases not covered by the general theorems of Kato, Kolyvagin, and Logachev. To give new evidence we would have to consider a modular abelian variety A_f with either f a newform in $S_2(\Gamma_0(N))$ and $\mathrm{ord}_{s=1} L(f, s) > 1$, or f a newform in $S_2(\Gamma_1(N))$ but not in $S_2(\Gamma_0(N))$ and $\mathrm{ord}_{s=1} L(f, s) > 0$. We would then show that $A_f(\mathbb{Q})$ is infinite, and more precisely that it has the rank predicted by Conjecture 3.1. In the above 2 cases the only known way to show that $A_f(\mathbb{Q})$ is infinite is to exhibit a point of infinite order in $A_f(\mathbb{Q})$, and this seems to require knowing equations for A_f. Also when $L(A_f, 1) = 0$, Conjecture 3.3 involves a regulator term, which we do not know how to compute without explicitly finding the points on a model for A_f. Thus we will focus on giving evidence for Conjecture 3.3 in the case when $L(f, 1) \neq 0$.

5 Modular symbols

In this section we describe how modular symbols are related to homology of modular curves, and illustrate how to compute with modular symbols in Magma. We also discuss computing decomposition of modular symbols spaces and, for efficiency reasons, computing in the $+1$ quotient.

Let N be a positive integer. The integral homology $H_1(X_0(N), \mathbb{Z})$ of the modular curve $X_0(N)$ is a free abelian group of rank equal to the genus of $X_0(N)$. The Hecke algebra $\mathbb{T} = \mathbb{Z}[T_1, T_2, T_3, \ldots]$ acts on a $H_1(X_0(N), \mathbb{Z})$ as a ring of homomorphisms and makes $H_1(X_0(N), \mathbb{Z})$ into a $\mathbb{T}$-module. This section is concerned with how to compute with this module using Magma. Section 12 contains a complete log of all Magma computations given below.[*]

Modular symbols provide a finite computable presentation for the homology of $X_0(N)$ along with the action of the Hecke algebra $\mathbb{T}$ on this homology. The relative rational homology $H_1(X_0(N), \mathbb{Q}, \text{cusps})$ is the rational homology of $X_0(N)$ relative to the cusps; it is the finitely generated free abelian group of homology equivalence classes of geodesic paths from α to β, where α and β lie in $\mathbb{P}^1(\mathbb{Q}) = \mathbb{Q} \cup \{\infty\}$. A finite presentation for $H_1(X_0(N), \mathbb{Q}, \text{cusps})$ can be found in [24]. For simplicity, we typically compute $H_1(X_0(N), \mathbb{Q}, \text{cusps})$ first, then find $H_1(X_0(N), \mathbb{Z})$ inside $H_1(X_0(N), \mathbb{Q}, \text{cusps})$ if it is needed. We now illustrate how Magma can compute a basis for $H_1(X_0(N), \mathbb{Q}, \text{cusps})$, and,

---

[*]See the Preface to this volume for style conventions regarding Magma code; code appearing in this book is available at http://magma.maths.usyd.edu.au/magma/.

given arbitrary α and β in $\mathbb{P}^1(\mathbb{Q})$, find an equivalent linear combination of basis elements.

M := ModularSymbols(389);
Basis(M);

The output of Basis(M) begins with the symbol $\{-1/337, 0\}$. Figure 2 illustrates how the expression $\{-1/337, 0\}$ represents the relative rational homology class determined by a geodesic path from $-1/337$ to 0 in the upper half plane. The cusps determined by $-1/337$ and 0 are equivalent by an element of $\Gamma_0(389)$, so the image of the geodesic path in the 32 holed torus $X_0(389)(\mathbb{C})$ is a closed loop.

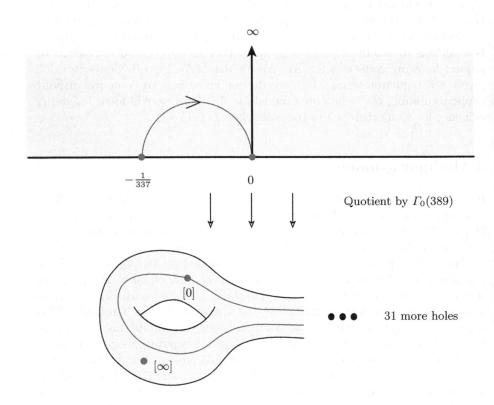

Fig. 2. The Modular Symbol $\{-1/337, 0\}$

The following Magma code illustrates how to find the image in the relative homology of an arbitrary path between cusps. The extra $<$ and $>$ are needed because we are considering modular symbols of weight $k = 2$; in general there is a coefficient which is a homogeneous polynomial of degree $k-2$, which is the first argument to the coercion. The Cusps()| part of the expression is needed so that the sequence is a sequence of cusps (this is not required if both cusps are rational numbers).

```
M ! <1, [Cusps() | −1/337, Infinity()] > ;
```

For more about computing with modular symbols, see [12, 13, 24, 37, 39].

Precise relationships between $H_1(X_0(N), \mathbb{Q})$ and $S_2(\Gamma_0(N))$, along with some linear algebra, make it possible for us to compute a basis of $S_2(\Gamma_0(N))$ from knowledge about $H_1(X_0(N), \mathbb{Q})$ as a $\mathbb{T}$-module. The following code, which computes a basis for $S_2(\Gamma_0(389))$, computes $H_1(X_0(389), \mathbb{Q})$ and uses it to deduce the basis.

```
S := CuspForms(389);
SetPrecision(S, 40);
Basis(S);
```

The SetPrecision command sets the output precision for q-expansions. The computed basis consists of q-expansions with coefficients in $\mathbb{Z}$.

Using NewformDecomposition, we find the submodules of $H_1(X_0(389), \mathbb{Q})$ that correspond to Galois-conjugacy classes of newforms. These in turn correspond to the modular abelian varieties A_f attached to newforms. Magma excels at dense linear algebra over $\mathbb{Q}$ and is highly optimized for computing these decompositions. The following commands compute a decomposition of the new subspace of $H_1(X_0(389), \mathbb{Q})$ corresponding to newforms.

```
M := ModularSymbols(389);
N := NewSubspace(CuspidalSubspace(M));
NewformDecomposition(N);
```

Since 389 is prime, the NewSubspace command is not necessary since everything is automatically new (there are no nonzero cusp forms of level 1 and weight 2). The decomposition consists of five factors of dimensions 2, 4, 6, 12, and 40; these correspond to newforms defined over fields of degrees 1, 2, 3, 6, and 20, respectively, which in turn correspond to abelian varieties over $\mathbb{Q}$ of dimensions 1, 2, 3, 6, and 20, respectively.

Remark 5.1 When information about the powers of 2 appearing in Conjecture 3.3 is not needed, we can instead do all computations in the "+1 quotient" of the space of modular symbols, which has half the dimension.

```
M := ModularSymbols(389, 2, +1); // the plus one quotient
```

6 Visibility theory

Mazur introduced the notion of visibility to unify diverse ideas for constructing elements of Shafarevich–Tate groups. In this section we define what it means for an element of the Shafarevich–Tate group to be visible, state a theorem that allows us to compute pieces of this visible subgroup in some cases, and illustrate the theorem with a 20 dimensional abelian variety of level 389.

Suppose $i : A \to J$ is an injective morphism of abelian varieties over $\mathbb{Q}$. Then the visible subgroup of $\mathrm{III}(A)$ is the kernel of the induced map $\mathrm{III}(A) \to \mathrm{III}(J)$.

Our interest in visibility in the present paper is that it allows us to obtain a provable divisor of $\#\mathrm{III}(A)$, which is useful in giving evidence for Conjecture 3.3. The following theorem is proved in [3, Thm. 3.1] for abelian varieties over number fields.

Theorem 6.1 *Let A and B be abelian subvarieties of an abelian variety J over $\mathbb{Q}$ such that $A(\overline{\mathbb{Q}}) \cap B(\overline{\mathbb{Q}})$ is finite. (Note that J need not be a Jacobian.) Let N be an integer divisible by the residue characteristics of primes of bad reduction for B (so if A and B are modular then N is the level). Suppose p is an odd prime and that*

$$p \nmid N \cdot \#(J/B)(\mathbb{Q})_{\mathrm{tor}} \cdot \#B(\mathbb{Q})_{\mathrm{tor}} \cdot \prod_p c_{A,p} \cdot c_{B,p},$$

where $c_{A,p} = \#\Phi_{A,p}(\mathbb{F}_p)$ (resp., $c_{B,p}$) is the Tamagawa number of A (resp., B) at p. Suppose furthermore that $B(\overline{\mathbb{Q}})[p] \subset A(\overline{\mathbb{Q}})$, where both are viewed as subgroups of $J(\overline{\mathbb{Q}})$. Then there is a natural map

$$\varphi : B(\mathbb{Q})/pB(\mathbb{Q}) \to \mathrm{III}(A)[p]$$

such that

$$\dim_{\mathbb{F}_p} \ker(\varphi) \leq \dim_{\mathbb{Q}} A(\mathbb{Q}) \otimes \mathbb{Q}.$$

In particular, if A has Mordell–Weil rank 0, then φ is injective.

Let A be the 20 dimensional quotient of $J_0(389)$ attached to a newform and B the elliptic curve quotient of $J_0(389)$. We use Magma to verify the hypothesis of Theorem 6.1 for $J = A^{\vee} + B^{\vee} \subset J_0(389)$ with $p = 5$, and hence deduce that $B(\mathbb{Q})/5B(\mathbb{Q}) = (\mathbb{Z}/5\mathbb{Z}) \times (\mathbb{Z}/5\mathbb{Z})$ injects into $\mathrm{III}(A)$.

Since A and B are quotients of $J_0(389)$, we have $N = 389$. Next we construct the corresponding spaces A and B of modular symbols.

```
M := ModularSymbols(389);
N := NewSubspace(CuspidalSubspace(M));
D := SortDecomposition(NewformDecomposition(N));
A := D[5]; B := D[1];
```

The command IntersectionGroup computes the group structure of the intersection of two abelian subvarieties. In our case these are the abelian varieties

$A^\vee$ and $B^\vee$, and we find that $A^\vee \cap B^\vee = (\mathbb{Z}/20\mathbb{Z}) \times (\mathbb{Z}/20\mathbb{Z})$. In particular, $B^\vee[5] = (\mathbb{Z}/5\mathbb{Z}) \times (\mathbb{Z}/5\mathbb{Z})$ is contained in $A^\vee$ as abelian subvarieties of $J_0(389)$.

IntersectionGroup(A, B);

Using the TorsionBound command (see Section 9 below), we obtain a multiple of the order of the torsion subgroup of B (it is 1) and of J/B (it is 97).

TorsionBound(A, 7);
TorsionBound(B, 7);

Neither torsion subgroup has order divisible by 5, so we can apply Theorem 6.1. The reason that TorsionBound($A,7$) is a multiple of the order of the torsion subgroup of J/B is because TorsionBound is an isogeny invariant and A is isogenous to J/B. (The kernel of the natural map from A to J/B is $A \cap B = (\mathbb{Z}/20\mathbb{Z}) \times (\mathbb{Z}/20\mathbb{Z})$, which is finite.)

Finally, we compute the Tamagawa numbers of A and B and obtain 97 and 1, respectively (see Section 8 below).

TamagawaNumber(A, 389);
TamagawaNumber(B, 389);

Putting everything together we see that $B(\mathbb{Q})/5B(\mathbb{Q})$ is a subgroup of $\text{III}(A)$. Finally, using the Rank command on the elliptic curve attached to B, we see that $B(\mathbb{Q})/5B(\mathbb{Q}) = (\mathbb{Z}/5\mathbb{Z}) \times (\mathbb{Z}/5\mathbb{Z})$.

E := EllipticCurve(B);
Rank(E);

Thus 25 divides $\#\text{III}(A)$, which gives evidence for Conjecture 3.3, as we will see in Section 10.

Frequently not all of $\text{III}(A)$ can be constructed using Theorem 6.1 and abelian subvarieties B of $J_0(N)$. One obstruction to visibility arises from a canonical homomorphism from $A^\vee$ to A. Jacobians of curves are canonically isomorphic to their dual abelian variety and the composition $A^\vee \to J_0(389)^\vee \cong J_0(389) \to A$ defines a homomorphism from $A^\vee$ to A. According to [3, §5.3], if p does not divide the kernel of $A^\vee \to A$, then no element of order p in $\text{III}(A)$ is visible in $J_0(N)$. The command ModularKernel computes the group structure of the kernel of $A^\vee \to A$.

G := ModularKernel(A);
Factorization($\#G$);

We find that the modular kernel has order $2^{24}5^2$, so any element of $\text{III}(A^\vee)$ that is visible in $J_0(389)$ has order divisible only by 2 and 5.

7 Computing special values of modular L-function

This section is about computing the quotient $L(A_f, 1)/\Omega_{A_f}$. We discuss the Manin constant and the LRatio command.

Let $A = A_f$ for some newform f and assume that $L(A, 1) \neq 0$. We can then rewrite Conjecture 3.3 as follows:

$$\frac{L(A, 1)}{\Omega_A} = \frac{\prod c_p \cdot \#\text{III}(A)}{\#A(\mathbb{Q})_{\text{tor}} \cdot \#A^\vee(\mathbb{Q})_{\text{tor}}}.$$

We do not know an algorithm, in general, to compute $L(A, 1)/\Omega_A$. However, we can compute $c_A \cdot L(A, 1)/\Omega_A$, where c_A is the Manin constant, which is defined in [1, §2.2]. We conjecture that $c_A = 1$, and prove in [1, §2.2.2] that if f is a newform on $\Gamma_0(N)$ then c_A is an integer divisible only by primes whose square divides $4N$. Moreover, if N is odd then $2^{\dim A}$ is the largest power of 2 that can divide c_A. See also [15] for results when A has dimension 1, and [10, §6.1.2] for a proof that c_A is an integer when $\Gamma_0(N)$ is replaced by $\Gamma_1(N)$.

The algorithm described in [10, §2.1.3], [2, §4] and [39, §3.10] to compute $c_A \cdot L(A, 1)/\Omega_A$ is implemented in Magma via the LRatio command

LRatio(A, 1);

For example, if A is as in Section 6, then $c_A \cdot L(A, 1)/\Omega_A = 2^{11} \cdot 5^2/97$.

8 Computing Tamagawa numbers

In this section we discuss computing Tamagawa numbers when $p \, || \, N$ and some bounds when $p^2 \, | \, N$. We also discuss issues that arise in going from the order of the component group to the Tamagawa number when $p \, || \, N$.

Let $A = A_f$ be a modular abelian variety attached to a newform $f \in S_2(\Gamma_0(N))$. When $p \, || \, N$, [11, §2.1] contains a computable formula for $\#\Phi_{A,p}(\overline{\mathbb{F}}_p)$ and for $c_p = \#\Phi_{A,p}(\mathbb{F}_p)$, where the latter formula is in some cases only valid up to a bounded power of 2. Also [20] is about how to compute these orders. Note that the Tamagawa number of A at p is the same as the Tamagawa number of $A^\vee$ at p.

When $p^2 \, | \, N$ the authors do not know an algorithm to compute c_p. However, in this case Lenstra and Oort proved in [22, Cor. 1.15] that

$$\sum_{\ell \neq p} (\ell - 1) \text{ord}_\ell(\#\Phi_{A,p}(\overline{\mathbb{F}}_p)) \leq 2 \dim(A_f),$$

so if $\ell \, | \, \#\Phi_{A,p}(\overline{\mathbb{F}}_p)$ then $\ell \leq 2 \cdot \dim(A_f) + 1$ or $\ell = p$. (Here $\text{ord}_\ell(x)$ denotes the exponent of the largest power of ℓ that divides x.)

Using [11], when $p \, || \, N$ we know how to compute the order of the component group over the algebraic closure, but not its structure as a group. The command ComponentGroupOrder computes the order of $\Phi_{A,p}(\overline{\mathbb{F}}_p)$. The command TamagawaNumber computes $c_p = \#\Phi_{A,p}(\mathbb{F}_p)$ when the subgroup of elements of $\Phi_{A,p}(\overline{\mathbb{F}}_p)$ fixed by the Galois group has order that does not depend on the underlying group structure. By computing the Atkin–Lehner involution on modular symbols, we can decide whether the Galois group acts trivially or by -1 on $\Phi_{A,p}(\overline{\mathbb{F}}_p)$ since the Atkin–Lehner involution acts as the negative of the canonical generator Frobenius of $\mathrm{Gal}(\overline{\mathbb{F}}_p/\mathbb{F}_p)$. We can thus compute $\#\Phi_{A,p}(\mathbb{F}_p)$ when the Galois group acts trivially. When the Galois group acts nontrivially, $\Phi_{A,p}(\mathbb{F}_p)$ is the 2-torsion subgroup of $\Phi_{A,p}(\overline{\mathbb{F}}_p)$, whose order we know as long as 4 does not divide $\#\Phi_{A,p}(\overline{\mathbb{F}}_p)$. It is an open problem to give an algorithm to compute the group structure of $\Phi_{A,p}(\overline{\mathbb{F}}_p)$ or the order of $\Phi_{A,p}(\mathbb{F}_p)$ in general.

Section 11 contains an example of an abelian variety of dimension 18 in which the author is only able to find the Tamagawa number up to a controlled power of 2.

9 Computing the torsion subgroup

In this section we describe how to compute a divisor and multiple of the order of the torsion subgroup and explain how knowing a divisor of $\#A_f(\mathbb{Q})_{\mathrm{tor}}$ yields a divisor of $\#A_f^\vee(\mathbb{Q})_{\mathrm{tor}}$.

The papers [2, §3.5–3.6] and [10, §2.1.1] contain discussions of how to compute a divisor and multiple of the order of the torsion subgroup $A_f(\mathbb{Q})_{\mathrm{tor}}$ of $A_f(\mathbb{Q})$, and likewise for $A_f^\vee(\mathbb{Q})_{\mathrm{tor}}$. (The multiple of $\#A_f^\vee(\mathbb{Q})_{\mathrm{tor}}$ is the same as for $A_f(\mathbb{Q})_{\mathrm{tor}}$, and the divisor can be computed as described below.) We compute the multiple by using that $A_f(\mathbb{Q})_{\mathrm{tor}}$ injects into $A_f(\mathbb{F}_p)$ for all p not dividing $2N$, and that $\#A_f(\mathbb{F}_p)$ is fairly easy to compute. We compute the lower bound by considering the subgroup of elements of $J_0(N)(\mathbb{Q})_{\mathrm{tor}}$ generated by rational cusps on $X_0(N)$ (see [40, §1.3]), and taking its image in $A_f(\mathbb{Q})_{\mathrm{tor}}$ or intersecting its image with $A_f^\vee(\mathbb{Q})_{\mathrm{tor}} \subset J_0(N)(\mathbb{Q})_{\mathrm{tor}}$. Note that there is no reason for the subgroup generated by rational cusps to equal the rational subgroup of the group generated by all cusps, and one might want to compute and work with this possibly larger group instead.

Let A and B be as in Section 6, where we showed that the torsion subgroup of B is trivial and the order of $B(\mathbb{Q})$ and $B^\vee(\mathbb{Q})$ divides 97. In Section 10, we give an example in which the divisor and multiple of the order of the torsion subgroup differ by a power of 2.

RationalCuspidalSubgroup(A); // subgroup of $A(\mathbb{Q})$

As mentioned in Section 6, there is a homomorphism $A^\vee \to A$ of degree $2^{24} \cdot 5^2$, which implies that 97 also divides $\#A_f^\vee(\mathbb{Q})_{\mathrm{tor}}$. Thus $\#A_f(\mathbb{Q})_{\mathrm{tor}} = \#A_f^\vee(\mathbb{Q})_{\mathrm{tor}} = 97$.

10 A divisor and multiple of the order of the Shafarevich–Tate group

In this section we substitute the values computed above into Conjecture 3.3 to obtain a conjectural divisor and multiple of the order of a Shafarevich–

Tate group. We then remark that the visibility computation of Section 6 gives evidence for Conjecture 3.3. This example is also discussed in [3, §4.2].

To obtain evidence for Conjecture 3.3, we consider an abelian variety A_f with $L(A_f, 1) \neq 0$ and combine the invariants whose computation is described above with Conjecture 3.3 to obtain a conjectural divisor and multiple of the order of $\text{III}(A_f)$. We then observe that this divisor and multiple is consistent with Conjecture 3.3.

We now combine the computations from the previous sections for the 20 dimensional quotient A of $J_0(389)$. Recall that Conjecture 3.3 asserts that

$$\frac{L(A, 1)}{\Omega_A} = \frac{\prod c_p \cdot \#\text{III}(A)}{\#A(\mathbb{Q})_{\text{tor}} \cdot \#A^\vee(\mathbb{Q})_{\text{tor}}}.$$

This equation becomes

$$\frac{2^n \cdot 2^{11} \cdot 5^2}{97} = \frac{97 \cdot \#\text{III}(A)}{97^2}$$

where $0 \leq n \leq 20$ (using the bound from [1, Thm. 2.7]). Thus the conjecture asserts that $\#\text{III}(A) = 5^2 \cdot 2^{11+n}$, and we have computed a conjectural divisor $5^2 \cdot 2^{11}$ and a conjectural multiple $5^2 \cdot 2^{31}$ of $\#\text{III}(A)$. Using visibility theory from Section 6 we have proved that $5^2 \mid \#\text{III}(A)$, which provides evidence for Conjecture 3.3.

11 An element of the Shafarevich–Tate group that becomes visible at higher level

We finish this paper by considering the 18-dimensional newform quotient A of $J_0(551)$. In this example, the level $551 = 19 \cdot 29$ is composite, the Shafarevich–Tate group is conjecturally nontrivial, and the methods of Section 6 do not produce nontrivial elements of the Shafarevich–Tate group at level 551.

This example is striking because it is, in some sense, the simplest known example of "visibility only at a higher level"; more precisely, the methods of Section 6 do produce a nontrivial element at the rather small level 1102. For a similar example, see [3, §4.3], where the levels involved are much larger.

We first compute the space of modular symbols corresponding to A:

```
M := ModularSymbols(551);
N := NewSubspace(CuspidalSubspace(M));
D := SortDecomposition(NewformDecomposition(N));
A := D[8];
```

Next we compute a divisor and a multiple of the order of the torsion subgroup of $A(\mathbb{Q})$ and $A^\vee(\mathbb{Q})$. Using odd primes $p \leq 7$ we obtain the multiple 160, and using the rational cuspidal subgroup we obtain the divisor 40.

```
TorsionBound(A, 7);
```

RationalCuspidalSubgroup(A);

Since the divisor and multiple are different, we try more finite fields. For $p \leq 29$ the multiple we obtain is still 160; however, for $p = 31$ the multiple is 80, which is where it appears to stabilize.

TorsionBound(A, 31);

We conclude that $40 \mid \#A(\mathbb{Q})_{\text{tor}} \mid 80$ and $5 \mid \#A^{\vee}(\mathbb{Q})_{\text{tor}} \mid 80$. We know that 5 divides $\#A^{\vee}(\mathbb{Q})_{\text{tor}}$ because, as we will see below, there is a homomorphism $A^{\vee} \to A$ of degree not divisible by 5.

Next we compute the modular kernel, which is of order $2^{44} \cdot 13^4$.

Factorization(#ModularKernel(A));

The only possible elements of $\text{III}(A)$ that we can construct using Theorem 6.1 at level 551 are of order 13.

The level 551 is not prime, so computation of the Tamagawa numbers involves certain relatively slow algorithms (a minute rather than seconds) that involve arithmetic in quaternion algebras. Also, in this example, we are unable to determine the exact power of 2 that divides the Tamagawa number at 19.

TamagawaNumber(A, 19); // takes over a minute; gives an error
TamagawaNumber(A, 29);

We find that $c_{29} = 40$. We also deduce that $c_{19} = 2$ or 4 by noting that the component group over $\overline{\mathbb{F}}_{19}$ has order $2^2 \cdot 13^2$ by using the command

ComponentGroupOrder(A, 19);

and noting that the Galois generator Frobenius acts as -1 because

AtkinLehnerOperator(A, 19)[1, 1];

returns 1. Finally note that the 2 torsion in any group of order $2^2 \cdot 13^2$ is a subgroup of order either 2 or 4.

Next we find that $L(A, 1)/\Omega_A = 2^n \cdot 2^2 \cdot 3^2/5$, with $0 \leq n \leq 18$, using the command

LRatio(A, 1);

and the fact that the Manin constant divides $2^{\dim A}$ (see [1, Thm. 2.7]).

Putting these computations together we find that Conjecture 3.3 asserts that

$$\frac{2^n \cdot 2^2 \cdot 3^2}{5} = \frac{2^m \cdot 40 \cdot \#\text{III}(A)}{40 \cdot 2^r \cdot 5 \cdot 2^s},$$

where $0 \leq n \leq 18$, $1 \leq m \leq 2$, $0 \leq r \leq 1$, and $0 \leq s \leq 4$. Solving for $\#\text{III}(A)$, we see that Conjecture 3.3 predicts that

$$\#\text{III}(A) = 2^t \cdot 3^2$$

with $2 \leq t \leq 24$.

Theorem 6.1 does not construct elements of order 2 (yet), so we do not consider the factor 2^t further. As mentioned above, we cannot construct any elements of $III(A)$ of order 3 using visibility at level 551. We can, however, consider the images of A in $J_0(2 \cdot 551)$ under various natural maps. These natural maps are the degeneracy maps δ_1 and δ_2, which correspond to the maps $f(q) \mapsto f(q)$ and $f(q) \mapsto f(q^2)$ from $S_2(\Gamma_0(551))$ to $S_2(\Gamma_0(2 \cdot 551))$.

We next compute the space of modular symbols that corresponds to the sum $C = \delta_1(A) + \delta_2(A)$ of the images of A at level $2 \cdot 551$ by the two degeneracy maps δ_1 and δ_2.

```
M := ModularSymbols(2*551, 2);
N := NewSubspace(CuspidalSubspace(M));
D := SortDecomposition(NewformDecomposition(N));
M_551 := ModularSymbols(M, 551);
N_551 := NewSubspace(CuspidalSubspace(M_551));
D_551 := SortDecomposition(NewformDecomposition(N_551));
A_551 := D_551[#D_551];
C := M !! A_551;  // sum of images under degeneracy maps
```

The sum C contains the 3-torsion of the rank 2 elliptic curve B defined by $y^2 + xy = x^3 + x^2 - 29x + 61$, as the following computation shows.

```
IntersectionGroup(C, D[1]);
B := EllipticCurve(D[1]); B;
```

It follows that $B[3]$ is contained in C. The following computation shows that the Tamagawa numbers of B are 2, 2, and 1 and $B(\mathbb{Q}) \cong \mathbb{Z} \times \mathbb{Z}$:

```
TamagawaNumber(B, 2);
TamagawaNumber(B, 19);
TamagawaNumber(B, 29);
MordellWeilGroup(B);
```

Theorem 6.1 implies that $B(\mathbb{Q})/3B(\mathbb{Q}) = \mathbb{Z}/3\mathbb{Z} \times \mathbb{Z}/3\mathbb{Z}$ is a subgroup of $III(C)$. By [27, §2], there is an isogeny φ from $A \times A$ to C whose kernel is isomorphic to the intersection of A with the Shimura subgroup of $J_0(551)$. The Shimura subgroup Σ is a subgroup of $J_0(N)$ that, according to [27, Prop. 2], is annihilated by $T_p - (p+1)$ for all primes $p \nmid 551$. Using Magma we find that $3 \nmid \det(T_3|_A - 4) = 12625812402998886400$, so the degree of φ is coprime to 3.

```
T_3 := HeckeOperator(A, 3);
d := Determinant(T_3−4);
Valuation(d, 3);
```

Since $3 \mid \#III(C)$ it follows that $3 \mid \#III(A)$. By [2, §5.3] the power of 3 that divides $\#III(A)$ is even, so $9 \mid \#III(A)$, as predicted by the Birch and Swinnerton-Dyer conjecture.

12 Complete Magma log

This is a complete log of using Magma V2.10-6 to do all of the computations discussed in this paper. The output has been edited slightly to save space.

```
>   M := ModularSymbols(389);
>   Basis(M);
        [
                {-1/337, 0}, {-1/237, 0},{-1/342, 0},{-1/266, 0},
                {-1/170, 0}, {-1/272, 0},{-1/333, 0},{-1/355, 0},
                {-1/270, 0},{-1/301, 0}, {-1/293, 0},{-1/87, 0},
                {-1/306, 0},{-1/205, 0},{-1/209, 0}, {-1/277, 0},
                {-1/383, 0},{-1/142, 0},{-1/178, 0},{-1/116, 0},
                {-1/61, 0},{-1/127, 0},{-1/235, 0},{-1/240, 0},
                {-1/93, 0}, {-1/121, 0},{-1/221, 0},{-1/199, 0},
                {-1/213, 0},{-1/370, 0}, {-1/282, 0},{-1/379, 0},
                {-1/100, 0},{-1/286, 0},{-1/165, 0}, {-1/158, 0},
                {-1/376, 0},{-1/228, 0},{-1/125, 0},{-1/72, 0},
                {-1/374, 0},{-1/140, 0},{-1/81, 0},{-1/186, 0},
                {-1/53, 0}, {-1/37, 0},{-1/175, 0},{-1/108, 0},
                {-1/183, 0},{-1/316, 0}, {-1/363, 0},{-1/250, 0},
                {-1/359, 0},{-1/162, 0},{-1/106, 0}, {-1/350, 0},
                {-1/216, 0},{-1/243, 0},{-1/111, 0},{-1/324, 0},
                {-1/311, 0},{-1/97, 0},{-1/259, 0},{-1/194, 0},
                {oo, 0}
        ]

>   M ! <1, [Cusps() | −1/337, Infinity()]>;
        {-1/337, 0} + -1*{oo, 0}

>   S := CuspForms(389);
>   SetPrecision(S, 40);
>   Basis(S);
        [
                q + 474049571*q^32 + 480335856*q^33 + 984946270*q^34
                + 1338756227*q^35 + 1246938503*q^36 - 29119245*q^37 +
                1504020580*q^38 - 2463550751*q^39 + O(q^40),
                . . .
        ]

>   M := ModularSymbols(389);
>   N := NewSubspace(CuspidalSubspace(M));
>   NewformDecomposition(N);
        [
                Modular symbols space for Gamma_0(389) of
                weight 2 and dimension 2 over Rational Field,
                Modular symbols space for Gamma_0(389) of
                weight 2 and dimension 4 over Rational Field,
```

```
                Modular symbols space for Gamma_0(389) of
                weight 2 and dimension 6 over Rational Field,
                Modular symbols space for Gamma_0(389) of
                weight 2 and dimension 12 over Rational Field,
                Modular symbols space for Gamma_0(389) of
                weight 2 and dimension 40 over Rational Field ]
```

> M := ModularSymbols(389, 2, +1);
> M := ModularSymbols(389);
> N := NewSubspace(CuspidalSubspace(M));
> D := NewformDecomposition(N);
> A := $D[5]$; B := $D[1]$;
> IntersectionGroup(A, B);

```
        Abelian Group isomorphic to Z/20 + Z/20
```

> TorsionBound(A, 7);

```
        97
```

> TorsionBound(B, 7);

```
        1
```

> TamagawaNumber(A, 389);

```
        97
```

> TamagawaNumber(B, 389);

```
        1
```

> E := EllipticCurve(B);
> Rank(E);

```
        2
```

> G := ModularKernel(A);

```
        Abelian Group isomorphic to Z/2 + Z/2 + Z/2 + Z/2 +
        Z/2 + Z/2 + Z/2 + Z/2 + Z/2 + Z/2 + Z/2 + Z/2 +
        Z/2 + Z/2 + Z/2 + Z/2 + Z/2 + Z/2 + Z/40 + Z/40
```

> Factorization(#G);

```
        [ <2, 24>, <5, 2> ]
```

> LRatio(A, 1);

```
        51200/97
```

> RationalCuspidalSubgroup(A);

```
        Abelian Group isomorphic to Z/97
```

```
>  M := ModularSymbols(551);
>  N := NewSubspace(CuspidalSubspace(M));
>  D := NewformDecomposition(N);
>  A := D[8];
>  TorsionBound(A, 7);
       160

>  RationalCuspidalSubgroup(A);
       Abelian Group isomorphic to Z/2 + Z/20

>  TorsionBound(A, 31);
       80

>  Factorization(#ModularKernel(A));
       [ <2, 44>, <13, 4> ]

>  TamagawaNumber(A, 19);
       No algorithm known to compute the Tamagawa number at 2.
       Use ComponentGroupOrder instead.

>  TamagawaNumber(A, 29);
       40

>  ComponentGroupOrder(A, 19);
       676

>  AtkinLehnerOperator(A, 19)[1,1];
       1

>  LRatio(A, 1);
       36/5

>  M := ModularSymbols(2*551, 2);
>  N := NewSubspace(CuspidalSubspace(M));
>  D := SortDecomposition(NewformDecomposition(N));
>  M₅₅₁ := ModularSymbols(M, 551);
>  N₅₅₁ := NewSubspace(CuspidalSubspace(M₅₅₁));
>  D₅₅₁ := NewformDecomposition(N₅₅₁);
>  A₅₅₁ := D₅₅₁[#D₅₅₁];
>  C := M !! A₅₅₁;
>  IntersectionGroup(C, D[1]);
       Abelian Group isomorphic to Z/32 + Z/32

>  B := EllipticCurve(D[1]); B;
       Elliptic Curve defined by y^2 + x*y = x^3 + x^2 - 29*x + 61
```

> TamagawaNumber(B, 2);

    2

> TamagawaNumber(B, 19);

    2

> TamagawaNumber(B, 29);

    1

> MordellWeilGroup(B);

    Abelian Group isomorphic to Z + Z

> MordellWeilGroup(B);

    Abelian Group isomorphic to Z + Z

> T_3 := HeckeOperator(A, 3);
> d := Determinant($T_3 - 4$);
> Valuation(d, 3);

    0

References

1. A. Agashe, W. A. Stein, *The Manin constant, congruence primes, and the modular degree*. Submitted.
2. A. Agashe, W. A. Stein, *Visible Evidence for the Birch and Swinnerton-Dyer Conjecture for Modular Abelian Varieties of Analytic Rank 0*, Math. Comp. **74**-249 (2005), 455–484.
3. A. Agashe, W. A. Stein, *Visibility of Shafarevich-Tate groups of abelian varieties*, J. Number Theory **97**-1 (2002), 171–185.
4. A. O. L. Atkin, J. Lehner, *Hecke operators on $\Gamma_0(m)$*, Math. Ann. **185** (1970), 134–160.
5. B. J. Birch, *Conjectures concerning elliptic curves*, pp. 106–112 in: *Proceedings of Symposia in Pure Mathematics, VIII*, Providence (R.I.): Amer. Math. Soc., 1965.
6. S. Bosch, W. Lütkebohmert, M. Raynaud, *Néron models*, Berlin: Springer-Verlag, 1990.
7. Wieb Bosma, John Cannon, Catherine Playoust, *The Magma algebra system I: The user language*, J. Symbolic Comput. **24** (1997), 235–265.
   See also the Magma home page at http://magma.maths.usyd.edu.au/magma/.
8. C. Breuil, B. Conrad, F. Diamond, R. Taylor, *On the modularity of elliptic curves over* $\mathbb{Q}$: *wild 3-adic exercises*, J. Amer. Math. Soc. **14**-4 (2001), 843–939 (electronic).
9. J. W. S. Cassels, *Diophantine equations with special reference to elliptic curves*, J. London Math. Soc. **41** (1966), 193–291.
10. B. Conrad, S. Edixhoven, W. A. Stein, $J_1(p)$ *Has Connected Fibers*, Documenta Mathematica **8** (2003), 331–408.

11. B. Conrad, W. A. Stein, *Component Groups of Purely Toric Quotients of Semi-stable Jacobians*, Math. Res. Letters **8**-5/6 (2001), 745-766.

12. J. E. Cremona, *Modular symbols for $\Gamma_1(N)$ and elliptic curves with everywhere good reduction*, Math. Proc. Cambridge Philos. Soc., **111**-2 (1992), 199-218.

13. J. E. Cremona, *Algorithms for modular elliptic curves*, Cambridge: Cambridge University Press, second edition, 1997.

14. F. Diamond, J. Im, *Modular forms and modular curves*, pp. 39-133 in: *Seminar on Fermat's Last Theorem*, Providence, RI, 1995.

15. B. Edixhoven, *On the Manin constants of modular elliptic curves*, pp. 25-39 in: G. van der Geer, F. Oort et al., (eds) *Arithmetic Algebraic Geometry*, Basel: Birkhäuser, Progress in Mathematics **89**, 1991.

16. E. V. Flynn, F. Leprévost, E. F. Schaefer, W. A. Stein, M. Stoll, J. L. Wetherell, *Empirical evidence for the Birch and Swinnerton-Dyer conjectures for modular Jacobians of genus 2 curves*, Math. Comp. **70**-236 (2001), 1675-1697 (electronic).

17. B. Gross, D. Zagier, *Heegner points and derivatives of L-series*, Invent. Math. **84**-2 (1986), 225-320.

18. Clay Mathematics Institute, *Millennium prize problems*, http://www.claymath.org/millennium_prize_problems/.

19. K. Kato, *p-adic Hodge theory and values of zeta functions of modular forms*, preprint, 244 pages.

20. D. R. Kohel, W. A. Stein, *Component Groups of Quotients of $J_0(N)$*, pp. 405-412 in: Wieb Bosma (ed.), *Proceedings of the 4th International Symposium (ANTS-IV), Leiden, Netherlands, July 2-7, 2000*, Springer: Berlin, 2000.

21. V. A. Kolyvagin, D. Y. Logachev, *Finiteness of the Shafarevich-Tate group and the group of rational points for some modular abelian varieties*, Algebra i Analiz **1**-5 (1989), 171-196.

22. H. W. Lenstra, Jr., F. Oort, *Abelian varieties having purely additive reduction*, J. Pure Appl. Algebra **36**-3 (1985), 281-298.

23. W-C. Li, *Newforms and functional equations*, Math. Ann. **212** (1975), 285-315.

24. J. I. Manin, *Parabolic points and zeta functions of modular curves*, Izv. Akad. Nauk SSSR Ser. Mat. **36** (1972), 19-66.

25. A. Néron, *Modèles minimaux des variétés abéliennes sur les corps locaux et globaux*, Inst. Hautes Études Sci. Publ.Math. No. **21** (1964), 128.

26. K. A. Ribet, *On modular representations Gal of $\mathrm{Gal}(\overline{\mathbf{Q}}/\mathbf{Q})$ arising from modular forms*, Invent. Math. **100**-2 (1990), 431-476.

27. K. A. Ribet, *Raising the levels of modular representations*, In: *Séminaire de Théorie des Nombres, Paris 1987-88*, Boston: Birkhäuser, 1990, pp. 259-271.

28. K. A. Ribet, *Abelian varieties over $\mathbb{Q}$ and modular forms*, pp. 53-79 in: *Algebra and topology 1992 (Taejŏn)*, Korea Adv. Inst. Sci. Tech., Taejŏn, 1992.

29. K. A. Ribet, W. A. Stein, *Lectures on Serre's conjectures*, pp. 143-232 in: *Arithmetic algebraic geometry (Park City, UT, 1999)*, IAS/Park City Math. Ser. **9**, Providence (R.I.): Amer. Math. Soc., 2001.

30. J-P. Serre, *Sur les représentations modulaires de degré 2 de $\mathrm{Gal}(\overline{\mathbf{Q}}/\mathbf{Q})$*, Duke Math. J. **54**-1 (1987), 179-230.

31. I. R. Shafarevich, *Exponents of elliptic curves*, Dokl. Akad. Nauk SSSR (N.S.) **114** (1957), 714-716.

32. G. Shimura, *On the factors of the jacobian variety of a modular function field*, J. Math. Soc. Japan **25**-3 (1973), 523-544.

33. G. Shimura, *Introduction to the arithmetic theory of automorphic functions*, Princeton (NJ): Princeton University Press, 1994. Reprint of the 1971 original, Kan Memorial Lectures **1**.

34. J. H. Silverman, *The arithmetic of elliptic curves*, New York: Springer-Verlag, 1992. Corrected reprint of the 1986 original.

35. J. H. Silverman, J. Tate, *Rational points on elliptic curves*, Undergraduate Texts in Mathematics, New York: Springer-Verlag, 1992.

36. W. A. Stein, *Modular Symbols*, Chapter 100 in: John Cannon, Wieb Bosma (eds.), *Handbook of Magma Functions*, Version 2.11, Volume **7**, Sydney, 2004, pp. 2947–3002.

37. W. A. Stein, *An introduction to computing modular forms using modular symbols*, to appear in an MSRI Proceedings.

38. W. A. Stein, *Shafarevich-Tate groups of nonsquare order*, in: J. Cremona, J.-C. Lario, J. Quer, K. Ribet (eds), Proceedings of MCAV 2002, Progress of Mathematics (to appear). *Proceedings of Modular Curves and Abelian Varieties*, Progress in Mathematics **224**, 2004.

39. W. A. Stein, *Explicit approaches to modular abelian varieties*, Ph.D. thesis, University of California, Berkeley, 2000.

40. G. Stevens, *Arithmetic on modular curves*, Boston Mass.: Birkhäuser, 1982.

41. J. Tate, *On the conjectures of Birch and Swinnerton-Dyer and a geometric analog*, pp. 415–440 in: Séminaire Bourbaki **9**, Exp. No. 306, Paris: Soc. Math. France, 1995.

42. R. Taylor, A. J. Wiles, *Ring-theoretic properties of certain Hecke algebras*, Ann. of Math. (2) **141**-3 (1995), 553–572.

43. A. J. Wiles, *Modular elliptic curves and Fermat's last theorem*, Ann. of Math. (2) **141**-3 (1995), 443–551.

Computing with the analytic Jacobian of a genus 2 curve

Paul B. van Wamelen

Department of Mathematics
Louisiana State University
Baton Rouge LA, USA
wamelen@math.lsu.edu

Summary. We solve two genus 2 curve problems using Magma. First we give examples of how Magma can be used to find the equation of a genus 2 curve whose Jacobian has prescribed Complex Multiplication. We treat 2 fields, one easy and one harder. Secondly we show how Magma can be used to find, and ultimately prove existence of, rational isogenies between the Jacobians of two genus 2 curves.

This paper illustrates computations that were done for the papers [14] and [16].

1 Introduction

In the last few years a lot of work has been done on the computational aspects of genus 2 curves and their Jacobians. In this paper we will demonstrate some of the analytic tools available in Magma[1] [1].

There are two computationally useful descriptions of the Jacobian of a genus g curve. We briefly recall them.

The first description is to think of the Jacobian as a certain group of divisors. Consider, Pic, the group of divisors on the curve modulo linear equivalence ([3, Chap 8], [5, I.2]). The subgroup of degree 0 elements, Pic^0, has the structure of an abelian variety. In fact Pic^0 can be thought of as a functor from the category of curves to that of abelian groups and the Jacobian is then a group scheme representing this functor, see [8, Theorem 1.1]. Our ground field is the complex numbers and in this case it can be shown that every element of Pic^0 can be represented by a formal sum $P_1 + P_2 + \cdots + P_g - gP_0$ where the P_i are complex points on the curve and P_0 is a fixed base point. This can be used to show that the Jacobian is birational to the g-th symmetric power of the curve. See [9, IIIa §1, §2] for an explicit description of how this works in the case where the curve is given by $y^2 = f(x)$, with $f(x)$ of odd degree (we will specialize to this case shortly) and [8] for the general case.

---

[1]You need Magma version 2.11 or later for the examples in this paper to work.

In the case of hyperelliptic curves with an odd degree model we will take the base point P_0 to be the unique point at infinity, ∞, on the curve. We will then often think of points on the Jacobian as unordered sets of g points on the curve. We call this description of the Jacobian the *algebraic* Jacobian. For a hyperelliptic curve (type CrvHyp) it can be computed with the Magma command Jacobian. Magma represents a point $P_1 + \cdots + P_g - gP_0$, with $P_i = (x_i, y_i)$, on an algebraic Jacobian by two polynomials a and b, where $a = \prod_{i=1}^{g}(x - x_i)$ and b is such that $b(x_i) = y_i$.

The second description of the Jacobian of a curve is as an abelian torus. We will call this the *analytic* Jacobian and it is constructed as follows. Suppose C is a curve of genus g defined over the complex numbers. We view the complex points on the curve, $C(\mathbb{C})$, as a compact Riemann surface of genus g. Then it is known that the dimension of the vector space of holomorphic differentials is equal to the genus of the curve. We let φ_i, $i = 1, 2, \ldots, g$ be a basis for this vector space and set $\overline{\varphi} = {}^{t}(\varphi_1, \ldots, \varphi_g)$, a vector of holomorphic 1-forms. For a hyperelliptic curve there is a natural choice of holomorphic differentials, namely $\varphi_i = x^{i-1}dx/dy$. Define Λ to be the image of the map from the first homology group of C, $H_1(C, \mathbb{Z})$, to $\mathbb{C}^g$ which sends a closed path, $\sigma \in H_1(C, \mathbb{Z})$, on the Riemann surface to $\int_\sigma \overline{\varphi}$. Λ turns out to be a lattice of $\mathbb{Z}$-rank $2g$ in $\mathbb{C}^g$, see [5, Theorem IV.3.3], [7, Lemma 11.1.1] or [4, 2.2]. We will call the complex torus $\mathbb{C}^g/\Lambda$ the analytic Jacobian.

To see how it is related to the algebraic Jacobian note that we can define a map Int : $C \to \mathbb{C}^g/\Lambda$ by

$$P \mapsto \int_P^{P_0} \overline{\varphi}.$$

By extending this mapping additively to the divisors of degree zero on C we get a map from Pic^0 to $\mathbb{C}^g/\Lambda$. The Abel part of the Abel–Jacobi theorem states that two divisors map to the same thing under Int if and only if they are linearly equivalent. The Jacobi part says that the map is surjective. See [7, Theorem 11.1.3], [10, §IIIa.5], [2, III.6], or [5, IV]. Therefore there is a bijection between the points of $\mathbb{C}^g/\Lambda$ and the points of the (algebraic) Jacobian.

By exhibiting a polarization we see that the torus $\mathbb{C}^g/\Lambda$ is an abelian manifold. We can find a polarization as follows.

Let $A_1, \ldots, A_g, B_1, \ldots, B_g$ be a symplectic homology basis. That is, the A_i and B_i are closed paths on the Riemann surface such that A_i intersects B_i with intersection number 1 for each i and all other intersection numbers are 0. Let ω_1 and ω_2 be the $g \times g$-matrices with entries

$$\omega_{1ij} = \int_{B_j} \varphi_i,$$

and

$$\omega_{2ij} = \int_{A_j} \varphi_i.$$

Then the columns of $P = (\omega_1, \omega_2)$ form a $\mathbb{Z}$-basis for the lattice $\Lambda = P\mathbb{Z}^{2g}$. The matrix P is called the *big period matrix*. The space of symmetric $g \times g$ complex matrices with positive definite imaginary part is called the Siegel upper-half space of degree g. Because we chose a symplectic basis for the homology it follows that $\tau = \omega_2^{-1}\omega_1$ is an element of Siegel upper half-space. We will refer to τ as the *small period matrix*. Set

$$J = \begin{pmatrix} 0 & 1_g \\ -1_g & 0 \end{pmatrix}.$$

If we define

$$E(Px, Py) = {}^t x J y,$$

for any $x, y \in \mathbb{R}^{2g}$, then E is a Riemann form for the torus $\mathbb{C}^g/\Lambda$, that is, $H(x, y) = E(ix, y) + iE(x, y)$ is a positive definite Hermitian form (see [7, Lemma 2.1.7]). The torus therefore acquires a polarization. For a torus, having a polarization is equivalent to it being embeddable into projective space as an algebraic variety over $\mathbb{C}$, see [12, Theorem A]. In our case the image variety is isomorphic to the (algebraic) Jacobian. See [9, Theorem II.2.1] and [5, VIII §1, §2] or [7, §11.1].

In Magma, for a polynomial f with complex coefficients, the analytic Jacobian of the corresponding hyperelliptic curve can be computed with AnalyticJacobian(f). The main task of this function is to find a suitable symplectic basis and to compute the period matrix. It also computes other technical data. These are all kept hidden in the AnHcJac data type.

These two descriptions of the Jacobian of a curve each have their strengths and weaknesses and it is therefore very useful to be able to translate between the two descriptions. In Magma the translations can be done with the functions FromAnalyticJacobian and ToAnalyticJacobian.

We will illustrate these tools by solving two specific genus 2 curve problems. The first concerns curves with Complex Multiplication. A curve of genus g is said to have Complex Multiplication if the endomorphism ring of its Jacobian is abelian and of free $\mathbb{Z}$-rank $2g$. In particular, a curve has CM if $\text{End}(J(C)) \otimes \mathbb{Q}$ contains a subfield of degree $2g$ (see [6, §1.3]). We will show how Magma can be used to find equations for genus 2 curves defined over the rationals and with Complex Multiplication. This problem was dealt with in the paper [14]. Here we show how the computations that were done for that paper can now be done in Magma. As an aside it should be noted that Murabayashi and Umegaki, [11], have now proved that the curves found in [14] represent all possible $\mathbb{C}$-isomorphism classes of genus 2 CM curves that can be defined over the rationals.

The second problem we consider is that of finding rational isogenies between genus 2 curves. That is, finding an surjective morphism defined over $\mathbb{Q}$ and with finite kernel between the Jacobians of the two curves. This work was developed in [16] and here we work through the details of a particular example using Magma. Similar ideas and techniques were also used in [15].

2 Finding genus 2 CM curves defined over the rationals

We give two examples* of how Magma can be used to find the equation of
a genus 2 curve whose Jacobian has Complex Multiplication by the maximal
order of a given CM field.

CM by the maximal order of $\mathbb{Q}(\sqrt{-2+\sqrt{2}})$

First we treat the field $F = \mathbb{Q}(\sqrt{-2+\sqrt{2}})$. This field can be defined in Magma
as follows:

```
>    Q := RationalField();
>    P<x> := PolynomialRing(Q);
>    R<s> := NumberField(x²−2);
>    PP<x> := PolynomialRing(R);
>    RF := NumberField(x²−(−2+s));
>    F<t> := AbsoluteField(RF);
>    O := MaximalOrder(F);
```

It is a CM field because it is a totally imaginary quadratic extension of a
totally real field, $R = \mathbb{Q}(\sqrt{2})$ in this case.

Recall that a CM type for a CM field is a maximal set of non-complex
conjugate embeddings of the field into $\mathbb{C}$. The ring of integers $\mathcal{O}$ in F is a
free, maximal rank, $\mathbb{Z}$-module in F and so for a CM type Φ we have that
$\mathbb{C}^2/\Phi(\mathcal{O})$ is a torus with Complex Multiplication by the maximal order of
our field $\mathbb{Q}(\sqrt{-2+\sqrt{2}})$ (see [6, Theorem 1.4.1.i]). In order to show that this
torus is isomorphic to the Jacobian of a curve of genus 2, we exhibit a principal
polarization. This can be done using Algorithm 2.1 in the next section. In this
case it turns out that a principal polarization is essentially just given by the
reciprocal of a generator of the different:

```
>    D := Different(O);
>    bool, gen := IsPrincipal(D);
>    ξ := 1/gen;
```

Let us check that ξ satisfies condition a) of [14, Theorem 3] for some type.

```
>    ξ in R;
```
```
         false
```

```
>    ξ² in R;
```
```
         true
```

```
>    Conjugates(ξ);
```
```
    [  0.E-57 − 18.9423467035944190772745406986641868034120390
```

---

*See the Preface to this volume for style conventions regarding Magma code; code
appearing in this book is available at http://magma.maths.usyd.edu.au/magma/.

```
06346606805251*i,   0.E-57 + 18.94234670359441907727454069861\
64186803412039030634660680525l*i,   0.E-57 -
0.0011665443177579766812784025947349111389906940129348043291\
2979*i,   0.E-57 + 0.0011665443177579766812784025947349111381\
99069401293480432929791*i ]
```

We can therefore use the type $\Phi = \{\phi_1, \phi_2\}$ consisting of embeddings 2 and 4. By Theorem 3 of [14] we see that, for $z, w \in \Phi(\mathcal{O})$,

$$E(z, w) = \sum_{j=1}^{2} \phi_j(\xi)(\overline{z}_j w_j - z_j \overline{w}_j),$$

defines a principal polarization. Recall that for $\alpha, \beta \in \mathcal{O}$, $E(\Phi(\alpha), \Phi(\beta)) = \mathrm{Tr}_{F/\mathbb{Q}}(\xi \overline{\alpha} \beta)$. We now find a Frobenius basis for our lattice with respect to the non-degenerate Riemann form given by ξ. That is, a basis of $\mathcal{O}$ with respect to which the Riemann form is given by the matrix $J = \left(\begin{smallmatrix} 0 & 1_2 \\ -1_2 & 0 \end{smallmatrix}\right)$. See [5, Chapter VIII] and [5, Lemma VI.3.1]. Given an alternating matrix E the Magma function FrobeniusForm computes a change of basis so that the corresponding form has matrix J. Once we have such a basis we will be able to compute the element of Siegel upper half-space corresponding to our abelian surface. We start by finding the automorphism of F corresponding to complex conjugation. We check that F has a cyclic Galois group and so complex conjugation is the unique Galois element of order 2.

```
>    G, Aut, τ := AutomorphismGroup(F);
>    G;

        Permutation group G acting on a set of cardinality 4
        Order = 4 = 2^2
        (1, 2, 4, 3)

>    exists(cc){τ(g) : g in G | Order(g) eq 2};

        true

>    Z := IntegerRing();
>    E := Matrix(Z, 4, 4, [Trace(ξ*cc(a)*b) : b in Basis(O), a in Basis(O)]);
>    D, C := FrobeniusForm(E); D;

        [ 0  0  1  0]
        [ 0  0  0  1]
        [-1  0  0  0]
        [ 0 -1  0  0]

>    newb := ElementToSequence( Matrix(O, C) * Matrix(O, 4, 1, Basis(O)) );
```

Let $\omega_1, \omega_2 \in M_2(\mathbb{C})$ be such that the columns of (ω_1, ω_2) form a Frobenius basis for the lattice. Then $\tau = \omega_2^{-1} \omega_1$ is the small period matrix of a curve with CM. We want to compute τ accurate to about 100 decimal places.

```
>    C := ComplexField(100);
```

```
>    SetKantPrecision(O, 100);
>    PMat := Matrix(C, 2, 4,
>        [ Conjugate(b, 2) : b in newb ] cat [ Conjugate(b, 4) : b in newb ]);
>    τ := Submatrix(PMat, 1, 3, 2, 2)^{-1} * Submatrix(PMat, 1, 1, 2, 2);
```

Finding a curve with this Jacobian is now essentially a matter of applying the Magma functions RosenhainInvariants and HyperellipticCurveFromIgusaClebsch. For a $g \times g$ (small) period matrix τ the function RosenhainInvariants returns a set S of $2g - 1$ complex numbers such that the hyperelliptic curve

$$y^2 = x(x - 1) \prod_{s \in S} (x - s),$$

has a Jacobian with small period matrix equivalent to τ.

```
>    S := RosenhainInvariants(τ);
>    KC<xc> := PolynomialRing(C);
>    f := xc*(xc−1)*&*{xc−a : a in S};
>    IC := IgusaClebschInvariants(f);
```

We want to check whether these invariants can be the invariants of a curve defined over the rationals. As the Igusa–Clebsch invariants live in weighted projective space they are not unique. If A' (the weight 2 Igusa–Clebsch invariant) is non-zero we can divide the invariants by appropriate powers of it in order to get a normalized set of invariants. These are rational numbers if there is a curve over $\mathbb{Q}$ with a Jacobian isomorphic to the constructed abelian manifold. We try to recognize them as such.

```
>    IC := [ 1, IC[2]/IC[1]^2, IC[3]/IC[1]^3, IC[4]/IC[1]^5 ];
>    ICp := [ BestApproximation(Re(r), 10^{50}) : r in IC ];
>    Maximum([ Abs(IC[i] − ICp[i]) : i in [1..#IC] ]);
```

```
        4.201141444535343702404324873935826042320688402002696\
        790295942549940288579655399942217127858421195e-95  3
```

```
>    ICp;
```

```
        [ 1, 5/324, 31/5832, 1/1836660096 ]
```

Unfortunately, it is not sufficient for these to be rational. The function HyperellipticCurveFromIgusaClebsch will decide whether a curve defined over the rationals with these invariants exists and if it does, return such a curve. The returned curve usually has huge coefficients but we can use the function ReducedModel to return a curve with smaller rational coefficients (and isomorphic over $\mathbb{C}$).

```
>    C₁ := HyperellipticCurveFromIgusaClebsch(ICp);
>    C₂ := ReducedModel(C₁ : AI := "Wamelen");
>    C₂;
```

```
        Hyperelliptic Curve defined by y^2 = -x^5 - 3*x^4 +
        2*x^3 + 6*x^2 - 3*x - 1 over Rational Field
```

We can use IsIsomorphicSmallPeriodMatrices to check that this curve's analytic Jacobian is isomorphic to the one corresponding to τ (it even returns the symplectic matrix relating the two elements of Siegel upper half-space).

```
>    f := Evaluate( HyperellipticPolynomials(C₂), xc );
>    A := AnalyticJacobian(f);
>    IsIsomorphicSmallPeriodMatrices( τ, SmallPeriodMatrix(A) );
```

```
        true
        [  0    0    7    3]
        [  0    0  -12   -5]
        [  5  -12   -1    0]
        [  3   -7    9    4]
```

We can use EndomorphismRing to check that the analytic Jacobian of this curve does have the correct CM. EndomorphismRing(τ) returns the endomorphism ring of the Jacobian with small period matrix τ as a matrix algebra.

```
>    MA := EndomorphismRing(SmallPeriodMatrix(A));
>    Dimension(MA);
```

```
        4
```

```
>    MAGens := SetToSequence(Generators(MA)); MAGens;
```

```
    [
        [ 1  -5   4  -7]
        [-1   9  -7  12]
        [ 9   6  -1   1]
        [ 6  -4   5  -9],

        [1 0 0 0]
        [0 1 0 0]
        [0 0 1 0]
        [0 0 0 1]
    ]
```

```
>    IsIsomorphic( F, NumberField(MinimalPolynomial(MAGens[1])) );
```

```
        true
```

CM by the maximal order of $\mathbb{Q}(\sqrt{-5 + \sqrt{5}})$

Now let us try the field $F = \mathbb{Q}(\sqrt{-5 + \sqrt{5}})$.

```
>    Q := RationalField();
>    P<x> := PolynomialRing(Q);
>    R<s> := NumberField(x²−5);
>    PP<x> := PolynomialRing(R);
>    RF := NumberField(x²−(−5+s));
>    F<t> := AbsoluteField(RF);
>    O := MaximalOrder(F);
```

```
>    ClassNumber(F);

        2
```

As the class number is not 1, finding all possible polarizations is not as trivial as it was for the example in the previous section. We use the following algorithm, see [14, Algorithm 1].

Algorithm 2.1 *To find all non-isomorphic principally polarized abelian varieties with CM by $\mathcal{O}(F)$.*

1. *Find all ideal classes $\mathfrak{A}$ such that $\mathfrak{A}\overline{\mathfrak{A}}$ is the ideal class of the codifferent $\mathfrak{D}_{F/\mathbb{Q}}^{-1}$.*
2. *Find a set of coset representatives of the units in $\mathcal{O}(R)$ modulo norms of units of $\mathcal{O}(F)$.*
3. *For each ideal class found in 1 pick an ideal $\mathfrak{a}$ and find a generator b of $\mathfrak{D}_{F/\mathbb{Q}}\mathfrak{a}\overline{\mathfrak{a}}$.*
4. *For each ideal class in step 3, if there exist a unit u in $\mathcal{O}(F)$ such that $ub = -\overline{ub}$, set $\xi_0 = (ub)^{-1}$ and go to the next step.*
5. *For each unit u^+ found in 2, choose a type Φ such that if $\xi = u^+\xi_0$ then $\mathrm{Im}(\phi_i(\xi)) > 0$ for each $\phi_i \in \Phi$.*
6. *For $z, w \in \Phi(\mathfrak{a})$, $E(z,w) = \sum_{j=1}^{2} \phi_j(\xi)(\overline{z}_j w_j - z_j \overline{w}_j)$, now defines a principal polarization of type Φ on $\mathbb{C}^g/\Phi(\mathfrak{a})$.*

First we find complex conjugation and compute the class group.

```
>    G, Aut, τ := AutomorphismGroup(F);
>    G;

        Permutation group G acting on a set of cardinality 4
        Order = 4 = 2^2
           (1, 2, 4, 3)

>    exists(cc){τ(g) : g in G | Order(g) eq 2};

        true


>    G, classmap := ClassGroup(F);
>    D := Different(O);
>    D @@ classmap;

        0
```

It follows that the different (and therefore also the codifferent) is in the principal class. We also check that for the non-principal class, $\mathfrak{A}$, we have that $\mathfrak{A}\overline{\mathfrak{A}}$ is the principal class.

```
>    l₁ := classmap(G.1);
>    IsPrincipal(l₁*cc(l₁));

        true
```

For step 1 of the algorithm we therefore find both classes.

We now find a set of coset representatives of the units in R modulo norms of units of F.

```
>    U, Umap := UnitGroup(O); U;

         Abelian Group isomorphic to Z/2 + Z
         Defined on 2 generators
         Relations:
            2*U.1 = 0

>    _, iso := IsIsomorphic(F, RF);
>    Uplus, Upmap := UnitGroup(R);
>    normUgens := { Norm(iso(F ! Umap(g)))@@Upmap :
>           g in Generators(U) };
>    subUplus := sub<Uplus | normUgens>;
>    cosetreps := Transversal(Uplus, subUplus);
>    ups := { (RF ! Upmap(cr))@@iso : cr in cosetreps };
>    ups;

         { 1,   1/2*(t^2 + 6),   1/2*(-t^2 - 6),   -1 }

>    uplus := 1/2*(t^2 + 6);
```

We therefore find $\pm 1, \pm(t^2 + 6)/2$. Note that by [14, Proposition 1] we only need to consider $u = 1, (t^2 + 6)/2$.

Next we have to, for each ideal class found in step 1, pick an ideal $\mathfrak{a}$ and find a generator b of $\mathfrak{D}_{F/\mathbb{Q}}\mathfrak{a}\bar{\mathfrak{a}}$. There is only one non-trivial class and we already have an ideal in it. The two b's are therefore just

```
>    _, b₁ := IsPrincipal(D); F ! b₁;

         t^3

>    _, b₂ := IsPrincipal( D*I₁*cc(I₁) ); F ! b₂;

         2*t^3
```

The following shows that in step 4 we can set $u = 1$.

```
>    b₁ eq −cc(b₁);

         true

>    b₂ eq −cc(b₂);

         true
```

We have now found 4 possible principally polarized tori (although some of them might turn out to be isomorphic). They are $\mathbb{C}^2/\Phi(\mathcal{O})$ with polarization given by $\xi := 1/b_1$ or $\xi := uplus/b_1$ and $\mathbb{C}^2/\Phi(I_1)$ with polarization given by $\xi := 1/b_2$ or $\xi := uplus/b_2$.

For each of these we want to compute the corresponding element of Siegel upper half-space. To do so, let us write a function that takes a ξ, the ideal, complex conjugation and a precision and then returns such an element to the given precision.

```
function xi2tau(ξ, la, cc, pres)
   O := MaximalOrder(Parent(ξ));
   E := Matrix(Z, 4, 4, [Trace(ξ*cc(a)*b) :
                b in Basis(la), a in Basis(la)]);
   _, C := FrobeniusForm(E);
   newb := ElementToSequence(
        Matrix(O, C)*Matrix(O, 4, 1, Basis(la)));
   C := ComplexField(pres);
   SetKantPrecision(O, pres);
   Φ := [i : i in [1..4] | Im(Conjugate(ξ, i)) gt 0];
   PMat := Matrix(C, 2, 4, [Conjugate(b, Φ[1]) : b in newb] cat
                            [Conjugate(b, Φ[2]) : b in newb]);
   return Submatrix(PMat, 1, 3, 2, 2)^{-1}*Submatrix(PMat, 1, 1, 2, 2);
end function;
```

We now compute the 4 τs.

```
>    τ1 := xi2tau(1/b1, O, cc, 100);
>    τ2 := xi2tau(uplus/b1, O, cc, 100);
>    τ3 := xi2tau(1/b2, l1, cc, 100);
>    τ4 := xi2tau(uplus/b2, l1, cc, 100);
```

Note that τ_1 and τ_2 correspond to isomorphic abelian surfaces. Similarly for τ_3 and τ_4:

```
>    IsIsomorphicSmallPeriodMatrices(τ1, τ2);

     true
     [-1   1   0   0]
     [ 0   0   1   1]
     [ 0   0   0   1]
     [-1   0   0   0]


>    bool := IsIsomorphicSmallPeriodMatrices(τ3, τ4); bool;

     true
```

We would now like to check whether τ_1 and τ_2 are the Jacobians of genus 2 curves defined over the rationals.

```
>    S := RosenhainInvariants(τ1);
>    f := xc * (xc−1) * &*{ xc−a : a in S };
>    IC := IgusaClebschInvariants(f);
>    IC := [ 1, IC[2]/IC[1]^2, IC[3]/IC[1]^3, IC[4]/IC[1]^5 ];
>    ICp := [ BestApproximation(Re(r), 10^{50}) : r in IC ];
>    Maximum([ Abs(IC[i] − ICp[i]) : i in [1..#IC] ]);
          3.68447393619862000602944199006783238987246924359791421\
          2280778631723613560610260453244233973797869119e-96 3
```

```
>   ICp;
```

>   [1, 1/49, 193/33075, 1/6202728393750]

```
>   C₁ := HyperellipticCurveFromIgusaClebsch(ICp);
>   C₂ := ReducedModel(C₁ : AI := "Wamelen"); C₂;
```

>   Hyperelliptic Curve defined by y^2 = 4*x^5 - 30*x^3 +
>   45*x - 22 over Rational Field

```
>   S := RosenhainInvariants(τ₃);
>   f := xc * (xc−1) * &*{ xc−a : a in S };
>   IC := IgusaClebschInvariants(f);
>   IC := [ 1, IC[2]/IC[1]², IC[3]/IC[1]³, IC[4]/IC[1]⁵ ];
>   ICp := [ BestApproximation(Re(r),10⁵⁰) : r in IC ];
>   ICp;
```

>   [1, 14641/4652649, 14641/13957947,
>   3138428376721/7091482188977263344 3750]

```
>   Maximum([ Abs(IC[i] − ICp[i]) : i in [1..#IC] ]);
```

>   3.24715556452396618483781950598769005811289036957 91015\
>   79261662125258622593240053919811159427642084149e-93 3

```
>   C₁ := HyperellipticCurveFromIgusaClebsch(ICp);
>   C₂ := ReducedModel(C₁ : AI := "Wamelen"); C₂;
```

>   Hyperelliptic Curve defined by y^2 = 8*x^6 - 52*x^5 +
>   250*x^3 - 321*x - 131 over Rational Field

3 Isogenies

We show how we can find candidates for rational isogenies between the Jacobians of genus 2 curves. We also give some idea of how these can then be shown to correspond to actual rational isogenies.

Let us consider the two curves

$$C_1 : y^2 = x^5 + 40x^4 + 136x^3 + 96x^2 + 16x,$$

and

$$C_2 : y^2 = x^6 - 8x^5 + 22x^4 - 16x^3 - 36x^2 + 64x - 24.$$

These are the fourth and fifth curves in the ninth isogeny class of [13]. We compute their analytic Jacobians to 100 decimal places and check whether they are isogenous.

```
>   K<x> := PolynomialRing(RationalField());
>   C<i> := ComplexField(100);
>   KC<xc> := PolynomialRing(C);
>   f₁ := x⁵ + 40*x⁴ + 136*x³ + 96*x² + 16*x;
```

```
>    f1C := Evaluate(f₁, xc);
>    A₁ := AnalyticJacobian(f1C);
>    f₂ := x⁶ − 8*x⁵ + 22*x⁴ − 16*x³ − 36*x² + 64*x − 24;
>    f2C := Evaluate(f₂, xc);
>    A₂ := AnalyticJacobian(f2C);
>    bool, M, α := IsIsogenous(A₁, A₂); bool;
         true
```

We see the curves are $\mathbb{C}$-isogenous, but we would like to know whether they are $\mathbb{Q}$-isogenous. We therefore look at the α matrix. Recall that α and M are two matrices $\alpha \in M_2(\mathbb{C})$ and $M \in M_4(\mathbb{Z})$ such that

$$\alpha P_1 = P_2 M,$$

where P_1 and P_2 are the period matrices for the curves C_1 and C_2, respectively. For an isogeny defined over $\mathbb{Q}$ we would have the entries of α in $\mathbb{Q}$.

```
>    α[1, 1];
```

```
     -3.553773974030037307344158953063146948164583499410307\
     8363326711483336752567887331027279378861117438 −
     0.643594252905582624735443437418209808924202742444 0076\
     5115615200093520748503218365195451342465953 95*i
```

Clearly this isogeny is not even defined over a real field. To find out whether our two curves are $\mathbb{Q}$-isogenous we fall back on **AnalyticHomomorphisms** so that we can look at *all* the isogenies between the two curves. **IsIsogenous** only looks for *an* isogeny (and makes a minimal attempt at finding a low degree isogeny, but ignores its field of definition). **AnalyticHomomorphisms** returns a basis for the $\mathbb{Z}$-module of isogenies between the Jacobians corresponding to the two given small period matrices.

```
>    Mlst := AnalyticHomomorphisms(
>         SmallPeriodMatrix(A₁), SmallPeriodMatrix(A₂) );
>    Mlst;
         [
             [ 1   2   1  -1]
             [-2  -3  -3   4]
             [-1  -1  -2   3]
             [ 1   1   2  -3],

             [ 1   1   0   1]
             [-3  -2   1  -3]
             [-3  -1  -2   3]
             [ 2   2  -2   4],
```

```
[ 1  0  1 -1]
[-4 -2 -2  2]
[-2 -2  2 -4]
[ 3  1  2 -3],

[ 1  0 -1  1]
[-3 -1  2 -3]
[ 6  4  0  0]
[ 5  3 -2  3]
]
```

For each of these four 'M' matrices let us find the corresponding α matrices.

```
>    P₁ := BigPeriodMatrix(A₁);
>    P₂ := BigPeriodMatrix(A₂);
>    alst := [ Submatrix(P₂* Matrix(C, M), 1, 1, 2, 2)
>                      * Submatrix(P₁, 1, 1, 2, 2)⁻¹ : M in Mlst ];
```

None of these are defined over $\mathbb{Q}$ either, but any $\mathbb{Z}$-linear combination of these matrices would also give an isogeny. We therefore want to find a linear combination of the above four matrices that is defined over $\mathbb{Q}$. One way of doing this is to find exact representations of the α matrices by recognizing their entries as algebraic numbers. This can be done in Magma through the use of the PowerRelation function. First we modify this function so that it will always return an irreducible polynomial.

```
function MyPowerRelation(a, d, pres)
    pol := PowerRelation( a, d : Al := "LLL", Precision := pres );
    if Degree(pol) le 0 or IsIrreducible(pol) then
      return pol;
    else
      fac := Factorization(pol);
      _, ind := Min([Abs(Evaluate(f[1], a)) : f in fac]);
      return fac[ind][1];
    end if;
end function;
```

It turns out that every entry of the four α matrices live in some field of degree at most 4.

```
>    AssertAttribute(FldPr, "Precision", 100);
>    Cp := ComplexField();
>    _<x> := PolynomialRing( IntegerRing() );
>    plst := [];
>    for α in alst do
>      for i in [1..2] do
>        for j in [1..2] do
>          pol := MyPowerRelation(Cp ! α[i, j], 4, 80);
>          if Degree(pol) gt 1 then
```

```
>          Append(~plst, pol);
>        end if;
>      end for;
>     end for;
>    end for;
>   end for;
>  plst;
        [
            x^4 - 8*x^3 + 20*x^2 - 16*x + 8,
            2*x^4 - 8*x^3 + 14*x^2 - 12*x + 5,
            x^4 - 16*x^3 + 88*x^2 - 192*x + 160,
            x^4 + 2*x^2 + 50,
            x^2 + 2,
            x^2 + 8,
            x^2 + 2,
            x^4 - 4*x^2 + 8,
            2*x^4 + 2*x^2 + 1,
            x^4 - 8*x^2 + 32,
            x^4 + 2*x^2 + 50,
            x^4 - 4*x^2 + 8,
            2*x^4 + 2*x^2 + 1,
            x^4 - 8*x^2 + 32,
            x^4 + 2*x^2 + 50
        ]
```

Now let us find one number field containing all of these entries. We do this using **CompositeFields**. We will be a bit optimistic by each time only taking the smallest degree extension returned by **CompositeFields**. As we will see later we get lucky and the field we get turns out to contain all the entries of all the α matrices.

```
>  F := NumberField(plst[1]);
>  for i in [2..#plst] do
>    F_2 := NumberField(plst[i]);
>    Flst := CompositeFields(F, F_2);
>    _, ind := Min([Degree(G) : G in Flst]);
>    F := Flst[ind];
>  end for;
>  DefiningPolynomial(F);
        x^8 - 16*x^7 + 112*x^6 - 448*x^5 + 1160*x^4 - 2112*x^3 +
            2624*x^2 - 1792*x + 784

>  O := MaximalOrder(F);
>  _, O_2 := OptimizedRepresentation(O);
>  pol := DefiningPolynomial(O_2); pol;
        x^8 + 4*x^7 + 12*x^6 + 20*x^5 + 24*x^4 + 20*x^3 +
            12*x^2 + 4*x + 1
```

Next we want to write every entry of the four α matrices as an element of the above degree 8 number field. To do this we use LinearRelation to write every such entry as a linear combination of a power basis of the number field. First we define a function to do this for a given basis and some element given to precision 100.

```
function inbase(basis, a)
    line := basis cat [Universe(basis) ! a];
    lst := LinearRelation(line:Al:="LLL", Precision := 80);
    if lst[#lst] eq 0 then
        return 0;
    else
        return [lst[i]/lst[#lst] : i in [1..#lst−1]];
    end if;
end function;
```

Now we compute the exact version of *alst*.

```
>   Kp<xp> := PolynomialRing(Cp);
>   aroot := Roots(Evaluate(pol, xp))[1][1];
>   basis := [aroot^i : i in [0..7]];
>   elst := [ [ inbase(basis, l) : l in ElementToSequence(α) ] : α in alst ];
>   elst;
    [
        [
            [ -13/7,  -44/7,   -95/7, -152/7,  -149/7, -94/7, -33/7,  -8/7 ],
            [  3/14,   25/7,   65/14,  38/7,   57/14,  20/7, 13/14,   2/7 ],
            [ -10/7,    6/7,   -30/7, -76/7,   -92/7, -54/7, -20/7,  -4/7 ],
            [  26/7,   53/7,    50/7,  38/7,    11/7,  13/7,   3/7,   2/7 ]
        ],
        [
            [    0,      0,       0,     0,       0,    0,      0,     0 ],
            [  20/7,   72/7,   137/7, 166/7,   156/7,  94/7,  33/7,   8/7 ],
            [ -40/7, -144/7,  -274/7,-332/7,  -312/7,-188/7, -66/7, -16/7 ],
            [  20/7,   72/7,   137/7, 166/7,   156/7,  94/7,  33/7,   8/7 ]
        ],
        [
            [  15/7,   12/7,     3/7, -26/7,   -37/7, -52/7, -19/7,  -8/7 ],
            [-25/14,  -52/7, -215/14,-130/7, -223/14, -64/7,-43/14,  -5/7 ],
            [ -10/7,  -92/7,  -212/7,-286/7,  -260/7,-180/7, -62/7, -18/7 ],
            [ -30/7, -150/7,  -321/7,-403/7,  -353/7,-218/7, -74/7, -19/7 ]
        ],
        [
            [  -1/7,   44/7,    95/7, 152/7,   149/7,  94/7,  33/7,   8/7 ],
            [-17/14,  -25/7,  -65/14, -38/7,  -57/14, -20/7,-13/14,  -2/7 ],
            [ -18/7,   -6/7,    30/7,  76/7,    92/7,  54/7,  20/7,   4/7 ],
            [ -26/7,  -53/7,   -50/7, -38/7,   -11/7, -13/7,  -3/7,  -2/7 ]
        ]
    ]
```

In order to find a rational isogeny we now simply want some $\mathbb{Z}$-linear combination of the four matrices so that the last 7 entries of each list above sum to zero. We do so by finding the nullspace of an appropriate matrix.

```
>     mat := Matrix(4, 28, [ &cat[ Remove(e[i], 1) : i in [1..4] ] ] : e in elst ]);
>     Nullspace(mat);

        Vector space of degree 4, dimension 1 over
        Rational Field
        Echelonized basis:
        (1 0 0 1)
```

We see that adding the first and fourth α matrices should give us a rational isogeny.

```
>     α := alst[1]+alst[4]; α;

        [1.999999999999999999999999999999999999999999999999999999\
         99999999999999999999999999999999999999999999999999 +
         0.000000000000000000000000000000000000000000000000000000\
         000000000000000000000000000000000000000000000000000153*i
         0.999999999999999999999999999999999999999999999999999999\
         99999999999999999999999999999999999999999999999459 +
         0.000000000000000000000000000000000000000000000000000000\
         000000000000000000000000000000000000000000000037476*i]
        [4.000000000000000000000000000000000000000000000000000000\
         000000000000000000000000000000000000000000000000003 +
         0.000000000000000000000000000000000000000000000000000000\
         000000000000000000000000000000000000000000000001*i
         -0.000000000000000000000000000000000000000000000000000000\
         000000000000000000000000000000000000000000000002]
```

We have therefore answered the original question: C_1 and C_2 do have $\mathbb{Q}$-isogenous Jacobians. As half our computations were inexact we could of course have been spectacularly unlucky and have found an incorrect isogeny (just because α *looks* rational to about 100 places does not prove that it is). Fortunately Magma didn't just tell us that the curves were $\mathbb{Q}$-isogenous but it also gave us an explicit representation of such an isogeny. We can use it, combined with more functionality in Magma, to actually prove that it is a $\mathbb{Q}$-isogeny. We can use α to map points on one *algebraic* Jacobian to points on the other. This in turn can be used to discover an algebraic equation for the isogeny. Through exact computations this can be verified to be a correct isogeny between the two curves.

We do this by using the two functions implementing the isomorphism between the analytic and algebraic Jacobians, called ToAnalyticJacobian and FromAnalyticJacobian. ToAnalyticJacobian maps the point (x, y) on the curve to the analytic Jacobian. The function returns a $g \times 1$ matrix which should be thought of as an element of $\mathbb{C}^g/\Lambda$ where Λ is the $\mathbb{Z}$-lattice generated by the columns of the big period matrix. By adding the images of g points on a genus g curve we get the map from the algebraic Jacobian to the analytic

Jacobian. FromAnalyticJacobian takes a $g \times 1$ complex matrix, z, (thought of as an element of $\mathbb{C}^g/\Lambda$) and returns a list of n (where $n \leq g$) pairs $P_i = \langle x_i, y_i \rangle$ satisfying $y^2 = f(x)$. The degree 0 divisor $P_1 + P_2 + \cdots + P_n - n\infty$ is then the image in the algebraic Jacobian of the point z in the analytic Jacobian.

First, as an example of the process, let us check that α maps rational points on the algebraic Jacobian of C_1 to rational points on the Jacobian of C_2. We find some rational points on the algebraic Jacobian of C_1 with RationalPoints. We pick a somewhat non-trivial one and translate it into an explicit degree two divisor by computing the two (x, y) coordinate pairs. We map each of these to the analytic Jacobian (with ToAnalyticJacobian) and simply add them to get our point on the analytic Jacobian of C_1. Now applying α corresponds to simply multiplying by the matrix α. Finally we use FromAnalyticJacobian to get back to a pair of points on the curve C_2. We check that this corresponds to a rational point on $J(C_2)$ by computing the polynomial with the two x-coordinates as roots.

```
>    C₁ := HyperellipticCurve(f₁);
>    J₁ := Jacobian(C₁);
>    pts₁ := RationalPoints(J₁:Bound:=500);
>    pts₁;

        {@ (1, 0, 0), (x, 0, 1), (x - 1, 17, 1), (x - 1, -17, 1),
          (x^2 + 36*x - 4, 16*x, 2), (x^2 + 36*x - 4, -16*x, 2), (x^2
          + 16*x + 4, 64*x + 16, 2), (x^2 + 16*x + 4, -64*x - 16,
          2), (x^2 + 8*x + 4, 32*x + 16, 2), (x^2 + 8*x + 4, -32*x -
          16, 2), (x^2 + 4*x - 4, 16*x, 2), (x^2 + 4*x - 4, -16*x,
          2), (x^2 - x, 17*x, 2), (x^2 - x, -17*x, 2), (x^2 - 12*x -
          4, 96*x + 32, 2), (x^2 - 12*x - 4, -96*x - 32, 2), (x^2 -
          364/9*x - 100/9, 10016/27*x + 2720/27, 2), (x^2 - 364/9*x
          - 100/9, -10016/27*x - 2720/27, 2) @}

>    P₁ := pts₁[6];
>    apol := ElementToSequence(P₁)[1];
>    bpol := ElementToSequence(P₁)[2];
>    divs₁ := [ r[1] : r in Roots(Evaluate(apol, xc)) ];
>    divs₁ := [ <d₁, Evaluate(bpol, d₁)> : d₁ in divs₁ ];
>    pt₁ := &+[ ToAnalyticJacobian(d[1], d[2], A₁) : d in divs₁ ];
>    P₂ := FromAnalyticJacobian(α*pt₁, A₂);
>    clst := Coefficients( (xc-P₂[1][1]) * (xc-P₂[2][1]) );
>    coefflst := [ BestApproximation(RealField() ! Re(c), 10^40) : c in clst ];
>    Maximum([ Abs(coefflst[i]-clst[i]) : i in [1..#clst] ]);

        1.17878540062786268333882653775943547641912909824783311\
        44854999813935149378524052865383696043861634 29e-94 2

>    xpol := K ! coefflst; xpol;

        x^2 + 4*x - 6
```

We can map any point $P = (x, y)$ on the curve C_1 to a point in the (algebraic) Jacobian of C_2 by first mapping P to the point $P + \infty$ in the algebraic Jacobian of C_1 and then mapping it to $J(C_2)$ by the isogeny found above. Let this image correspond to the point $P_1 + P_2$ in the symmetric square of C_2 and let x_i be the x-coordinate of P_i, $i = 1, 2$. Then it can be shown that $s_1 = x_1 + x_2$ and $s_2 = x_1 x_2$ are both rational functions of x. These rational functions determine an algebraic description of the isogeny, and once they are known it can be checked by an exact computation that it gives an isogeny. For a given x we can compute the values of s_1 and s_2 in Magma by going through the analytic Jacobian as was done in the previous paragraph.

In this particular case we know from [16] the rational functions giving s_1 and s_2. We check this by computing them for a 'random' point on C_1.

```
>    pntx := Pi(C)+i*67/109;
>    pnty := Sqrt( Evaluate(f1C,pntx) );
>    pt := ToAnalyticJacobian(pntx, pnty, A₁);
>    P₂ := FromAnalyticJacobian(α*pt, A₂);
>    (8*(2+3*pntx))/(4+8*pntx+pntx²) − (P₂[1][1]+P₂[2][1]);
```

```
    -0.0000000000000000000000000000000000000000000000000000000000\
    00000000000000000000000000000000000000000000000000111 -
    0.0000000000000000000000000000000000000000000000000000000000\
    000000000000000000000000000000000000000000000000079091*i
```

```
>    (8*(2+3*pntx))/(4+8*pntx+pntx²) − (P₂[1][1]*P₂[2][1]);
```

```
    -0.0000000000000000000000000000000000000000000000000000000000\
    0000000000000000000000000000000000000000000000000147 -
    0.0000000000000000000000000000000000000000000000000000000000\
    000000000000000000000000000000000000000000000026197*i
```

Note that s_1 being a rational function in x implies that for any x there is a linear relation of the form

$$1 + a_1 x + a_2 x^2 + \cdots + a_n x^n + b_0 s_1 + b_1 s_1 x + \cdots + b_m s_1 x^m = 0,$$

for some n and m. This shows that if we compute s_1 (and similarly for s_2) for enough x values we can *discover* the coefficients a_i and b_i by computing an appropriate nullspace.

References

1. Wieb Bosma, John Cannon, Catherine Playoust, *The Magma algebra system I: The user language*, J. Symbolic Comput. **24** (1997), 235–265.
   See also the Magma home page at http://magma.maths.usyd.edu.au/magma/.
2. H. M. Farkas, I. Kra, *Riemann surfaces*, Graduate Texts in Mathematics **71**, New York: Springer-Verlag, 1992 (second edition).

3. William Fulton, *Algebraic curves. An introduction to algebraic geometry*, New York-Amsterdam: W. A. Benjamin, Inc., 1969. Notes written with the collaboration of Richard Weiss, Mathematics Lecture Notes Series.

4. Phillip Griffiths and Joseph Harris, *Principles of algebraic geometry*, Wiley Classics Library, New York: John Wiley & Sons Inc., 1994. Reprint of the 1978 original.

5. S. Lang, *Introduction to Algebraic and Abelian Functions*, Springer-Verlag, 1982.

6. S. Lang, *Complex Multiplication*, Springer-Verlag, 1983.

7. H. Lange, Ch. Birkenhake, *Complex Abelian Varieties*, Springer-Verlag, 1992.

8. J. S. Milne, *Jacobian varieties*, in: G. Cornell, J. H. Silverman (editors), *Arithmetic Geometry*, Springer-Verlag, 1986.

9. D. Mumford, *Tata Lectures on Theta I*, Progr. Math. **28**, Birkhäuser, 1983.

10. D. Mumford, *Tata Lectures on Theta II*, Progr. Math. **43**, Birkhäuser, 1984.

11. Naoki Murabayashi, Atsuki Umegaki, *Determination of all $\mathbb{Q}$-rational CM-points in the moduli space of principally polarized abelian surfaces*, J. Algebra **235**-1 (2001), 267–274.

12. Michael Rosen, *Abelian varieties over $\mathbb{C}$*, pp. 79–101 in: *Arithmetic geometry (Storrs, Conn., 1984)*, New York: Springer, 1986.

13. N. P. Smart, *S-unit equations, binary forms and curves of genus 2*, Proc. London Math. Soc. (3) **75**-2 (1997), 271–307.

14. Paul van Wamelen, *Examples of genus two CM curves defined over the rationals*, Math. Comp. **68**-225 (1999), 307–320.

15. Paul van Wamelen, *Proving that a genus 2 curve has complex multiplication*, Math. Comp. **68**-228 (1999), 1663–1677.

16. Paul van Wamelen, *Poonen's question concerning isogenies between Smart's genus 2 curves*, Math. Comp. **69**-232 (2000), 1685–1697.

Graded rings and special K3 surfaces

Gavin Brown

Institute of Mathematics, Statistics and Actuarial Science
University of Kent
Canterbury, CT2 7NF, UK
gdb@kent.ac.uk

Summary. Many recent constructions of varieties, including the lists of K3 surfaces in Magma, use graded ring methods. We show how to apply the method using Magma and, as an application, construct 27 families of K3 surfaces that appear as degenerate cases of surfaces in the usual lists. These are displayed in Tables 1–3 and include both standard degenerations and new examples.

1 Introduction

Projective varieties, graded rings and Hilbert series

A projective variety $X \subset \mathbb{P}^n$ has an associated graded ring, its homogeneous coordinate ring $R(X)$. For example

$$X \colon (x^4 + y^4 + z^4 + w^4 = 0) \subset \mathbb{P}^3 \quad \text{has} \quad R(X) = \frac{k[x, y, z, w]}{(x^4 + y^4 + z^4 + w^4)},$$

a ring that is graded by the homogeneous degree of polynomials. We set up the variety in Magma [3] as follows*

```
>    P3<x, y, z, w> := ProjectiveSpace(Rationals(), 3);
>    X := Scheme(P3, x^4 + y^4 + z^4 + w^4); X;
        Scheme over Rational Field defined by
        x^4 + y^4 + z^4 + w^4
```

As we recount in Section 3, the *Hilbert series* $P_{R(X)}(t)$ records the vector space structure of $R(X)$: if r_n denotes the dimension of the vector space of homogeneous polynomials of degree n in R, then

$$P_{R(X)}(t) = 1 + r_1 t + r_2 t^2 + \cdots = \sum_{n \geq 0} r_n t^n.$$

---

*See the Preface to this volume for style conventions regarding Magma code; code appearing in this book is available at http://magma.maths.usyd.edu.au/magma/.

In Section 2, we look at the standard method for realising such power series as rational functions of t. Calculating in Magma (where the usual assignment $P := \ldots$ is replaced by $P<t> := \ldots$ as a quick way to name the variable t):

```
>    R := CoordinateRing(X);
>    P<t> := HilbertSeries(R);
>    P;
```

$$(-t\char94 3 - t\char94 2 - t - 1)/(t\char94 3 - 3*t\char94 2 + 3*t - 1)$$

Viewing $P(t)$ as a power series

```
>    S<s> := PowerSeriesRing(Rationals());
>    S ! P;
```

$$1 + 4*s + 10*s\char94 2 + 20*s\char94 3 + 34*s\char94 4 + 52*s\char94 5 + \ldots$$

which is the familiar count of monomials of given degree in four variables, with the twist that the equation of X gives a relation among the 35 quartics, $x^4 = -(y^4 + z^4 + w^4)$ say, and this defect propagates into higher degree.

Families of K3 surfaces having constant Hilbert series

The coordinate ring of any variety defined by an equation of degree 4 in $\mathbb{P}^3$ has the same Hilbert series as that of X defined above. We regard X as being a member of this family of surfaces, and denote a general member by $X_4 \subset \mathbb{P}^3$.

The general $X_4 \subset \mathbb{P}^3$ is a (projective) K3 surface. A *K3 surface* is a simply-connected, nonsingular surface with trivial canonical class. (In Section 5 we allow mild singularities on X, as is common.) This definition will not occupy us again since it is already implicit in our calculations.

There is a database of K3 surfaces in Magma, discussed in Section 5. It includes the surface $X_6 \subset \mathbb{P}^3(1,1,1,3)$. This expression denotes a hypersurface defined by a degree 6 equation of the form $y^2 = f_6(x_1, x_2, x_3)$ in variables x_1, x_2, x_3, y of weights $1, 1, 1, 3$ respectively. In Section 6 we recall the well-known K3 surface $Y_{2,6} \subset \mathbb{P}^4(1,1,1,2,3)$ defined by two equations. These two K3 surfaces, X_6 and $Y_{2,6}$, have identical Hilbert series, but they lie in quite different spaces and so have different graded rings. In this paper, we use Magma to find other examples of such behaviour, in which the Hilbert series cannot distinguish between two K3 surfaces having different graded rings.

Overview of the chapter

As a standard method, one uses the Riemann–Roch theorem (RR) to make Hilbert series, followed by calculations that construct corresponding graded rings. In Sections 2–4, we sketch this method of graded rings and an elementary application of it to subcanonical curves. To use these methods, one must work in weighted projective spaces (wps) from the start: both Fletcher [10] and Dolgachev [8] give introductory tutorials in wps while Reid [16] discusses graded rings.

In Section 5, we look at K3 surfaces and introduce Magma's K3 database. This is a list of families of K3 surfaces embedded in wps. (The word 'database' does not have the connotation of being exhaustive.) The first such list is Reid's 'famous 95', a list of 95 K3 weighted hypersurfaces written down in 1979 by hand. We look for degenerations of varieties in the K3 database, that is, special fibres of families of K3 surfaces (in a precise and limited sense described below). It is easy to find degenerations of the famous 95, and Tables 1 and 2 of Section 6 lists some families of hypersurfaces which have K3 surfaces in codimension 2 as special members. Finally, in Sections 7–8, we discuss unprojection and show how, with more careful book-keeping, one can use it to discover degenerations in higher codimension. Table 3 lists such examples in codimension 3.

Related projects

The method of graded rings was initiated in its modern form by Reid [17], in part inspired by Enriques' construction of surfaces of general type such as $X_{10} \subset \mathbb{P}^3(1,1,2,5)$, a surface defined by an equation of the form

$$z^2 = y^5 + a_4(x_1,x_2)y^3 + b_6(x_1,x_2)y^2 + c_8(x_1,x_2)y + d_{10}(x_1,x_2)$$

in variables x_1, x_2, y, z of weights $1, 1, 2, 5$ respectively. Since then, the method has been used in several different contexts: K3 surfaces and Fano 3-folds [10], [1], [22], Calabi–Yau 3-folds [23], [7], [6], del Pezzo surfaces [20], log Enriques surfaces and Fano–Enriques 3-folds (work in progress by Jorge Caravantes). Computer algebra is used both for systematic calculations and for experiments by many of these authors. Many of these results are available at [4].

The case study in this paper is small, involving searches of only a few hundred cases to find examples with suitable behaviour. Nevertheless, the paradigm it illustrates—first making easy calculations in low codimension and then lifting them through a structured database using unprojection— will apply in many other situations. For example, the construction of the K3 database itself, as described in [5], takes this approach. A more serious example would be to relate the deformation properties of K3 surfaces before and after projection as a part of the classification of Fano 3-folds. This would be the 3-fold analogue of a familiar story for surfaces: the classification of del Pezzo surfaces into 9 families using descriptions of elliptic curves polarised by low degree divisors, as found in [13], Section III.3 (especially Theorem 3.5).

The significance of such databases of varieties is an issue; after all, these lists often constitute results having no theorems. The 'famous 95' are distinguished by proofs that they are the only families of hypersurfaces in wps that contain K3 surfaces; a recent proof is Johnson–Kollár [11], Corollary 11. This is not true of the existing lists of K3 surfaces in higher codimension. In this paper, we present examples of K3 surfaces that do not occur in the existing lists. Some of these examples are well known, but most are not. We show how these examples were discovered using Magma. Their real significance will eventually be as a small piece in the final (finite) classification of Fano 3-folds, a major current project in birational geometry. The constructions here also invite theorems that we do not discuss: a proof in the style of [11] that codimension 2 is exhausted; the statement and proof that unprojection—Type I in particular—can be made in families, proving existence of the complicated K3 surfaces that would arise if these computer experiments were taken further.

2 Elementary example

Describing a family of graded rings with a given Hilbert series

We illustrate the naive method of constructing a graded ring given a Hilbert series. Consider the following power series $P = P(t)$:

```
>    T<t> := PowerSeriesRing(Rationals() : Precision:=50);
>    P := 1 + t + t² + t³ + 2*t⁴ + 3*t⁵ + 4*t⁶
>                    + &+[ (n−3)*tⁿ : n in [7..49] ] + O(t⁵⁰);
```

Multiplying P by $1 - t$ cancels some small powers of t in the expansion, in the sense that

```
>    (1−t) * P;
         1 + t^4 + t^5 + t^6 + t^8 + t^9 + t^10 +  ...   + O(t^50)
```

We attempt to write P as a rational function by further cancellation. The next nontrivial power of t in the expansion is t^4, so we multiply by $1 - t^4$.

```
>    (1−t⁴)*(1−t) * P;
         1 + t^5 + t^6 + t^11 + O(t^50)
```

That completes the task (at least up to the given precision) since the last display can be rewritten as

$$P = \frac{1 + t^5 + t^6 + t^{11}}{(1 - t^4)(1 - t)}.$$

One can continue the procedure, working on the t^5 and t^6 terms.

```
>    (1−t⁵)*(1−t⁴)*(1−t) * P;
         1 + t^6 - t^10 - t^16 + O(t^50)

>    (1−t⁶)*(1−t⁵)*(1−t⁴)*(1−t) * P;
         1 - t^10 - t^12 + t^22 + O(t^50)
```

Although it could have been calculated from the previous expression, this display shows clearly that

$$P = \frac{1 - t^{10} - t^{12} + t^{22}}{(1 - t^6)(1 - t^5)(1 - t^4)(1 - t)}.$$

Expressed in this way, the power series P is visibly the Hilbert series of any variety

$$X \colon (f_{10} = g_{12} = 0) \subset \mathbb{P}^3(1, 4, 5, 6)$$

where f, g is a regular sequence of polynomials of the indicated degrees, 10 and 12 respectively, in variables x, y, z, w of weights $1, 4, 5, 6$.

Working with a particular family member

The calculation above described a family of varieties. Since Magma works with individual varieties (or schemes) defined by equations, we study a single member of the family by choosing particular equations f_{10} and g_{12}.

Lemma 2.1. *For suitable choice of equations, the variety*

$$X \colon (f_{10} = g_{12} = 0) \subset \mathbb{P}^3(1, 4, 5, 6)$$

is a nonsingular curve of genus 4 polarised by a subcanonical divisor D with $\deg D = 1$ and $6D = K_X$ where K_X is a canonical divisor.

This is clear by the standard methods of wps, but we demonstrate it using Magma on the example defined by the equations

$$f = wy + z^2 + x^{10}, \quad g = w^2 + y^3 + (xz)^2.$$

Certainly the variety defined by these equations is quasi-smooth (in the sense of wps [10], [8]): X only has orbifold singularities at points where the stabiliser subgroup of the $\mathbb{C}^\star$ action jumps up.

```
>    P3<x, y, z, w> := ProjectiveSpace(Rationals(), [1, 4, 5, 6]);
>    f := w*y + z² + x¹⁰;
>    g := w² + (x*z)² + y³;
>    X := Scheme(P3, [f, g]);
>    IsNonsingular(X);
         true
```

The intrinsic **IsNonsingular** computes the Jacobian ideal and returns true if, in the projective case, that ideal is supported on the irrelevant maximal ideal. In other words, the equations of X do not lead to any singularities on X.

To use Magma's RR machinery, we must rewrite X as a plane curve and identify the divisor $D \subset X$. This is not automated so we work interactively as follows. The polarising divisor D is $(x = 0) \cap X$. This divisor is supported on a point p_0 of X, and has degree 1. We identify the point p_0. (Intrinsics to compute this are not available yet for wps so we are lucky to be able to spot the solution: intersection simply concatenates the equations of two schemes.)

```
>    p0 := P3 ! [0, −1, 1, 1];
>    p0 in Intersection(X, Scheme(P3, x));
         true (0 : -1 : 1 : 1)
```

The point p_0 was constructed initially as a point of $\mathbb{P}^3(1, 4, 5, 6)$. The second return value above is the point regarded as lying on the intersection.

We will map X to the plane in two steps: first we map to an ordinary projective 3-space using the complete linear system of degree 6. (We could have used LinearSystem(P_3, 6) to identify this map.)

```
>    P<a,b,c,d> := ProjectiveSpace(Rationals(),3);
>    φ := map< P_3 → P | [x^6, x^2*y, x*z, w] >;
>    Y := φ(X);
>    Y;
```

```
          Scheme over Rational Field defined by
          a*c^2 + a*d^2 + b^3,
          a^2 + b*d + c^2
```

```
>    p_1 := φ(p_0);
>    p_1;
```

```
          (0 : 0 : 0 : 1)
```

```
>    p_1 in Y;
```

```
          true (0 : 0 : 0 : 1)
```

Second we map to the plane by projection from the point $(1 : 0 : 0 : 0)$ (which is the default for the Projection intrinsic).

```
>    P_2<r,s,t> := ProjectiveSpace(Rationals(), 2);
>    ψ := Projection(P, P_2);
>    p_2 := ψ(p_1);
>    Z := ψ(Y);
>    Z;
```

```
          Scheme over Rational Field defined by
          r^6 + r*s^4*t + 2*r*s^2*t^3 + r*t^5 + s^6 +
          2*s^4*t^2 + s^2*t^4
```

```
>    p_2 in Z;
```

```
          true (0 : 0 : 1)
```

At this stage, Z is regarded as a scheme. To use the general curve machinery—so that we can compute its genus, or RR for its divisors—we must change its type to a curve.

```
>    IsCurve(Z);
```

```
          true
```

```
>    _,C := IsCurve(Z);
>    C;
```

```
          Curve over Rational Field defined by
          r^6 + r*s^4*t + 2*r*s^2*t^3 + r*t^5 + s^6 +
          2*s^4*t^2 + s^2*t^4
```

Now we can use the curve machinery.

```
>    Genus(C);

        4
```

We also realise p_2 as a point on the curve C.

```
>    p := C ! p₂;
>    IsNonsingular(p);

        true
```

```
>    GapNumbers(p);

        [ 1, 2, 3, 7 ]
```

The gap numbers of a point record the increments np to $(n+1)p$ which do not increase the number of sections, and these show that the degree 1 divisor supported at p embeds C as $X \subset \mathbb{P}(1,4,5,6)$. It is easier to see this by considering the dimensions of RR spaces of the multiples of p (regarded as a divisor using the Place(p) command):

```
>    [ Dimension(RiemannRochSpace(n∗Place(p))) : n in [0..8] ];

        [ 1, 1, 1, 1, 2, 3, 4, 4, 5 ]
```

In fact, this is what the GapNumbers intrinsic computes: there are no new basis elements when we step up to p, $2p$, $3p$, $7p$.

```
>    IsCanonical(6∗Place(p));

        true
```

These results are just as the claims we made about X: we have made an explicit curve with the given genus containing a point dividing the canonical class with the given gap sequence.

3 Graded rings of polarised varieties

The natural language for our calculations is graded rings and Hilbert series. This standard material is explained at greater length in [1].

Let X be a projective variety and D an ample (and in particular $\mathbb{Q}$-Cartier) divisor on X. Define the graded ring

$$R(X, D) = \bigoplus_{i \geq 0} \mathrm{H}^0(X, iD)$$

which is finitely generated (since D is ample) and graded by $i \in \mathbb{N}$. There is a natural embedding

$$X \hookrightarrow \mathrm{Proj}\, R(X, D).$$

If $X \subset \mathbb{P}^n$ and D is a hyperplane section of X, then $R(X, D)$ is the homogeneous coordinate ring of $X \subset \mathbb{P}^n$. The example of the last section constructed

$$C \hookrightarrow X = \operatorname{Im}(C) \subset \operatorname{Proj} R(C, p) = \mathbb{P}^3(1, 4, 5, 6).$$

For a graded ring A, denote the i-th graded piece by A_i. It is well known that the graded ring $A = R(X, D) = \oplus_{i \geq 0} A_i$ has a *Hilbert series*

$$P_A = \sum_{i \geq 0} (\dim A_i) t^i$$

and that this is equal to a rational function. Indeed, there are finitely many polynomials $f_{A,1}, \ldots, f_{A,r}$ so that, for all sufficiently large i, the coefficient $\dim A_i$ is equal to $f_{A,j}(i)$ where $j = i \bmod r$. In the case $r = 1$, one usually writes $f_A = f_{A,1}$ and refers to it as the *Hilbert polynomial*. In our context, this occurs when D is a Cartier divisor.

4 Subcanonical curves

These calculations originated in Reid's 'homework' [16] and a University of Warwick MMath project by Adam Keenan. They are discussed in detail in the ISSAC2003 tutorial [4].

The setup

We want to write down examples of a curve C polarised by a *subcanonical* divisor D, that is, a divisor satisfying $K_C = kD$ for some integer k. So the basic input is a pair of integers (g, d), where $g \geq 2$ is a genus and d divides $2g - 2$, and the output is a description of the ring $A = R(C, D)$ for some curve C of genus g admitting a subcanonical divisor D of degree d.

It is easy to see from RR that the Hilbert polynomial $f_A(n)$ is equal to $nd + 1 - g$, and moreover vanishing implies that the Hilbert polynomial determines the coefficient of t^n in the Hilbert series whenever $nd > 2g - 2$. So nominating a pair (g, d) determines a Hilbert series except for the coefficients of a few initial terms, for which the Hilbert polynomial $f_A(n)$ gives a lower bound for the coefficient of p_n of t^n in $P_A(t)$ when $1 < n < k$ rather than determining it exactly. There will be many possibilities for different coefficients in this range corresponding to different gap sequences for a divisor D on a curve C. Of course, there are some rules that limit the possibilities for the initial coefficients, a few of which are:

1. The coefficient p_k of t^k in $P_A(t)$ is the genus g.
2. Given the values of $p_1, \ldots, p_{\lfloor k/2 \rfloor}$, Serre duality computes $p_{\lceil k/2 \rceil}, \ldots, p_{k-1}$.
3. If $p_n > 0$, then one sees by restriction to nD that $p_{m+n} - p_m \leq nd$ for any $m > 0$.
4. The divisors nD in the range $1 < n < k$ are special (in the sense of [9], Section IV.1) so from Clifford's theorem one knows that $p_n - 1 \leq (n/2)d$ whenever $p_n \neq 0$.

5. The resulting ring A should be a ring without zero-divisors, so, for example, if $p_{2n} = 0$ then $p_n = 0$.

Rather than absorbing these points in one go, it is easy to try out some numbers and cross-check when the result seems to be nonsense: typically one of these rules has been violated. Observe that point 5 (which is more of a hint about other trivial restrictions than a rule) does not imply that the sequence of p_n is monotonic unless $p_1 > 0$.

An example: $g = 7$, $d = 6$

Consider curves of genus $g = 7$. In this case, $2g - 2 = 12$ so d must divide 12. Try $d = 6$ in Magma: before starting we need to name the variable t used in the polynomial numerators

```
>    R<t> := PolynomialRing(Rationals());
```

The syntax used to create a curve with prescribed leading terms $P(t) = p_0 + p_1 t + \cdots + p_s t^s + \cdots$ is

```
SubcanonicalCurve(g, d, [p_0,p_1,...p_s])
```

for some $s \geq \lceil k/2 \rceil$ (although any p_i with i strictly above this threshold are ignored). Taking p_1 to be 3 (so that the Hilbert series starts $P(t) = 1+3t+\cdots$) gives the curve

```
>    SubcanonicalCurve(7, 6, [1,3]);

        Subcanonical curve C, D of genus 7, K_C = 2*D,
        deg(D) = 6, codim = 2 with
          Weights: [ 1, 1, 1, 2 ]
          Numerator: t^7 - t^4 - t^3 + 1
```

We could hope to build a real example of C, D by writing down a pair of polynomials of degrees 3,4 in four variables x, x_1, x_2, y of degrees $1,1,1,2$ as in Section 2 but we don't do that yet.

Trying leading terms $P(t) = 1 + 2t + \cdots$ instead gives

```
>    SubcanonicalCurve(7, 6, [1,2]);

        Subcanonical curve C, D of genus 7, K_C = 2*D,
        deg(D) = 6, codim = 4 with
          Weights: [ 1, 1, 2, 2, 2, 2 ]
          Numerator: t^12 - 9*t^8 + 16*t^6 - 9*t^4 + 1
```

Writing down equations for this curve is more complicated. According to the numerator, there appear to be 9 equations with 16 syzygies between them. Such 9×16 formats do exist in codimension 4: Papadakis [14] has made a study of two families of codimension 4 equation formats that he calls Tom&Jerry, and it is possible that these are the only two such formats.

Experimentation may persuade you that these are the only two possible Hilbert series with $g = 7$, $d = 6$, although a proof would need more detailed geometry of curves. Other values of d give many more examples when $g = 7$.

5 K3 database

Magma contains the applicable RR formula for several different contexts. In each case, given appropriate RR data that could be associated to a polarised variety X, A, there are intrinsics that will compute a Hilbert series and then use differencing to find a possible model for the graded ring $R(X, A)$. The basic differencing calculation is generic and can be made on any geometric power series, but experience has suggested many small modifications to the basic game to suit different contexts.

One such RR formula due to Altınok [2], [1] applies to polarised K3 surfaces. Such surfaces are allowed to have Gorenstein (cyclic) quotient singularities which we will describe.

Quotient singularities

Let $\mathbb{Z}/r\mathbb{Z}$ act on the plane $\mathbb{C}^2$ in coordinates u, v by

$$\zeta \cdot u = \zeta^a u, \quad \zeta \cdot v = \zeta^{r-a} v$$

where $\zeta \in \mathbb{Z}/r\mathbb{Z}$ is a complex rth root of unity and a is coprime to r. We can form the quotient singularity

$$\mathbb{C}^2/(\mathbb{Z}/r\mathbb{Z})(a, r-a)$$

as the surface whose coordinate ring is the invariant subring of $\mathbb{C}[u, v]$ by this action. This singularity is denoted by $\frac{1}{r}(a, r-a)$. (As usual, to say that the surface may have quotient singularities means that it may have finitely many points which have an analytic neighbourhood isomorphic to an analytic neighbourhood of the origin in such a model quotient singularity.) The integer $r > 1$ is called the *index* of the singularity $\frac{1}{r}(a, r-a)$.

In the tables, we write the equivalent form $\frac{b(r-b)}{r}$, where $ab = 1 \bmod r$, in place of $\frac{1}{r}(a, r-a)$: this rational number is exactly the contribution that the singularity makes to the degree of the surface.

Numerical type

The input to Altınok's RR formula is the *numerical type* of a K3 surface. That is data as follows:

> *genus* : an integer $g \geq -1$, and
> *basket* : a list $\mathcal{B}$ of quotient singularities $\frac{1}{r}(a, r-a)$.

In Magma, the basket $\mathcal{B}$ is denoted by a sequence comprising sequences of coprime integers $[r, a]$ with $r > a$. Following Barlow–Fletcher–Reid, [18], baskets of singularities are common for this type of RR formula. A basket lists quotient singularities that will make the same contribution to RR as the actual singularities of the variety. In fact, it is known that in the case of K3 surfaces, a general surface we make will have exactly these singularities.

Building a single K3 surface

The intrinsic K3Surface(g, B) takes the two parts of a numerical type as its arguments and returns a description of a K3 surface. For example

```
>    g := 0;
>    B := [ [2,1], [2,1], [3,1], [4,1] ];
>    X := K3Surface(g, B);
>    X;

          Codimension 1 K3 surface with data
            Weights: [ 1, 2, 3, 4 ]
            Numerator: -t^10 + 1
            Basket: [ 2, 1 ], [ 2, 1 ], [ 3, 1 ], [ 4, 1 ]
```

The intrinsic K3Surface first calculates a Hilbert series using Altınok's formula with the data g, B, and then runs a generic differencing routine to compute weights and numerator.

Interpreting the numerator as usual, we write a variety

$$X \colon (f_{10} = 0) \subset \mathbb{P}^3(1, 2, 3, 4).$$

One can check as before that if the degree 10 polynomial f is sufficiently general, then X is a K3 surface bearing exactly the four quotient singularities listed in the basket.

Finiteness of baskets

It is a standard fact of the geometry of K3 surfaces [1], Section 4.8, that if $r_1, \ldots, r_k$ are the indexes of the singularities of the basket, then

$$\sum r_i - 1 \le 19.$$

So the number of possible baskets is finite a priori. (They can be computed using the intrinsic K3Baskets, although it does nothing beyond simple book-keeping.) Of course, this is not a bound on the number of families of K3 surfaces since the genus g can be arbitrarily large.

The K3 database

The K3 database in Magma is a list of families of K3 surfaces. The families are in 1-1 correspondence with a subset of the possible pairs g, B for small g.

```
>    D := K3Database();
>    #D;
          24099
```

The 391 K3 surfaces in this list of codimension at most 4, including the famous 95 hypersurfaces, were first calculated by Reid, Fletcher and Altınok. There are several ways to extract K3 surfaces from the database.

```
>    K3Surface(D, [1,2,3,4]);
```

```
K3 surface no.272, genus 0, in codimension 1 with data
  Weights: [ 1, 2, 3, 4 ]
  Basket: 2 x 1/2(1,1), 1/3(1,2), 1/4(1,3)
  Degree: 5/12          Singular rank: 7
  Numerator: -t^10 + 1
  Projection to codim 1 K3 no.267 -- type I from 1/4(1,3)
  Projection to codim 1 K3 no.253 -- type I from 1/3(1,2)
  Unproj'n from codim 2 K3 no.273 -- type I from 1/7(3,4)
  Unproj'n from codim 2 K3 no.274 -- type I from 1/5(2,3)
  Unproj'n from codim 2 K3 no.276 -- type I from 1/5(1,4)
  Unproj'n from codim 2 K3 no.309 -- type I from 1/4(1,3)
  Unproj'n from codim 2 K3 no.360 -- type I from 1/3(1,2)
  Unproj'n from codim 3 K3 no.821 -- type II_1 from 1/2(1,1)
```

As already mentioned, if $X_d \subset \mathbb{P}(a_1, a_2, a_3, a_4)$ is a K3 surface, then it is a member of one of these 95 hypersurfaces. In higher codimension, however, the same statement is not true; the authors of the original lists were well aware that they overlooked some degenerate cases, and we discuss those next.

6 Simple degenerations of the famous 95

Simple degenerations

Recall the well-known monogonal degeneration of K3 surfaces:

$$X = X_6 \subset \mathbb{P}^3(1,1,1,3) \quad \text{degenerates to} \quad Y = Y_{2,6} \subset \mathbb{P}^4(1,1,1,2,3).$$

In coordinates x_0, x_1, x_2, u, y of weights $1, 1, 1, 2, 3$, this is obtained by sending $\lambda \to 0$ in the equations

$$(\lambda u - g_2(x_0, x_1, x_2) = f_6(x_0, x_1, x_2, u, y) = 0) \subset \mathbb{P}^4(1,1,1,2,3)$$

for sufficiently general f, g of degrees $2, 6$ respectively. When $\lambda \neq 0$ one eliminates u from the equations to see a double plane $X_6 \subset \mathbb{P}^3(1,1,1,3)$. But for $\lambda = 0$ the equations define $Y_{2,6}$ of the same numerical type as a double plane, but spanning the $\mathbb{P}^4$. For general equations f, g, both X and Y are nonsingular since the equation f_6 involves u^3 and y^2, so in particular neither X nor Y contains the index 2 singularity that arises from the new variable u of weight 2. We want to write down more examples of degenerations of K3 surfaces having fixed numerical type but lying in different ambient spaces.

This example is the first dichotomy for linear systems on K3 surfaces that are not very ample (away from -2-curves): a basis x_0, x_1, x_2 of $\mathrm{H}^0(X, D)$ maps the general X surjectively $2 : 1$ to $\mathbb{P}^2$ while sections of $\mathrm{H}^0(Y, D)$ describe the special Y as an elliptic surface over a plane conic $g_2(x_0, x_1, x_2) = 0$. Such examples are not quasi-smooth since Y meets the locus $\mathbb{P}(2,3)$ with a node given by $g_2 = 0$, projection from which is the elliptic pencil. A degeneration to a quasi-smooth K3 surface is given by the hyperelliptic degeneration

$$X = X_4 \subset \mathbb{P}^3(1,1,1,1) \quad \text{degenerates to} \quad Y = Y_{2,4} \subset \mathbb{P}^4(1,1,1,1,2)$$

in which a general such Y is a $2:1$ cover of a quadric in $\mathbb{P}^3$. This is an example of the second dichotomy: see Saint-Donat [21] or Reid [19] Chapter 3. Both cases are of interest to us, although the definition of simple degeneration below restricts us to the hyperelliptic case.

Let $X \colon (f(x_0, \ldots, x_n) = 0) \subset \mathbb{P}$ be an irreducible hypersurface in wps $\mathbb{P}$. A *simple degeneration* of $X \subset \mathbb{P}$ is

1. a wps $\mathbb{P}' \supset \mathbb{P}$ containing $\mathbb{P}$ as a linear hyperplane $u = 0$;
2. a variety

$$\mathcal{X} \colon (F(x_0, \ldots, x_n, u, \lambda) = G(x_0, \ldots, x_n, u, \lambda) = 0) \subset \mathbb{P}' \times \mathbb{C}$$

defined by equations satisfying, for $g = g(x_0, \ldots, x_n)$,

$$F(x_0, \ldots, x_n, g, 1) = f, \qquad G(x_0, \ldots, x_n, u, \lambda) = \lambda u - g$$

and that every fibre $\mathcal{X}_\lambda \subset \mathbb{P}'$ is quasi-smooth (including at $\lambda = 0$).

If $\mathcal{X}$ is such a simple degeneration, the fibre $\mathcal{X}_\lambda$ above $\lambda \in \mathbb{C}^\times$ is isomorphic to $X \subset \mathbb{P}$ by a projective linear isomorphism of $\mathbb{P}'$. We are only interested in simple degenerations that have special fibre $Y = \mathcal{X}_0 \subset \mathbb{P}'$ a K3 surface that spans $\mathbb{P}'$ (and so, in particular, is not the same polarised variety as X).

In a neighbourhood of any singular point of $Y \subset \mathbb{P}'$, either u is eliminated near P or u is a polarising axis of P and some other variable is eliminated. This puts numerical conditions on simple degenerations. In the following lemma, the weights of variables are not ordered in any particular way: the numerical conditions apply to a singularity at any coordinate point of $\mathbb{P}$.

Lemma 6.1. *Let $X_m \subset \mathbb{P} = \mathbb{P}^3(a, b, c, d)$ be one of the famous 95 having a singularity $\frac{1}{d}(a, b)$. If $Y_{m,n} \subset \mathbb{P}' = \mathbb{P}^4(a, b, c, d, n)$ is a simple degeneration of X variable then $d < n < m$ and n divides m. Moreover, either d divides $m - n$ or d divides one of $n - a$ and $n - b$.*

The lemma follows by considering the polarisation of the singular point. Fix coordinates x, y, z, t, u on $\mathbb{P}^4$. In a neighbourhood of P_t, the singularity in question, there are two equations in an ambient 4-space. To be quasi-smooth, both equations must be nonsingular at the origin, each used to implicitise a single variable there. Either u is an implicit function of the other variables or it is not. If it is, the equation of degree m contains a term $t^r u$ and so d divides $m - n$. And moreover, some other variable must be made implicit by the equation of degree n, and so $d < n$ for a similar monomial reason. On the other hand, if u is not implicit, then either x or y must be and this is realised by the equation of weight n. This shows that d divides $n - a$ or $n - b$.

Table 1. Special codimension 2 K3 surfaces among the famous 95

No.	general K3/basket $\mathcal{B}$	special K3	deg u
1	$X_4 \subset \mathbb{P}(1,1,1,1)$ no singularities	$Y_{2,4} \subset \mathbb{P}(1,1,1,1,2)$	2
2	$X_6 \subset \mathbb{P}(1,1,2,2)$ $3 \times \frac{1}{2}$	$Y_{3,6} \subset \mathbb{P}(1,1,2,2,3)$	3
3	$X_8 \subset \mathbb{P}(1,2,2,3)$ $4 \times \frac{1}{2} + \frac{2}{3}$	$Y_{4,8} \subset \mathbb{P}(1,2,2,3,4)$	4
4	$X_{10} \subset \mathbb{P}(1,2,3,4)$ $2 \times \frac{1}{2} + \frac{2}{3} + \frac{3}{4}$	$Y_{5,10} \subset \mathbb{P}(1,2,3,4,5)$	5
5	$X_{12} \subset \mathbb{P}(1,2,4,5)$ $3 \times \frac{1}{2} + \frac{4}{5}$	$Y_{6,12} \subset \mathbb{P}(1,2,4,5,6)$	6
6	$X_{12} \subset \mathbb{P}(2,3,3,4)$ $3 \times \frac{1}{2} + 4 \times \frac{2}{3}$	$Y_{6,12} \subset \mathbb{P}(2,3,3,4,6)$	6
7	$X_{14} \subset \mathbb{P}(2,3,4,5)$ $3 \times \frac{1}{2} + \frac{2}{3} + \frac{3}{4} + \frac{2 \cdot 3}{5}$	$Y_{7,14} \subset \mathbb{P}(2,3,4,5,7)$	7
8	$X_{18} \subset \mathbb{P}(3,4,5,6)$ $\frac{1}{2} + 3 \times \frac{2}{3} + \frac{3}{4} + \frac{4}{5}$	$Y_{9,18} \subset \mathbb{P}(3,4,5,6,9)$	9

Automated search through the Magma *K3 database*

We apply this lemma to the famous 95 using Magma. We want to add a new variable u and a new equation of the same degree to the graded ring so that we do not change its Hilbert series. The elementary ingredients of the proof form the tests used in the following code samples. If input to Magma in this order, this code will generate the contents of Table 1.

The new equation must not be reducible or linear, so must involve at least 3 variables of lower degrees and not have any factors. In particular, we cannot add a variable or equation of degree 1. Given a sequence of weights of available variables Q and the target degree d, the following code works out monomials of the given degree.

```
/*
   Input:     Q seq of integers; q,n both integers.
   Procedure: update Q to include (multiple occurrences of)
      q in every elt S of Q for which (sum of S) + q <= n.
*/
procedure add_q(~Q, q, n)
   Q := &cat[ [ S cat [q:i in [1..p]]:
               p in [0..Floor((n-&+S)/q)]] : S in Q];
end procedure;
```

```
/*
    Input:    Q seq of integers, n an integer.
    Output:   seqs of elts of Q (with repeats) that sum to n.
*/
function sum_to(Q, n)
    Qout := [ [Integers() | ] ];
    for q in Q do
        add_q(~Qout, q, n);
    end for;
    return [ S : S in Qout | &+S eq n ];
end function;
```

We collect together the numerical information needed to find a degeneration in a single function.

```
/*
    Input:    codimension 1 K3 surface X.
    Output:   a seq of integers for which there may be an
              irreducible   simple degeneration of X.
*/
function specialk3s(X)
    d := ApparentEquationDegrees(X)[1];
    W := Weights(X);  // i.e. X is (f_d = 0) in P(W)
    R<x, y, z, t> := PolynomialRing(Rationals(), W);
    // Find suitable degrees that divide d.
    E := [ e : e in [2..d-1] | IsDivisibleBy(d, e) and not e in W ];
    data := [ <d, e, MRe> : e in E | #MRe ge 2
            where MRe is MonomialsOfWeightedDegree(R, e) ];
    return [ [x[1], x[2]] : x in data | TotalDegree(GCD(x[3])) eq 0
            and #[R.i : i in [1..Rank(R)] | IsDivisibleBy(&*x[3], R.i)] ge 3];
end function;
```

Finally, given a sequence D of K3 surfaces, we run through those that are in codimension 1 and determine whether they satisfy the numerical conditions. For each one that does satisfy these conditions, we consider how the polarisation of the largest singularity would degenerate. This is bound up in a single function.

```
/*
    Return sequence of simple degenerations of codimension 1
    K3 surfaces in the sequence of K3 surfaces D.
*/
function degenerations(D)
    degens := [ ];
    for X in D do
        if Codimension(X) ne 1 then continue X; end if;
        W := Weights(X);
        B := Basket(X);
```

```
        for S in specials(X) do
          d, e := Explode(S);
          if #B eq 0 then
            Append(~degens, <X, e, d>);
          else
            // check polarisation of the biggest singularity
            r := Maximum([ p[1] : p in B ]);
            if r lt e and
              &or[ (d−a) mod r eq 0 : a in W cat [e] ] and
              &or[ (e−a) mod r eq 0 : a in W ] then
              Append(~degens, <X, e, d>);
            end if;
          end if;
        end for;
      end for;
      return degens;
    end function;
```

To run the code on all 95 hypersurfaces

```
>   D := K3Database();
>   K3s := [ K3SurfaceWithCodimension(D, 1, i) : i in [1..95] ];
>   E := degenerations(K3s);
>   #E;
        9
```

Finally, we check as before that a general element of each family is quasi-smooth to complete the calculation: with luck, the following code returns a series of 'true' values. (In fact, this output includes the monogonal degeneration described above which is detected here since it is not quasi-smooth.)

```
>   for e in E do
>     P := ProjectiveSpace(Rationals(), Weights(e[1]) cat [e[2]]);
>     Le := LinearSystem(P, e[2]);
>     Ld := LinearSystem(P, e[3]);
>     f := &+[ Random([1..5])*m : m in Sections(Ld) ];
>     g := &+[ Random([1..5])*m : m in Prune(Sections(Le)) ];
>     X := Scheme(P, [f, g]);
>     IsNonsingular(X);
>   end for;
```

The intrinsic LinearSystem(P, d) returns a vector space based by monomials of degree d on the wps P, and the Prune command removes the final monomial which is the linear term u that must not appear in the degenerate small equation. This code generates Table 1.

Table 2. Monogonal degenerations among the famous 95 in codimension 2

general K3/basket $\mathcal{B}$	special K3	deg u
$X_6 \subset \mathbb{P}(1,1,1,3)$ no singularities	$Y_{2,6} \subset \mathbb{P}(1,1,1,2,3)$	2
$X_{12} \subset \mathbb{P}(1,2,3,6)$ $2 \times \frac{1}{2} + 2 \times \frac{2}{3}$	$Y_{4,12} \subset \mathbb{P}(1,2,3,4,6)$	4
$X_{18} \subset \mathbb{P}(1,3,5,9)$ $2 \times \frac{2}{3} + \frac{2 \cdot 3}{5}$	$Y_{6,18} \subset \mathbb{P}(1,3,5,6,9)$	6
$X_{18} \subset \mathbb{P}(2,3,4,9)$ $4 \times \frac{1}{2} + 2 \times \frac{2}{3} + \frac{3}{4}$	$Y_{6,18} \subset \mathbb{P}(2,3,4,6,9)$	6
$X_{24} \subset \mathbb{P}(3,4,5,12)$ $2 \times \frac{2}{3} + 2 \times \frac{3}{4} + \frac{2 \cdot 3}{5}$	$Y_{8,24} \subset \mathbb{P}(3,4,5,8,12)$	8
$X_{30} \subset \mathbb{P}(4,5,6,15)$ $2 \times \frac{1}{2} + \frac{2}{3} + \frac{3}{4} + 2 \times \frac{4}{5}$	$Y_{10,30} \subset \mathbb{P}(4,5,6,10,15)$	10

Other examples

The conditions applied in the code were found experimentally by applying the function repeatedly, strengthening the conditions each time a false example arose. Omitting the conditions at some singularities gives the monogonal degenerations of Table 2—these only fail to be simple in that they have contracted the section of the elliptic pencil, a -2-curve, to a singular point.

Other more complicated examples include the degeneration

$$Y_{6,12} \subset \mathbb{P}^4(2,2,3,5,6) \quad \text{of} \quad X_{12} \subset \mathbb{P}^3(2,2,3,5).$$

The surface Y is not quasi-smooth since the equation of degree 6 cannot contain a linear term near the weight 5 coordinate point P_z (the z-axis). In fact, this point is an elliptic Gorenstein singularity which is the cyclic hyperquotient

$$(y^2 = x^3 + xt^4 + t^6) \subset \frac{1}{5}(2,3,1).$$

This can be written as the maximal Pfaffians of a skew-symmetric 5×5 matrix of $\mathbb{Z}/5\mathbb{Z}$-invariant functions, as the Buchsbaum–Eisenbud Theorem suggests.

7 Unprojection

The geometric notion of projection of a variety X from a point $p \in X$ often corresponds to the algebraic notion of elimination of a variable of the homogeneous coordinate ring $R[X]$, that is, finding the subring of $R[X]$ generated by the other variables: the projection ψ in Section 2 is an example.

Gorenstein projection is a more precise notion applied to projectively Gorenstein polarised varieties X, A that constructs Gorenstein subrings of $R(X, A)$. The best-understood example was invented by Kustin and Miller [12] and recast by Papadakis and Reid [15]. It is known as *Type I* Gorenstein projection. It too is typically the simple elimination of a variable, but a variable that is chosen carefully. The key property of Gorenstein projection is that the projection is reversible given some data about its image: in particular, one must specify a divisor on the image known as the *unprojection divisor* that is the image of the centre of projection. The reverse procedure—recovering X, A from the image of projection Z together with $D \subset Z$, the unprojection divisor—is called *unprojection*. The papers mentioned above work out this unprojection.

As an aside, the algebra of unprojection can be carried out at the level of free resolutions, as detailed in Kustin–Miller's original paper, or more recently by Papadakis [14]. It is, therefore, ideally suited to calculation in a system like Magma, although not implemented at the time of writing.

If a K3 surface X admits a Gorenstein projection from a singularity $p \in X$ to another K3 surface Z, then any special surface Y of the same numerical type as X will also admit a Gorenstein projection. The image of this projection might be some special element of the family containing Z, or it might be a member of a special family.

We use unprojection as a construction of a new variety. The main point in calculations is that although the some member X of the general family may contain a suitable unprojection divisor, the low degree equation of a simple degeneration Y imposes many more conditions, usually preventing this divisor from appearing on Y.

Let $X \subset \mathbb{P}$ be a K3 surface and $Y \subset \mathbb{P}'$ a simple degeneration lying in an ambient space with exactly one extra variable u. Suppose that X contains an unprojection divisor D described as the image of an embedding

$$\varphi \colon \mathbb{P}(a, b) \hookrightarrow D \subset X \subset \mathbb{P}.$$

We say that X and Y *have the same unprojection divisor* if Y contains the image of the map

$$\varphi' \colon \mathbb{P}(a, b) \hookrightarrow \mathbb{P}'$$

defined as an extension of φ by $\varphi'^*(u) = 0$: in other words, $\mathbb{P}(a, b) \times \mathbb{C} \subset \mathcal{X}$ independently of $\lambda \in \mathbb{C}$.

We work only with Type I projections in which the map φ embeds $\mathbb{P}(a, b)$ as a coordinate line. (There are other types, but their theory is not yet complete.) We will use the K3 database to find such $\mathbb{P}(a, b)$ in the surfaces X of Table 1 and then use elementary calculations to discover in which cases the curve extends to the corresponding Y. Then we use unprojection to make simple degenerations of the unprojections of X.

8 Special K3 surfaces in Fletcher's 84

In [1], there is a calculus that predicts the behaviour of RR data under pro-jection: given a projection $X \to Z$ from a point $\frac{1}{r}(a, r - a) \in X$,

- $g(X) = g(Z)$
- $[r, a]$ in the basket of X contributes $\{[a, r], [r - a, r]\}$ to the basket of Z

taking $r \bmod a$ or $r \bmod r - a$ in the basket calculation, and ignoring any singularity $[s, b]$ with $s = 1$. Surfaces in the K3 database are related to one another by projection, which is calculated using this calculus. Recall the K3 surface $X_{10} \subset \mathbb{P}^3(1, 2, 3, 4)$ of Section 2. It has six unprojections, of which we look at two here. Consider first

> K3Surface(D, 0, 274) : Minimal;

```
K3 surface (g=0, no.274) in P^4(1,2,3,4,5)
   Basket: 1/2(1,1), 1/4(1,3), 1/5(2,3)
   Numerator: t^15 - t^9 - t^6 + 1
```

(where we use abbreviated printing, as indicated by : Minimal). The projection of this from the point $\frac{1}{5}(2, 3)$ results in a surface X_{10} containing the unpro-jection divisor $\mathbb{P}(2, 5 - 2) = \mathbb{P}(2, 3)$. It is possible to find such a K3 surface containing this curve, of course, since

$$X_{10} \subset \mathbb{P}^3(1, 2, 3, 4)$$

and $\mathbb{P}(2, 3)$ lies linearly in this space inside X_{10} as long as y^5 is not in the defining equation of X_{10}. But the degeneration is

$$Y_{5,10} \subset \mathbb{P}^4(1, 2, 3, 4, 5).$$

The small degree equation may be $xt = yz$ or $x^5 = yz$. In either case, the equation does not vanish on the linear $\mathbb{P}(2, 3)$ and so this curve is not contained in Y. We cannot lift the degeneration of X_{10} to the unprojected variety by a Type I unprojection of the whole family.

As a second case, consider

> K3Surface(D, 0, 273) : Minimal;

```
K3 surface (g=0, no.273) in P^4(1,2,3,4,7)
   Basket: 2 x 1/2(1,1), 1/7(3,4)
   Numerator: t^17 - t^9 - t^8 + 1
```

To see the image of projection from $\frac{1}{7}(3, 4)$ we must find X_{10} containing a linear $\mathbb{P}(3, 7 - 3) = \mathbb{P}(3, 4)$; this is easy. This time the linear $\mathbb{P}(3, 4)$ is also contained in the degeneration $Y_{5,10}$ so we can make the Type I unprojection of Y. The result is number 7 of Table 3: the second column records the projection from $\frac{1}{7}(3, 4) \in X$ to a member of family number 4 of Table 1.

We sketch code that will reproduce the contents of Table 3 following this idea. First we identify possible unprojection divisors $\mathbb{P}(r, s)$ for a given basket B according to the calculus at the beginning of this section.

```
function unproj_divs(B)
    // easy cases: 1/r(1, r−1) could be on ℙ(1, r)
    result := [ [1, 1] ] cat [ [r, 1] : r in [1.. max_r] | [r, 1] in B ]
            where max_r is Maximum([ b[1] : b in B ] cat [ 0 ]);
    // hard cases: 1/r(s, −s)+1/s(r, −r) could be on ℙ(r, s)
    Bred := SetToSequence(SequenceToSet(B));
    for i in [1..#Bred] do
      p₁ := Bred[i]; r := p₁[1]; a := p₁[2];
      for j in [i+1..#Bred] do
        p₂ := Bred[j]; s := p₂[1]; b := p₂[2];
        if ((r mod s eq b) or (r mod s eq s−b))
              and ((s mod r eq a) or (s mod r eq r−a)) then
          Append(∼result, [r, s]);
        end if;
      end for;
    end for;
    return result;
end function;
```

To find other degenerations, we run through possible unprojections of the families of Table 1 (as recorded in Section 6 as the Magma sequence E) and check whether the low degree equation can be chosen to vanish on the unprojection divisor while remaining irreducible.

```
res := [ ];
for i in [1..#E] do
  X := E[i][1];
  W := Weights(X);
  P := ProjectiveSpace(Rationals(), W);
  e := E[i][2]; // the degree of the small equation
  for p₁ in unproj_divs(Basket(X)) do
    i₁ := Index(W, p₁[1]); // this is 0 if p₁[1] not in W
    if p₁ eq [1, 1] then
      inds := [ i : i in [1..#W] | W[i] eq 1 ];
      case #inds ge 2:
        when true: i₂ := inds[2];
        else continue p₁;
      end case;
    else
      i₂ := Index(W, p₁[2]);
    end if;
    if i₁*i₂ ne 0 then
      line := Scheme(P, [P.i : i in [1.. Length(P)] | i notin [i₁, i₂]]);
      Le := LinearSystem(P, e);
      M := Sections(LinearSystem(Le, line));
      if TotalDegree(GCD(M)) eq 0 and
```

```
#[ P.i : i in [1..Length(P)] | IsDivisibleBy(&*M, P.i) ] ge 3 then
   Append(~res, <p₁, W>);
     end if;
   end if;
 end for;
end for;
```

Printing the result *res* would present the contents of Table 3. (The irre-

Table 3. Special K3 surfaces among Fletcher's 84 with projections to Table 1

No./proj'n	general K3/basket $\mathcal{B}$	special K3	deg u
1	$X_{3,3} \subset \mathbb{P}(1,1,1,1,2)$	$Y \subset \mathbb{P}(1,1,1,1,2,2)$	2
$\frac{1}{2} \mapsto$ No.1	$\frac{1}{2}$	$Y = Y_{2,3,3,4,4}$	
2	$X_{4,4} \subset \mathbb{P}(1,1,2,2,2)$	$Y \subset \mathbb{P}(1,1,2,2,2,3)$	3
$\frac{1}{2} \mapsto$ No.2	$4 \times \frac{1}{2}$	$Y = Y_{3,4,4,5,6}$	
3	$X_{4,5} \subset \mathbb{P}(1,1,2,2,3)$	$Y \subset \mathbb{P}(1,1,2,2,3,3)$	3
$\frac{2}{3} \mapsto$ No.2	$2 \times \frac{1}{2} + \frac{2}{3}$	$Y = Y_{3,4,5,6,6}$	
4	$X_{5,6} \subset \mathbb{P}(1,2,2,3,3)$	$Y \subset \mathbb{P}(1,2,2,3,3,4)$	4
$\frac{2}{3} \mapsto$ No.3	$3 \times \frac{1}{2} + 2 \times \frac{2}{3}$	$Y = Y_{4,5,6,7,8}$	
5	$X_{6,7} \subset \mathbb{P}(1,2,2,3,5)$	$Y \subset \mathbb{P}(1,2,2,3,4,5)$	4
$\frac{2 \cdot 3}{5} \mapsto$ No.3	$3 \times \frac{1}{2} + \frac{2 \cdot 3}{5}$	$Y = Y_{4,6,7,8,9}$	
6	$X_{6,8} \subset \mathbb{P}(1,2,3,4,4)$	$Y \subset \mathbb{P}(1,2,3,4,4,5)$	5
$\frac{3}{4} \mapsto$ No.4	$2 \times \frac{1}{2} + 2 \times \frac{3}{4}$	$Y = Y_{5,6,8,9,10}$	
7	$X_{8,9} \subset \mathbb{P}(1,2,3,4,7)$	$Y \subset \mathbb{P}(1,2,3,4,5,7)$	5
$\frac{2 \cdot 5}{7} \mapsto$ No.4	$2 \times \frac{1}{2} + \frac{2 \cdot 5}{7}$	$Y = Y_{5,8,9,10,12}$	
8	$X_{6,7} \subset \mathbb{P}(1,2,3,3,4)$	$Y \subset \mathbb{P}(1,2,3,3,4,5)$	5
$\frac{2}{3} \mapsto$ No.4	$\frac{1}{2} + 2 \times \frac{2}{3} + \frac{3}{4}$	$Y = Y_{5,6,7,8,10}$	
9	$X_{7,8} \subset \mathbb{P}(1,2,3,4,5)$	$Y \subset \mathbb{P}(1,2,3,4,5,6)$	6
$\frac{2}{3} \mapsto$ No.5	$2 \times \frac{1}{2} + \frac{2}{3} + \frac{4}{5}$	$Y = Y_{6,7,8,9,12}$	
10	$X_{8,9} \subset \mathbb{P}(2,3,3,4,5)$	$Y \subset \mathbb{P}(2,3,3,4,5,6)$	6
$\frac{2 \cdot 3}{5} \mapsto$ No.6	$2 \times \frac{1}{2} + 3 \times \frac{2}{3} + \frac{2 \cdot 3}{5}$	$Y = Y_{6,8,9,11,12}$	
11	$X_{9,10} \subset \mathbb{P}(2,3,4,5,5)$	$Y \subset \mathbb{P}(2,3,4,5,5,7)$	7
$\frac{2 \cdot 3}{5} \mapsto$ No.7	$2 \times \frac{1}{2} + \frac{3}{4} + 2 \times \frac{2 \cdot 3}{5}$	$Y = Y_{7,9,10,12,14}$	
12	$X_{10,12} \subset \mathbb{P}(2,3,4,5,8)$	$Y \subset \mathbb{P}(2,3,4,5,7,8)$	7
$\frac{3 \cdot 5}{8} \mapsto$ No.7	$3 \times \frac{1}{2} + \frac{3}{4} + \frac{3 \cdot 5}{8}$	$Y = Y_{7,10,12,14,15}$	
13	$X_{12,13} \subset \mathbb{P}(3,4,5,6,7)$	$Y \subset \mathbb{P}(3,4,5,6,7,9)$	9
$\frac{2 \cdot 5}{7} \mapsto$ No.8	$2 \times \frac{1}{2} + 2 \times \frac{2}{3} + \frac{4}{5} + \frac{2 \cdot 5}{7}$	$Y = Y_{9,12,13,16,18}$	

ducibility test above was too coarse, and three entries of *res* are omitted from Table 3.) One can make further checks as before that the listed examples do exist as claimed.

And so we could continue. There are probably a few other examples to be added to Table 3 in codimension 3: we might be able to make more complicated unprojections—so-called Type II$_1$—of the degenerations of Table 1. In codimension 4, this method will offer many possible examples that might be analysed by methods such as those of Papadakis [14].

References

1. Selma Altınok, Gavin Brown, Miles Reid, *Fano 3-folds, K3 surfaces and graded rings*, pp. 25–53 in: *Topology and geometry: commemorating SISTAG*, Contemp. Math. **314**, Providence (RI): Amer. Math. Soc., 2002.

2. Selma Altınok, *Hilbert series and applications to graded rings*, Int. J. Math. Math. Sci. **7** (2003), 397–403.

3. Wieb Bosma, John Cannon, Catherine Playoust, *The Magma algebra system I: The user language*, J. Symbolic Comput. **24** (1997), 235–265.
   See also the Magma home page at http://magma.maths.usyd.edu.au/magma/.

4. Gavin Brown, *A database of polarized K3 surfaces*, to appear in Exp. Math. (2006). See also www.kent.ac.uk/ims/grdb/

5. Gavin Brown, *Datagraphs in algebraic geometry*, in: F. Winkler, U. Langer (eds.), *Symbolic and Numerical Scientific Computing – Proc. of SNSC'01*, LNCS **2630**, Heidelberg: Springer-Verlag, 2003.

6. Anita Buckley, Balázs Szendrői, *Orbifold Riemann-Roch for threefolds with an application to Calabi-Yau geometry*, J. Algebraic Geom. **14**-4 (2005), 601–622.

7. Anita Buckley, *Orbifold Riemann-Roch for 3-folds and applications to Calabi-Yaus*, PhD thesis, University of Warwick, 2003, viii+119pp.

8. Igor Dolgachev, *Weighted projective varieties*, pp. 34–71 in: *Group actions and vector fields (Vancouver, B.C., 1981)*, Lecture Notes in Math. **956**, Berlin: Springer, 1982.

9. Robin Hartshorne, *Algebraic geometry*, Graduate Texts in Mathematics **52**, New York: Springer-Verlag, 1977.

10. A. R. Iano-Fletcher, *Working with weighted complete intersections* pp. 101–173 in: *Explicit birational geometry of 3-folds*, London Math. Soc. Lecture Note Ser. **281**, Cambridge: Cambridge Univ. Press, 2000.

11. Jennifer M. Johnson, János Kollár, *Fano hypersurfaces in weighted projective 4-spaces*, Experiment. Math. **10**-1 (2001), 151–158.

12. Andrew R. Kustin, Matthew Miller, *Constructing big Gorenstein ideals from small ones*, J. Algebra **85**-2 (1983), 303–322.

13. János Kollár, *Rational curves on algebraic varieties*, Ergebnisse der Mathematik und ihrer Grenzgebiete (3. Folge) **32**, Berlin: Springer-Verlag, 1996.

14. Stavros Papadakis, *Kustin-Miller unprojection with complexes*, J. Algebraic Geom. **13**-2 (2004), 249–268.

15. Stavros Papadakis, Miles Reid, *Kustin–Miller unprojection with complexes*, J. Algebraic Geom. **13**-3 (2004), 563–577.

16. Miles Reid, *Graded rings and varieties in weighted projective space*. Links from `www.maths.warwick.ac.uk/~miles/surf/`

17. Miles Reid, *Surfaces with $p_g = 0$, $K^2 = 1$*, J. Fac. Sci. Univ. Tokyo Sect. IA Math. **25**-1 (1978), 75–92.

18. Miles Reid, *Young person's guide to canonical singularities*, pp. 345–414 in: *Algebraic geometry, Bowdoin, 1985 (Brunswick, Maine, 1985)*, Proc. Sympos. Pure Math. **46**, Providence (RI): Amer. Math. Soc., 1987.

19. Miles Reid, *Chapters on algebraic surfaces*, pp. 3–159 in: Complex algebraic geometry (Park City, UT, 1993), IAS/Park City Math. Ser. **3**, Providence (RI): Amer. Math. Soc., 1997.

20. Miles Reid, Kaori Suzuki, *Cascades of projections from log del Pezzo surfaces*, pp. 227–249 in: *Number theory and algebraic geometry – to Peter Swinnerton-Dyer on his 75th birthday*, Cambridge: Cambridge University Press, 2003.

21. B. Saint-Donat, *Projective models of K3 surfaces*, Amer. J. Math. **96** (1974), 602–639.

22. Kaori Suzuki, *On Fano indices of $\mathbb{Q}$-Fano 3-folds*, Manuscripta Math. **114**-2 (2004), 229–246.

23. Balázs Szendrői, *Calabi-Yau threefolds with a curve of singularities and counterexamples to the Torelli problem*, Internat. J. Math. **11**-3 (2000), 449–459.

Constructing the split octonions

Donald E. Taylor

School of Mathematics and Statistics
The University of Sydney
Sydney, Australia
D.Taylor@maths.usyd.edu.au

Summary. In this chapter we construct the split octonion algebra over a ring, first using the structure constant machinery of **Magma** and then using Lie algebras. The second method provides an explicit representation of the algebraic group of type G_2 over a field F as the automorphism group of the split octonions $\mathbb{O}(F)$; the Lie algebra of type G_2 is realized as the algebra of derivations of $\mathbb{O}(F)$. Finally, we look at the action of $G_2(F)$ and show that the subgroups $\mathrm{SU}(3, F)$ and $\mathrm{SL}(3, F)$ occur naturally as stabilizers of elements of $\mathbb{O}(F)$.

1 Introduction

A *composition algebra* is a vector space A over a field F with a bilinear multiplication $A \times A \to A : (x, y) \mapsto xy$ and a non-degenerate quadratic form $Q : A \to F$ such that for all $x, y \in A$,

$$Q(xy) = Q(x)Q(y).$$

We assume that A has an identity element, denoted by 1, but we do not assume that the multiplication is associative. (It was shown by Jacobson [11] that when the dimension of A is finite it is possible to ensure that A has an identity element by a slight modification of the product.)

The *polar form* of Q is the symmetric bilinear form $\langle x, y \rangle$ defined by

$$\langle x, y \rangle := Q(x + y) - Q(x) - Q(y).$$

The condition for Q to be non-degenerate is that $Q(x) = 0$ and $\langle x, y \rangle = 0$ for all $y \in A$ should imply $x = 0$. The quadratic form Q is called the *norm* of A.

For $x \in A$, the *trace* of x is $\mathrm{Tr}(x) := \langle x, 1 \rangle$ and the *conjugate* of x is $\bar{x} := \mathrm{Tr}(x)\, 1 - x$. Every element $x \in A$ satisfies the quadratic equation $x^2 - \mathrm{Tr}(x)x + Q(x)\, 1 = 0$ and therefore $x\bar{x} = \bar{x}x = Q(x)\, 1$. The map $A \to A : x \mapsto \bar{x}$ is an *involution*, namely a linear transformation such that $\bar{\bar{x}} = x$ and $\overline{xy} = \bar{y}\,\bar{x}$.

When F is the field of real numbers it was proved by Hurwitz [10] in 1898 that the dimension of a composition algebra is 1, 2, 4 or 8 (see Rost [14] for another approach). Jacobson [11] extended this result to all fields of characteristic other than 2 and in 1963 Springer showed that when the form $\langle -, - \rangle$ is non-degenerate it holds for all fields without restriction. An excellent account of this result can be found in the book by Springer and Veldkamp [16].

Given an associative composition algebra A and a non-zero element $\lambda \in F$, a new composition algebra B of twice the dimension can be constructed by means of the *Cayley–Dickson* process. As a vector space we have $B = A \oplus A$ and the multiplication is defined by

$$(a_1, a_2) \times (b_1, b_2) := (a_1 b_1 + \lambda \bar{b}_2 a_2, b_2 a_1 + a_2 \bar{b}_1).$$

The norm $\hat{Q}$ of B is given by $\hat{Q}(a, b) := Q(a) - \lambda Q(b)$, where Q is the norm of A. It follows that $\overline{(a, b)} = (\bar{a}, -b)$. It can be shown [16] that every composition algebra over F can be constructed by repeated applications of this process: beginning with the field F, or in the case of fields of characteristic 2, with $F \oplus F$ or a quadratic extension of F.

The four-dimensional composition algebras are known as *quaternion* algebras. For example, the algebra $M_2(F)$ of 2×2 matrices over F is a quaternion algebra for which the quadratic form is the determinant. The eight-dimensional composition algebras are known as *octonion* algebras and they are intimately connected with the exceptional groups and Lie algebras of type G_2 (see [4] [8] and [16]).

In fact, the only composition algebras over F are the field F itself, $F \oplus F$, quadratic extensions of F, quaternion algebras, octonion algebras and, in the case of fields of characteristic 2, certain purely inseparable extensions of F (see [18] or [12]). The octonion algebras are *alternative*; that is, they are not associative but the identities $a^2 b = a(ab)$ and $(ab)b = ab^2$ hold for all $a, b \in A$.

If $Q(a) \neq 0$ for all non-zero elements $a \in A$, then A is a division algebra; i.e., given $a, b \in A$ where $a \neq 0$ there are unique elements x and $y \in A$ such that $ax = b$ and $ya = b$. On the other hand, if there is an element $a \neq 0$ such that $Q(a) = 0$, then A has zero divisors and the Witt index of Q is half the dimension of A. In this case, A is uniquely determined up to isomorphism by its dimension and it is known as the *split composition algebra* over F. For example, the split composition algebra of dimension 4 over F is the matrix algebra $M_2(F)$. The split octonion algebra over F will be denoted by $\mathbb{O}(F)$.

In this chapter we present **Magma** [3] code to construct the split octonions. The split octonions will be obtained from a construction beginning with the Lie algebra of type E_6 and involving the Lie algebras of types D_4 and G_2. The Lie algebra of type G_2 is its algebra of derivations and the Chevalley group of type G_2 is its automorphism group. A similar construction for the 27-dimensional exceptional Jordan algebra and the Lie algebra of type F_4 can be found in [15].

When F is the field of real numbers there are two octonion algebras: the classical *Cayley division algebra* $\mathcal{C}$ corresponding to the positive definite quadratic form $\sum_{i=1}^{8} x_i^2$ and the split octonions $\mathbb{O}(\mathbb{R})$ with quadratic form $\sum_{i=1}^{8}(-1)^i x_i^2$. We define the multiplication for the Cayley division algebra in the next section and use Magma to realize it as a structure constant algebra.

Over the reals, the simple Lie algebra of type G_2 is often defined as the algebra of derivations of the Cayley algebra $\mathcal{C}$. The recent expository article by Baez [2] explores this connection in some detail. An earlier article by van der Blij [17] is also worth consulting.

2 Structure constant algebras

It is possible to use Magma to construct the various octonion algebras from their known multiplication tables and then obtain the 14-dimensional algebra G_2 as a Lie algebra of derivations. We illustrate this approach in this section and in later sections we reverse the process and use the Lie algebra machinery [13] of Magma to first construct G_2 and then derive the split octonions from the Lie algebra.

The Magma constructor Algebra<R, n | T> creates the *structure constant algebra* with standard basis $e_1, e_2, \ldots, e_n$ over the ring R. The sequence T contains quadruples $< i, j, k, c_{ij}^k >$ giving the non-zero structure constants c_{ij}^k such that $e_i \circ e_j = \sum_k c_{ij}^k e_k$. All other structure constants are defined to be 0.

A construction for the Cayley algebra $\mathcal{C}$ can be found in the Structure Constants chapter [5] of the Magma Handbook. In this example there is a basis $1 = e_1, e_2, \ldots, e_8$ such that for $i \geq 2$ we have $e_i^2 = -1$ and for $i, j \geq 2$ and $i \neq j$ we have $e_i e_j = \pm e_k$, where the triples $\{i, j, k\}$ form the lines of the 7-point projective plane (also known as the *Fano* plane) on the points $\{2, 3, \ldots, 8\}$. The signs are determined by setting $e_2 e_3 = e_5$ and using the fact that for $i, j \geq 2$ and $i \neq j$, the elements e_i and e_j generate an associative algebra such that $e_i e_j = e_k$ implies $e_{i+1} e_{i+1} = e_{k+1}$ (subscripts modulo 7).

To carry out this construction* in Magma we begin with the Fano plane on $\{0, 1, \ldots, 6\}$

```
>    fano := { <n, (n+1) mod 7, (n+3) mod 7> : n in [0..6] };
```

and then act with the symmetric group Sym(3) to obtain the structure constants. (In Magma the infix operator *cat* concatenates two sequences.)

```
>    sc := [ <f[1^g]+2, f[2^g]+2, f[3^g]+2, Sign(g)> : g in Sym(3), f in fano];
>    sc cat:= [ <i, i, 1, −1> : i in [2..8] ]
>        cat [ <1, i, i, 1> : i in [1..8] ] cat [<i, 1, i, 1> : i in [2..8] ];
```

Since the structure constants are integers, they may be used to define an algebra over any ring. Over a finite field F every quadratic form has a non-trivial zero and hence the only octonion algebra over F is the split algebra $\mathbb{O}(F)$. Thus the following function constructs the Cayley division algebras

---

*See the Preface to this volume for style conventions regarding Magma code; code appearing in this book is available at http://magma.maths.usyd.edu.au/magma/.

over the rational and real numbers and, by necessity, it produces the split octonions over any finite field.

```
>    Cayley := func< R | Algebra< R, 8 | sc > >;
```

Using this function we can construct the Cayley algebra over a polynomial ring and then calculate with indeterminates.

```
>    R<u1, u2, u3, u4, u5, u6, u7, u8, v1, v2, v3, v4, v5, v6, v7, v8> :=
>                 PolynomialRing(Rationals(), 16);
>    A := Cayley(R);
>    u := A ! [u1, u2, u3, u4, u5, u6, u7, u8];
>    v := A ! [v1, v2, v3, v4, v5, v6, v7, v8];
>    u*v;
```

```
    (u1*v1-u2*v2-u3*v3-u4*v4-u5*v5-u6*v6-u7*v7-u8*v8
    u1*v2+u2*v1+u3*v5+u4*v8-u5*v3+u6*v7-u7*v6-u8*v4
    u1*v3-u2*v5+u3*v1+u4*v6+u5*v2-u6*v4+u7*v8-u8*v7
    u1*v4-u2*v8-u3*v6+u4*v1+u5*v7+u6*v3-u7*v5+u8*v2
    u1*v5+u2*v3-u3*v2-u4*v7+u5*v1+u6*v8+u7*v4-u8*v6
    u1*v6-u2*v7+u3*v4-u4*v3-u5*v8+u6*v1+u7*v2+u8*v5
    u1*v7+u2*v6-u3*v8+u4*v5-u5*v4-u6*v2+u7*v1+u8*v3
    u1*v8+u2*v4+u3*v7-u4*v2+u5*v6-u6*v5-u7*v3+u8*v1)
```

The quadratic form Q for the Cayley algebra is the standard one, namely $Q(x) = \sum_{i=1}^{8} x_i^2$. Thus the output for $u * v$ combined with the fact that $Q(uv) = Q(u)Q(v)$ provides an explicit expression of the product of two sums of eight squares as a sum of eight squares.

```
>    Q := func< x | InnerProduct(x, x) >;
>    Q(u), Q(u*v) eq Q(u)*Q(v);
```

```
    u1^2 + u2^2 + u3^2 + u4^2 + u5^2 + u6^2 + u7^2 + u8^2
    true
```

A linear transformation T of an algebra A is a *derivation* if

$$(u\,v)T = (uT)\,v + u\,(vT)$$

for all $u, v \in A$. If $e_1, e_2, \ldots, e_n$ is a basis for A and if $e_i\,e_j = \sum_k c_{ij}^k e_k$, then this condition becomes

$$\sum_{\ell} c_{ij}^{\ell} a_{\ell k} = \sum_{\ell} c_{\ell j}^{k} a_{i\ell} + \sum_{\ell} c_{i\ell}^{k} a_{j\ell}$$

where (a_{ij}) is the matrix of T. Thus finding a basis for the vector space of derivations of A is simply a matter of solving these n^3 linear equations for the a_{ij}. This is encapsulated in the following Magma function. The rows of the $n^2 \times n^3$ matrix M are indexed by pairs (i, j) and the columns are indexed by triples (i, j, k). The returned value is a basis matrix of the null space of M, namely a matrix N with rank(M) linearly independent rows such that $NM = 0$.

```
DerivationBasis := function(A)
    F := BaseRing(A);
    n := Dimension(A);
    M := Zero(KMatrixSpace(F,n²,n³));
    for i := 1 to n do
      for j := 1 to n do
        p := n²*(i−1)+n*(j−1);
        e := A.i*A.j;
        for k := 1 to n do
          pk := p+k; qq := n*(k−1);
          qi := qq+i; qj := qq+j;
          r := n*qq + n*(j−1);
          s := n²*(i−1)+qq;
          for l := 1 to n do
          f := e[l];
            M[n*(l−1)+k,pk] +:= f;
            M[qi,r+l] −:= f;
            M[qj,s+l] −:= f;
          end for;
        end for;
      end for;
    end for;
    return KernelMatrix(M);
  end function;
```

The space $\mathrm{Der}(A)$ of derivations of an algebra A is a Lie algebra with multiplication $[a, b] := ab − ba$. In the case of a composition algebra, the dimension of $\mathrm{Der}(A)$ is 14. We verify this for the Cayley algebra over the field of 5 elements.

```
>   Cay₅ := Cayley(GF(5));
>   Der₅ := DerivationBasis(Cay₅);
>   Nrows(Der₅);
      14
```

To conclude this section we define a function that computes the structure constants for an algebra of derivations and returns the resulting Lie algebra. This function illustrates several useful features of Magma, namely error checking and the use of a local function (in this case **comm**) that is defined within LieAlg. Notice that the sequence K consists of the rows of D converted to $n \times n$ matrices.

```
LieAlg := function( D )
    sq, n := IsSquare(Ncols(D));
    error if not sq, "the number of columns must be a square";
    m := Nrows(D);
    F := BaseRing(D);
    K := [ MatrixAlgebra(F,n) ! Eltseq(D[i]) : i in [1..m] ];
```

```
comm := func< a, b | a*b − b*a >;
struct := [ Eltseq( Solution(D, V ! Eltseq(comm(K[i], K[j]))) ) :
              i,j in [1..m] ] where V is VectorSpace(F, n²);
return LieAlgebra< F, m | struct >;
end function;
```

This enables us to check that the Lie algebra Der_5 of derivations of the Cayley algebra Cay_5 constructed above is of type G_2.

```
>    SemiSimpleType(LieAlg(Der₅));

     G2
```

3 Lie algebras of type D_4 and E_6

We have just seen (in the case of the field F of 5 elements) that the Lie algebra of derivations of the split octonions $\mathbb{O}(F)$ is the simple Lie algebra of type G_2 over F. The group of automorphisms of $\mathbb{O}(F)$ is the algebraic group $G_2(F)$, the exceptional Lie group of smallest dimension. However, instead of using Magma to construct G_2 directly from the octonions we shall reverse the process and construct the octonions from a Lie algebra on which G_2 acts. (See [1] for similar constructions of the Cayley numbers and the exceptional Jordan algebra over the real and complex numbers.)

We use Chevalley bases (see [9]) to define the various simple Lie algebras we encounter. As a result we shall see that there is a natural $\mathbb{Z}$-form carried by the octonions and hence the construction can be transferred to an arbitrary field.

Let $\mathcal{L}$ be a complex semisimple Lie algebra with root system Φ, simple roots Δ, Cartan subalgebra $\mathcal{H}$ and Cartan decomposition

$$\mathcal{L} = \mathcal{H} \oplus \bigoplus_{\alpha \in \Phi} \mathcal{L}_\alpha.$$

For each $\alpha \in \Phi$ choose $e_\alpha \in \mathcal{L}_\alpha$ so that $\{e_\alpha \mid \alpha \in \Phi\}$ together with the elements $h_\alpha = [e_{-\alpha} e_\alpha]$ for $\alpha \in \Delta$ is a Chevalley basis for $\mathcal{L}$. If $\alpha, \beta \in \Phi$ and $\alpha + \beta \in \Phi$, then write

$$[e_\alpha, e_\beta] = N_{\alpha\beta} e_{\alpha+\beta}.$$

For a Chevalley basis, the *structure constants* $N_{\alpha\beta}$ are integers, which are determined up to a sign by the root system (see [9]).

We shall use Magma to construct an 8-dimensional vector space which is a module for the Lie algebra of type D_4 and then define a multiplication which turns it into an alternative algebra. This construction will use an embedding of D_4 in E_6 and the 8-dimensional alternative algebra will turn out to be the split octonions. By construction, the elements of the simple Lie algebra of type G_2 will act as derivations on it and the corresponding Chevalley group of type G_2 will be its group of automorphisms. In addition to Magma's general machinery for sets, modules, and algebras we shall use the built-in functions that construct a root system of a given type, its Cartan matrix, and the Lie structure constants of the associated Lie algebra.

From now on let Φ be a root system of type E_6 and let $\Delta = \{\alpha_1, \ldots, \alpha_6\}$ be the set of simple roots labelled according to the Dynkin diagram

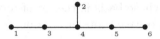

This is the default labelling used by Magma, as can be checked by typing DynkinDiagram(*RD*) after entering the command on the next line.

>   RD := RootDatum("E6");
> Φ := Roots(RD);

There are 36 positive roots and 24 of these lie outside the subsystem (of type D_4) spanned by α_2, α_3, α_4 and α_5. These 24 roots are partitioned into 3 subsets of size 8 according to the values of their first and last coordinates. We label these sets X_{10}, X_{01} and X_{11}.

> X_{10} := Reverse([x : x in Φ | x[1] *eq* 1 *and* x[6] *eq* 0]);
> X_{01} := Reverse([x : x in Φ | x[1] *eq* 0 *and* x[6] *eq* 1]);
> X_{11} := Reverse([x : x in Φ | x[1] *eq* 1 *and* x[6] *eq* 1]);
>   #Φ, #X_{10}, #X_{01}, #X_{11};

     72 8 8 8

The output above confirms that there are 72 roots and that each of the sets X_{ij} contains 8 elements.

We have reversed the order of the roots so that, in each case, they are ordered from highest to lowest. For example, the highest root of X_{10} is $X_{10}[1]$:

> $X_{10}[1]$;

     (1 1 2 2 1 0)

Let Φ^+ be the set of positive roots and Φ^- the set of negative roots. For $\alpha \in \Phi^+$, write $f_\alpha = e_{-\alpha}$ and let $\mathfrak{d}_4$ be the Lie algebra generated by the root vectors e_{α_i}, f_{α_i}, $2 \leq i \leq 5$. Then $\mathfrak{d}_4$ is a Lie algebra of type D_4 and its root system is

> X_{00} := {@ x : x in Φ | x[1] *eq* 0 *and* x[6] *eq* 0 @};

In Magma, vectors are represented as rows and matrices act on the right. Thus for $x, y \in \mathcal{L}$ we define ad(y) to be the linear transformation that sends x to $[x, y]$. The resulting homomorphism ad : $\mathcal{L} \to \text{End}(\mathcal{L})$ is known as the *adjoint representation* of $\mathcal{L}$.

If α belongs to X_{00}, $\beta \in X_{ij}$ and $\alpha + \beta$ is a root, then $\alpha + \beta \in X_{ij}$. Therefore the vectors e_α for $\alpha \in X_{ij}$ span a $\mathfrak{d}_4$-module V_{ij} and the Lie bracket induces a $\mathfrak{d}_4$-invariant bilinear pairing $V_{10} \times V_{01} \to V_{11}$.

For $\beta \in X_{00}$, the entries in the matrices representing the action of $\mathrm{ad}(e_\beta)$ on V_{ij} (with respect to the basis indexed by X_{ij}) are the structure constants $N_{\alpha\beta}$, where $\alpha \in X_{ij}$.

The Magma function LieConstant_N(R, i, j) returns the structure constant $N_{\alpha\beta}$, where α is the i-th and β the j-th root of the root datum R. Thus the following function returns the matrix of $\mathrm{ad}(e_\alpha)$ acting (on the right) on the $\mathfrak{d}_4$-invariant vector space whose basis is the set of root vectors indexed by X, where X is a sequence of positive roots or a sequence of negative roots of R.

```
LieRootMatrix := function(R, α, X)
  rts := Roots(R);
  n := #X;
  a := Index(rts, α);
  ad := Zero(MatrixRing(Integers(), n));
  for j := 1 to n do
    c := LieConstant_N( R, Index(rts, X[j]), a );
    if c ne 0 then
      ad[j][Index(X, α+X[j])] := c;
    end if;
  end for;
  return ad;
end function;
```

(The matrix of $\mathrm{ad}(e_\alpha)$ acting on the left is the negative of the transpose of the matrix returned here.)

We shall use $d4e_{11}[i]$ to denote the matrix representing the action of the root vector of the i-th simple root of $\mathfrak{d}_4$ on V_{11} and $d4f_{11}$ will be the sequence of matrices representing the actions of the root vectors of the negatives of the simple roots. Similarly, $d4e_{10}$, $d4f_{10}$, $d4e_{01}$ and $d4f_{01}$ will be the corresponding sequences of matrices for the action of $\mathfrak{d}_4$ on V_{10} and V_{01}. Because we use a Chevalley basis, all of the matrices have integer entries and we work with modules over $\mathbb{Z}$ whenever possible.

```
>   d4e₁₀ := [ LieRootMatrix(RD, Φ[i], X₁₀) : i in [2..5] ];
>   d4f₁₀ := [ LieRootMatrix(RD, -Φ[i], X₁₀) : i in [2..5] ];
>   d4e₀₁ := [ LieRootMatrix(RD, Φ[i], X₀₁) : i in [2..5] ];
>   d4f₀₁ := [ LieRootMatrix(RD, -Φ[i], X₀₁) : i in [2..5] ];
>   d4e₁₁ := [ LieRootMatrix(RD, Φ[i], X₁₁) : i in [2..5] ];
>   d4f₁₁ := [ LieRootMatrix(RD, -Φ[i], X₁₁) : i in [2..5] ];
```

It turns out that the matrix of $\mathrm{ad}(e_{-\alpha})$ is the transpose of $\mathrm{ad}(e_\alpha)$. We can use Magma to check this:

```
>   forall{ i : i in [1..4] | Transpose(d4e₁₀[i]) eq d4f₁₀[i] };
        true
```

The Cartan subalgebra $\mathcal{H}$ of $\mathfrak{d}_4$ has basis $\{h_2, h_3, h_4, h_5\}$, where $h_i = h_{\alpha_i}$, and its dual has basis $\{\alpha_2, \alpha_3, \alpha_4, \alpha_5\}$. Moreover, for $h \in \mathcal{H}$ and $\alpha \in X$ we have

$[e_\alpha, h] = \alpha(h)e_\alpha$ and so the weights of the module V_{ij} are simply the roots $\alpha \in X_{ij}$ restricted to $\mathcal{H}$.

The Cartan matrix of Φ has entries $\alpha_i(h_j)$ and since this expression is linear in the first variable, the values of $\alpha(h_j)$ are given by the product of the row vector α by the matrix C. That is, αC gives the coefficients of α as a sum of fundamental weights. Thus the following Magma calculation shows that the highest weights of the $\mathfrak{d}_4$-modules V_{10}, V_{01} and V_{11} have multiplicity one and correspond to the end nodes of the D_4 Dynkin diagram. In particular, the modules are irreducible and pairwise non-isomorphic.

```
>    C := CartanMatrix(RD);
>    X10[1]*C, X01[1]*C; X11[1]*C;
        ( 0  0  1  0  0 -1)
        (-1  0  0  0  1  0)
        ( 0  1  0  0  0  0)
```

(Note that restricting to $\mathcal{H}$ means ignoring the first and last coefficients.)

4 Triality

The Dynkin diagram of the root system of type D_4 is

The $\mathfrak{d}_4$-module V_{10} has highest weight λ_2, V_{01} has highest weight λ_4, and V_{11} has highest weight λ_1, where $\lambda_1, \lambda_2, \lambda_3, \lambda_4$ are the fundamental weights of $\mathfrak{d}_4$, and where the Dynkin diagram for $\mathfrak{d}_4$ is labelled as above.

The graph automorphisms of this diagram comprise the symmetric group on $\{1,2,4\}$ generated by the permutations $\sigma = (1,2)$ and $\tau = (1,4)$ of the simple roots. Each permutation π of the simple roots extends to an automorphism (also denoted by π) of $\mathfrak{d}_4$ such that $h_\alpha^\pi = h_{\alpha^\pi}$ and $e_\alpha^\pi = \pm e_{\alpha^\pi}$ for all roots α of $\mathfrak{d}_4$, where the sign is positive if α is a simple root (see [6]). The elements $\sigma\tau$ and $\tau\sigma$ are the *triality* automorphisms of $\mathfrak{d}_4$, of order 3.

```
>    σ := Sym(4) ! (1,2);
>    τ := Sym(4) ! (1,4);
```

From the $\mathfrak{d}_4$-module V_{11} we may define a new $\mathfrak{d}_4$-module V_{11}^σ by declaring the action of $x \in \mathfrak{d}_4$ on $v \in V_{11}$ to be vx^σ. Since $[e_\alpha, h_\beta^\sigma] = \alpha^\sigma(h_\beta)e_\alpha$ we see that V_{11}^σ has highest weight λ_2; that is, $V_{11}^\sigma \simeq V_{10}$.

Similarly we define the $\mathfrak{d}_4$-module V_{11}^τ via the action $(v, x) \mapsto vx^\tau$. The highest weight of V_{11}^τ is λ_4 and so $V_{11}^\tau \simeq V_{01}$. In Magma, the matrices representing the action of $\mathfrak{d}_4$ on the modules V_{11}^σ and V_{11}^τ are:

```
>    d4es11 := [ d4e11[i^σ] : i in [1..#d4e11] ];
```

```
>    d4fs₁₁ := [ d4f₁₁[iᵒ] : i in [1..#d4f₁₁] ];
>    d4et₁₁ := [ d4e₁₁[iᵀ] : i in [1..#d4e₁₁] ];
>    d4ft₁₁ := [ d4f₁₁[iᵀ] : i in [1..#d4f₁₁] ];
```

In order to establish explicit isomorphisms $V_{11}^\sigma \simeq V_{10}$ and $V_{11}^\tau \simeq V_{01}$ we proceed as follows. A vector v^+ is said to be *maximal* if it is sent to 0 by the action of all root elements e_β, where β is positive. (Recall that root elements act on the spaces V_{ij} via the adjoint representation.) For the modules V_{11}, V_{01} and V_{10}, the first vector in the given basis is a maximal vector. From highest weight theory each irreducible module has a basis consisting of vectors of the form $v^+ f_{\beta_1}^{i_1} \cdots f_{\beta_m}^{i_m}$, where $\{\beta_1, \ldots, \beta_m\}$ is the set of positive roots of $\mathfrak{d}_4$. To set this up in Magma we need a generic space V of dimension 8 and the ring M of 8×8 matrices.

```
>    V := RSpace(Integers(), 8);
>    M := MatrixRing(Integers(), 8);
```

To obtain a basis $v_1, v_2, \ldots, v_8$ for one of our modules it is enough to take v_1 to be a maximal vector (namely $V.1$) and to give a sequence of pairs $s = [(i_1, j_1), \ldots, (i_7, j_7)]$ such that the basis vector v_{k+1} is obtained by applying the j_k-th negative root element to the i_k-th basis vector v_{i_k}. We shall carry out this construction several times and so we define a Magma function to perform the task. In the application of this function, L will be a sequence of matrices giving the action of the negative root elements on the module. (The sequence of basis vectors is built recursively and the built-in function Self refers to the sequence under construction.)

```
>    basisSeq := func< s, L |
>        [ i eq 0 select V.1 else Self(s[i][1]) * L[s[i][2]] : i in [0..#s] ] >;
```

We claim that the sequences

```
>    d4s₁₀ := [ [1,2],[2,3],[3,4],[3,1],[5,4],[6,3],[7,2] ];
>    d4s₀₁ := [ [1,4],[2,3],[3,1],[3,2],[5,1],[6,3],[7,4] ];
```

define bases for V_{10} and V_{01}.

```
>    d4B₁₀ := basisSeq(d4s₁₀, d4f₁₀);
>    d4B₀₁ := basisSeq(d4s₀₁, d4f₀₁);
```

This can be seen directly.

```
>    d4B₁₀;
        [  (1  0  0  0  0  0  0  0),
           (0 -1  0  0  0  0  0  0),
           (0  0  1  0  0  0  0  0),
           (0  0  0  0  1  0  0  0),
           (0  0  0 -1  0  0  0  0),
           (0  0  0  0  0 -1  0  0),
           (0  0  0  0  0  0 -1  0),
           (0  0  0  0  0  0  0 -1)  ]
```

```
>    d4B01;
        [  (1  0  0  0  0  0  0  0),
           (0 -1  0  0  0  0  0  0),
           (0  0  1  0  0  0  0  0),
           (0  0  0 -1  0  0  0  0),
           (0  0  0  0 -1  0  0  0),
           (0  0  0  0  0  1  0  0),
           (0  0  0  0  0  0 -1  0),
           (0  0  0  0  0  0  0  1)  ]
```

The same sequences must define bases for V_{11}^σ and V_{11}^τ.

```
>    d4Bs11 := basisSeq(d4s10, d4fs11);
>    d4Bt11 := basisSeq(d4s01, d4ft11);
```

If P_{10} and P_{01} are the matrices relating the bases just defined we can check directly that P_{10} conjugates d4Be11s to d4e10, that P_{01} conjugates d4et11 to d4e01 and thus $V_{11}{}^\sigma \simeq V_{10}$ and $V_{11}{}^\tau \simeq V_{01}$.

```
>    P10 := (M ! d4Bs11)^-1 * M ! d4B10;
>    P01 := (M ! d4Bt11)^-1 * M ! d4B01;
>    [ P10^-1 * e * P10 : e in d4es11 ] eq d4e10;

        true

>    [ P01^-1 * e * P01 : e in d4et11 ] eq d4e01;

        true
```

We shall return to these isomorphisms after we have defined an algebra structure on V_{11}.

5 The Lie algebra of type G_2

The Lie algebra $\mathfrak{g}_2$ of type G_2 may be regarded as the subalgebra of $\mathfrak{d}_4$ generated by $e_A = e_{\alpha_2} + e_{\alpha_3} + e_{\alpha_5}$, $e_B = e_{\alpha_4}$, $e_{-A} = e_{-\alpha_2} + e_{-\alpha_3} + e_{-\alpha_5}$ and $e_{-B} = e_{-\alpha_4}$. These elements are fixed by the automorphisms σ and τ and so the spaces V_{10}, V_{01} and V_{11} are isomorphic as $\mathfrak{g}_2$-modules.

The matrices for the $\mathfrak{g}_2$ action on V_{10} are

```
>    e10 := [ d4e10[1]+d4e10[2]+d4e10[4], d4e10[3] ];
>    f10 := [ d4f10[1]+d4f10[2]+d4f10[4], d4f10[3] ];
```

Similarly the actions on V_{01} and V_{11} are given by

```
>    e01 := [ d4e01[1]+d4e01[2]+d4e01[4], d4e01[3] ];
>    f01 := [ d4f01[1]+d4f01[2]+d4f01[4], d4f01[3] ];
>    e11 := [ d4e11[1]+d4e11[2]+d4e11[4], d4e11[3] ];
>    f11 := [ d4f11[1]+d4f11[2]+d4f11[4], d4f11[3] ];
```

It turns out that the matrices representing the actions of $\mathfrak{g}_2$ on V_{01} and V_{11} are identical:

```
>    (e_01 eq e_11) and (f_01 eq f_11);
         true
```

Over any field of characteristic different from 2, the $\mathfrak{g}_2$-modules V_{10}, V_{01} and V_{11} are completely reducible and split into the sum of an irreducible 7-dimensional submodule and a 1-dimensional complement.

Writing $x_1, x_2, \ldots, x_8$ to denote the standard basis vectors of V_{10} (y_1, $y_2, \ldots, y_8$ for the standard basis of V_{01}, and $z_1, z_2, \ldots, z_8$ for the standard basis of V_{11}), the 1-dimensional $\mathfrak{g}_2$-submodules of V_{10}, V_{01} and V_{11} are spanned by $x_4 + x_5$, $y_4 - y_5$ and $z_4 - z_5$, respectively.

```
>    [ (V.4+V.5) * e : e in (e_10 cat f_10) ];
         [
             (0 0 0 0 0 0 0 0),
             (0 0 0 0 0 0 0 0),
             (0 0 0 0 0 0 0 0),
             (0 0 0 0 0 0 0 0)
         ]

>    [ (V.4-V.5) * e : e in (e_11 cat f_11) ];
         [
             (0 0 0 0 0 0 0 0),
             (0 0 0 0 0 0 0 0),
             (0 0 0 0 0 0 0 0),
             (0 0 0 0 0 0 0 0)
         ]
```

The irreducible 7-dimensional submodule of V_{10} has maximal vector x_1 and a basis *Bx*, where

```
>    ξ := [ [1,1],[2,2],[3,1],[4,1],[5,2],[6,1] ];
>    Bx := basisSeq(ξ, f_10);
```

We obtain compatible bases for V_{01} and V_{11} by applying f_{01} and f_{11} to the sequence ξ.

```
>    By := basisSeq(ξ, f_01);
>    Bz := basisSeq(ξ, f_11);
>    By eq Bz;
         true
```

Here are the basis vectors:

```
>    Bx;
         [ (1  0  0  0  0  0  0  0),
           (0 -1  0  0  0  0  0  0),
           (0  0  1  0  0  0  0  0),
           (0  0  0 -1  1  0  0  0),
           (0  0  0  0  0 -2  0  0),
           (0  0  0  0  0  0 -2  0),
           (0  0  0  0  0  0  0 -2) ]
```

> Bz;

```
[ (1  0  0  0  0  0  0  0),
  (0 -1  0  0  0  0  0  0),
  (0  0  1  0  0  0  0  0),
  (0  0  0 -1 -1  0  0  0),
  (0  0  0  0  0  2  0  0),
  (0  0  0  0  0  0 -2  0),
  (0  0  0  0  0  0  0  2) ]
```

From these bases and the given basis vectors for the 1-dimensional submodules we may define $\mathfrak{g}_2$-module isomorphisms $\varphi : V_{11} \to V_{10}$ and $\psi : V_{11} \to V_{01}$ with matrices P and Q, where

$$P = \begin{pmatrix} 1\,0\,0 & 0 & 0 & 0\,0\,0 \\ 0\,1\,0 & 0 & 0 & 0\,0\,0 \\ 0\,0\,1 & 0 & 0 & 0\,0\,0 \\ 0\,0\,0 & \frac{1}{2}(\lambda+1) & \frac{1}{2}(\lambda-1) & 0\,0\,0 \\ 0\,0\,0 & -\frac{1}{2}(\lambda-1) & -\frac{1}{2}(\lambda+1) & 0\,0\,0 \\ 0\,0\,0 & 0 & 0 & -1\,0\,0 \\ 0\,0\,0 & 0 & 0 & 0\,1\,0 \\ 0\,0\,0 & 0 & 0 & 0\,0\,-1 \end{pmatrix} \quad \text{and}$$

$$Q = \begin{pmatrix} 1\,0\,0 & 0 & 0 & 0\,0\,0 \\ 0\,1\,0 & 0 & 0 & 0\,0\,0 \\ 0\,0\,1 & 0 & 0 & 0\,0\,0 \\ 0\,0\,0 & \frac{1}{2}(\mu+1) & -\frac{1}{2}(\mu-1) & 0\,0\,0 \\ 0\,0\,0 & -\frac{1}{2}(\mu-1) & \frac{1}{2}(\mu+1) & 0\,0\,0 \\ 0\,0\,0 & 0 & 0 & 1\,0\,0 \\ 0\,0\,0 & 0 & 0 & 0\,1\,0 \\ 0\,0\,0 & 0 & 0 & 0\,0\,1 \end{pmatrix}$$

where λ and μ are non-zero scalars chosen so that $(z_4 - z_5)^\varphi = \lambda(x_4 + x_5)$ and $(z_4 - z_5)^\psi = \mu(y_4 - y_5)$.

6 The split octonions

If $\alpha \in X_{10}$, $\beta \in X_{01}$ and $\alpha + \beta$ is a root, then $\alpha + \beta \in X_{11}$. Thus the Lie bracket defines a bilinear "multiplication" $V_{10} \times V_{01} \to V_{11} : (x, y) \mapsto [x, y]$. The matrix of products $[x_i, y_j]$ of the basis vectors x_i and y_j is

$$\begin{pmatrix} 0 & 0 & 0 & 0 & z_1 & z_2 & z_3 & z_5 \\ 0 & 0 & -z_1 & -z_2 & 0 & 0 & z_4 & z_6 \\ 0 & z_1 & 0 & -z_3 & 0 & -z_4 & 0 & z_7 \\ 0 & z_2 & z_3 & 0 & z_4 & 0 & 0 & z_8 \\ z_1 & 0 & 0 & z_5 & 0 & z_6 & z_7 & 0 \\ z_2 & 0 & -z_5 & 0 & -z_6 & 0 & z_8 & 0 \\ -z_3 & -z_5 & 0 & 0 & z_7 & z_8 & 0 & 0 \\ z_4 & z_6 & z_7 & z_8 & 0 & 0 & 0 & 0 \end{pmatrix}$$

This can be seen by inspecting *coeffmat* and *ndxmat* after running the following code

```
coeffmat := M ! 0; ndxmat := M ! 0;
for i := 1 to 8 do
  a := Index(Φ, X₁₀[i]);
  for j := 1 to 8 do
    c := LieConstant_N( RD, a, Index(Φ, X₀₁[j]) );
    if c ne 0 then
      ndxmat[i,j] := Index(X₁₁, X₁₀[i]+X₀₁[j]);
      coeffmat[i,j] := c;
    end if;
  end for;
end for;
```

Using the isomorphisms between V_{10}, V_{01} and V_{11} we define a product on V_{11} by

$$u \star v = [u^\varphi, v^\psi]$$

and we choose λ, μ and ξ so that $\xi(z_4 - z_5)$ is an identity element for V_{11}.

From the table of products given above we see that

$$\xi(z_4 - z_5) \star z_4 = -\tfrac{1}{2}\xi\lambda(\mu - 1)z_4 + \tfrac{1}{2}\xi\lambda(\mu + 1)z_5.$$

Hence for $\xi(z_4 - z_5)$ to be an identity element (on the left) we must have $\mu = -1$ and $\xi\lambda = 1$, whence $\xi(z_4 - z_5) \star z_i = z_i$ for all i. Similarly

$$z_4 \star \xi(z_4 - z_5) = \tfrac{1}{2}(\lambda + 1)z_4 - \tfrac{1}{2}(\lambda - 1)z_5$$

and so $\lambda = \xi = 1$.

With this choice of values, the identity element of V_{11} is $z_4 - z_5$ and the revised values of P and Q are

$$P = \begin{pmatrix} 1 & 0 & 0 & 0 & 0 & 0 & 0 & 0 \\ 0 & 1 & 0 & 0 & 0 & 0 & 0 & 0 \\ 0 & 0 & 1 & 0 & 0 & 0 & 0 & 0 \\ 0 & 0 & 0 & 1 & 0 & 0 & 0 & 0 \\ 0 & 0 & 0 & 0 & -1 & 0 & 0 & 0 \\ 0 & 0 & 0 & 0 & 0 & -1 & 0 & 0 \\ 0 & 0 & 0 & 0 & 0 & 0 & 1 & 0 \\ 0 & 0 & 0 & 0 & 0 & 0 & 0 & -1 \end{pmatrix} \quad Q = \begin{pmatrix} 1 & 0 & 0 & 0 & 0 & 0 & 0 & 0 \\ 0 & 1 & 0 & 0 & 0 & 0 & 0 & 0 \\ 0 & 0 & 1 & 0 & 0 & 0 & 0 & 0 \\ 0 & 0 & 0 & 0 & 1 & 0 & 0 & 0 \\ 0 & 0 & 0 & 1 & 0 & 0 & 0 & 0 \\ 0 & 0 & 0 & 0 & 0 & 1 & 0 & 0 \\ 0 & 0 & 0 & 0 & 0 & 0 & 1 & 0 \\ 0 & 0 & 0 & 0 & 0 & 0 & 0 & 1 \end{pmatrix}$$

It turns out that these are the matrices P_{10} and P_{01} determined earlier.

We can check that the representations of $\mathfrak{g}_2$ on V_{10}, V_{01} and V_{11} are isomorphic:

```
>   [ P10 * x * P10 ⁻¹ : x in e10 ] eq e11;
        true

>   [ P10 * x * P10 ⁻¹ : x in f10 ] eq f11;
        true

>   [ P01 * x * P01 ⁻¹ : x in e01 ] eq e11;
        true

>   [ P01 * x * P01 ⁻¹ : x in f01 ] eq f11;
        true
```

The product $z_i \star z_j$ is cz_k, where c is $coeff[i,j]$ and k is $ndx[i,j]$. The code to compute this is a slight modification of that used previously to calculate the Lie products $[x_i, y_j]$. The effect of the matrix P_{10} is encoded by the 8-tuple sgn and the effect of P_{01} is encoded by the permutation π.

```
coeff := M ! 0; ndx := M ! 0;
π := Sym(8) ! (4,5);
sgn := [1, 1, 1, 1, −1, −1, 1, −1];
for i := 1 to 8 do
  a := Index(Φ, X10[i]);
  for j := 1 to 8 do
    c := LieConstant_N( RD, a, Index(Φ, X01[jᵖ]) );
    if c ne 0 then
      ndx[i,j] := Index(X11, X10[i]+X01[jᵖ]);
      coeff[i,j] :− c ∗ sgn[i];
    end if;
  end for;
end for;
```

The resulting table of products $z_i \star z_j$ is

$$
\begin{pmatrix}
0 & 0 & 0 & z_1 & 0 & z_2 & z_3 & z_5 \\
0 & 0 & -z_1 & 0 & -z_2 & 0 & z_4 & z_6 \\
0 & z_1 & 0 & 0 & -z_3 & -z_4 & 0 & z_7 \\
0 & z_2 & z_3 & z_4 & 0 & 0 & 0 & z_8 \\
-z_1 & 0 & 0 & 0 & -z_5 & -z_6 & -z_7 & 0 \\
-z_2 & 0 & z_5 & z_6 & 0 & 0 & -z_8 & 0 \\
-z_3 & -z_5 & 0 & z_7 & 0 & z_8 & 0 & 0 \\
-z_4 & -z_6 & -z_7 & 0 & -z_8 & 0 & 0 & 0
\end{pmatrix}
$$

With the arrays *coeff* and *ndx* at our disposal we are able to define a Magma function that creates a structure constant algebra over a ring R. We shall prove that the algebra is the split octonions over R.

```
SplitOctonions := func< R |
    Algebra< R, 8 | [ < i, j, ndx[i,j], coeff[i,j] > : i,j in [1..8]
                     | ndx[i,j] ne 0 ] > >;
O := SplitOctonions(Integers());
```

The Lie algebra $\mathfrak{g}_2$ acts on E_6 as derivations and this carries over to its action on the algebra $\mathbb{O} = \mathbb{O}(\mathbb{Z})$. In order to check this we need a way to act on an element of a structure constant algebra with a matrix. In the case of the octonion algebra $\mathbb{O}$ just constructed it would be enough to coerce the elements of $\mathbb{O}$ into the $\mathbb{Z}$-module V and then act on the right with the matrix. But for future use we take this opportunity to define a more generic method of applying a matrix to an algebra element; this is embodied in the following function.

```
>    apply := function(u, g)
>       P := Parent(u);
>       return P ! Eltseq(Matrix(B, u) * Matrix(B, g))
>         where B is BaseRing(P);
>    end function;
```

The element g is a derivation if

$$(u \star v)g = (ug) \star v + u \star (vg)$$

for all $u, v \in \mathbb{O}$. The Lie algebra $\mathfrak{g}_2$ is generated by the matrices in e_{11} and f_{11} and every element of $\mathfrak{g}_2$ is a derivation of $\mathbb{O}$, as can be seen from the following calculation.

```
>    forall{ g : g in (e11 cat f11) | forall{ <i,j> : i,j in [1..8] |
>       apply(O.i*O.j, g) eq apply(O.i, g) * O.j + O.i * apply(O.j, g) } };
        true
```

We claim that the algebra just constructed satisfies the alternative laws:

$$x^2 y = x(xy) \quad \text{and} \quad xy^2 = (xy)y.$$

The linearized versions of these laws are

$$(x, y, z) + (y, x, z) = 0 \quad \text{and} \quad (x, y, z) + (x, z, y) = 0,$$

where the *associator* (x, y, z) is defined to be $(xy)z - x(yz)$.

We check the linearized alternative laws on the basis elements:

```
>    assoc := func< x, y, z | (x * y) * z - x * (y * z) >;
>    forall{ <i,j,k> : i,j,k in [1..8] |
>       assoc(O.i, O.j, O.k) + assoc(O.j, O.i, O.k) eq O ! 0 };
        true
```

```
>    forall{ <i,j,k> : i,j,k in [1..8] |
>       assoc(O.i,O.j,O.k) + assoc(O.i,O.k,O.j) eq O!0 };
```

true

Thus the algebra $\mathbb{O}$ is alternative but not associative:

```
>    exists(i,j,k){<i,j,k> : i,j,k in [1..8] |
>       assoc(O.i,O.j,O.k) ne O!0 };
```

true

```
>    i,j,k;
```

1 2 7

The algebra is not a division algebra since it contains non-zero elements whose square is zero; for example, z_1.

7 The quadratic form

In order to complete the proof that $\mathbb{O}$ is a $\mathbb{Z}$-form of the split octonions we define a quadratic form q that preserves the multiplication; i.e.,

$$q(u \star v) = q(u)q(v)$$

for all $u,v \in \mathbb{O}$.

We shall obtain q via the Killing form on E_6. First observe that the vector space V_z spanned by $\{\,e_{-\alpha} \mid \alpha \in X_{11}\,\}$ has highest weight λ_1 as a $\mathfrak{d}_4$-module and therefore it is isomorphic to V_{11}. To obtain the action of $\mathfrak{d}_4$ on this module we first obtain its basis

```
>    Z := {@ x : x in Φ | x[1] eq −1 and x[6] eq −1 @};
```

and then construct matrices as before

```
>    d4eZ := [ LieRootMatrix(RD,Φ[i],Z) : i in [2..5] ];
>    d4fZ := [ LieRootMatrix(RD,−Φ[i],Z) : i in [2..5] ];
>    eZ := [ d4eZ[1]+d4eZ[2]+d4eZ[4], d4eZ[3] ];
>    fZ := [ d4fZ[1]+d4fZ[2]+d4fZ[4], d4fZ[3] ];
```

The matrices of the action of $\mathfrak{d}_4$ on V_z are the negatives of the matrices for the action on V_{11}:

```
>    forall{ i: i in [1..4] | d4e11[i] eq −d4eZ[i] and d4f11[i] eq −d4fZ[i] };
```

true

In both V_{11} and V_z the first vector in the basis is a maximal vector and so we obtain a "standard basis" as follows

```
>    d4sz := [ [1,1],[2,3],[3,2],[3,4],[5,2],[6,3],[7,1] ];
>    d4Bz := basisSeq(d4sz, d4f11);
>    d4Bz;
```

```
[ (1  0  0  0  0  0  0  0),
  (0 -1  0  0  0  0  0  0),
  (0  0  1  0  0  0  0  0),
  (0  0  0 -1  0  0  0  0),
  (0  0  0  0 -1  0  0  0),
  (0  0  0  0  0  1  0  0),
  (0  0  0  0  0  0 -1  0),
  (0  0  0  0  0  0  0  1) ]
```

The corresponding basis for V_z constructed from the sequence *d4sz* and matrix representation *d4fZ* turns out to be the standard one.

```
>    d4Bz := basisSeq(d4sz, d4fZ);
>    d4Bz eq Basis(V);
        true
```

The Killing form, namely $\kappa(x, y) = \mathrm{Tr}(\mathrm{ad}(x)\,\mathrm{ad}(y))$, for E_6 defines a non-degenerate $\mathfrak{d}_4$-invariant pairing

$$\kappa : V_z \times V_{11} \to \mathbb{C}$$

and the following Magma code returns the value of the Killing form on the root elements e_α and e_β.

```
KForm := function( RD, α, β )
  rtn := 0;
  rts := Roots(RD);
  if α eq −β then
    a := Index(rts, α);
    aa := Index(rts, −α);
    for γ in rts do
      if γ ne −α then
        c := LieConstant_N( RD, Index(rts, γ), a );
        if c ne 0 then
          rtn +:= c * LieConstant_N( RD, ga, aa )
              where ga is Index(rts, γ+α);
        end if;
      else
        rtn +:= 2;
      end if;
    end for;
    rtn +:= 2; // contribution from the Cartan subalgebra
  end if;
  return rtn;
end function;
```

Thus we obtain a $\mathfrak{d}_4$-invariant bilinear form β on V_{11} by setting $\beta(u, v) = \kappa(u^\eta, v)$, where $\eta : V_{11} \to V_z$ is the $\mathfrak{d}_4$-isomorphism obtained above.

Let b be the 8×8 matrix defined by $b_{ij} = \beta(z_i, z_j)$ for $i < j$ and with all other entries 0. We can check directly that $\beta(z_i, z_i) = 0$ for all i and so the quadratic form whose polar form is β is given by $q(u) = ubu^t$. The corresponding polar form is $(u, v) \mapsto q(u + v) - q(u) - q(v) = u(b + b^t)v^t$. We scale the form so that it takes the value 1 at the identity element $z_4 - z_5$.

```
>   η := [1, −1, 1, −1, −1, 1, −1, 1];
>   b := M ! 0;
>   for i := 1 to 7 do
>     for j := i+1 to 8 do
>       b[i,j] := η[i] * KForm( RD, Z[i], X₁₁[j] )/24;
>     end for;
>   end for;
>   b;
        [ 0  0  0  0  0  0  0  1]
        [ 0  0  0  0  0  0 -1  0]
        [ 0  0  0  0  0  1  0  0]
        [ 0  0  0  0 -1  0  0  0]
        [ 0  0  0  0  0  0  0  0]
        [ 0  0  0  0  0  0  0  0]
        [ 0  0  0  0  0  0  0  0]
        [ 0  0  0  0  0  0  0  0]
```

```
>   q := func< u | InnerProduct( uu * Matrix(B,b), uu )
>           where uu is Matrix(B,u) where B is BaseRing(Parent(u)) >;
>   q(V.4−V.5);
        1
```

The polar form of q is the bilinear form β.

```
>   β := func< u, v |
>     InnerProduct( Matrix(B,u)*Matrix(B,b+Transpose(b)), Matrix(B,v) )
>       where B is BaseRing(Parent(u)) >;
```

The Lie algebra $\mathfrak{d}_4$ preserves this form; namely, for all $u, v \in V_{11}$ and $x \in \mathfrak{d}_4$ we have $\beta(ux, v) + \beta(u, vx) = 0$, as can be checked in Magma.

```
>   forall{ <i,j,e> : i,j in [1..8], e in (d4e₁₁ cat d4f₁₁) |
>           β( apply(O.i, e), O.j ) + β( O.i, apply(O.j, e) ) eq 0 };
        true
```

We check that q preserves the multiplication by using the polynomial ring R in the indeterminates $u_1, \ldots, u_8$ and $v_1, \ldots, v_8$ that was defined earlier.

```
>   OR := SplitOctonions(R);
>   u := OR ! [u₁, u₂, u₃, u₄, u₅, u₆, u₇, u₈];
>   v := OR ! [v₁, v₂, v₃, v₄, v₅, v₆, v₇, v₈];
>   q(u∗v) eq q(u)∗q(v);
        true
```

With this machinery at hand we have convenient formulas both for the product and for the quadratic form.

```
>    u * v;

                (u1*v4 - u2*v3 + u3*v2 - u5*v1
                 u1*v6 - u2*v5 + u4*v2 - u6*v1
                 u1*v7 - u3*v5 + u4*v3 - u7*v1
                 u2*v7 - u3*v6 + u4*v4 - u8*v1
                 u1*v8 - u5*v5 + u6*v3 - u7*v2
                 u2*v8 - u5*v6 + u6*v4 - u8*v2
                 u3*v8 - u5*v7 + u7*v4 - u8*v3
                 u4*v8 - u6*v7 + u7*v6 - u8*v5)
```

```
>    q(u);

                 u1*u8 - u2*u7 + u3*u6 - u4*u5
```

For each element $x \in \mathfrak{d}_4$, the elements x^σ and x^τ satisfy

$$(u \star v)x = ux^\sigma \star v + u \star vx^\tau$$

for all $u, v \in V$. This is the *Principle of Local Triality* (see [19]).

```
>    forall{ i : i in [1..4] |
>        apply(u, d4es11[i])*v + u*apply(v, d4et11[i]) eq apply(u*v, d4e11[i])
>        and
>        apply(u, d4fs11[i])*v + u*apply(v, d4ft11[i]) eq apply(u*v, d4f11[i])
>    };

        true
```

8 The Chevalley groups of type G_2

Given a root $\alpha \in \Phi$ of the Lie algebra $\mathcal{L}$, the linear transformation $E_\alpha = \mathrm{ad}(e_\alpha)$ is nilpotent and therefore if F is a field and $t \in F$, the *root element* $x_\alpha(t) = \exp(tE_\alpha)$ is well-defined provided the divided powers $E^k/k!$ of E are computed over $\mathbb{Z}$ before coercing to the field F. Furthermore, the matrices in e_{11} and f_{11} are nilpotent and the corresponding root elements generate a representation of the group $G_2(F)$ on the split octonions.

Magma has an extensive suite of functions to compute with Lie algebras, root systems and groups of Lie type (see [7]). In particular, given F we can construct the group $G_2(F)$ as the group of Lie type whose elements are words in the Steinberg generators. To illustrate this we take F to be the field of 5 elements.

```
>    F := GF(5);
>    G := GroupOfLieType("G2", F);
>    Random(G);

        x1(2) x2(3) x4(1) (1 2) n1 n2 n1 n2 n1 x1(2) x3(2) x5(2)
```

The output shows a typical element of the group as a "Steinberg word". The Magma notation for a root element $x_i(t)$ of G is the root term **elt**$<G \mid <i, t>>$, where t belongs to the field. Note that if N is the number of positive roots and $i \leq N$, then **elt**$< G \mid <i+N, t> >$ represents a negative root term. In the output above, $(1, 2)$ refers to a torus element, which can be input as **elt**$< G \mid$ VectorSpace$(F, 2)$! $[1, 2] >$. The elements n_1 and n_2 represent generators of the Weyl group and they can be input as **elt**$< G \mid 1 >$ and **elt**$< G \mid 2 >$, respectively.

We are now able to construct an explicit representation of this group on the split octonions using the matrices obtained above. In order to use the built-in function Representation(G, pos, neg) we need matrices for all the root elements of the Lie algebra of type G_2. In computing these matrices we must divide by the structure constants and therefore it is essential that they are calculated in the $\mathbb{Z}$-form of the complex Lie algebra before coercing them into the field F.

```
>    G2RD := RootDatum("G2");
>    comm := func< a, b | a*b - b*a >;
>    for sum in [ 3 .. NumPosRoots(G2RD) ] do
>        r, s := ExtraspecialPair( G2RD, sum );
>        c := LieConstant_N( G2RD, r, s );
>        e11[sum] := (1/c)*comm(e11[r], e11[s]);
>        f11[sum] := -(1/c)*comm(f11[r], f11[s]);
>    end for;
```

An *extraspecial pair* for a positive root γ is a pair of positive roots (α, β) where α is a simple root and $\gamma = \alpha + \beta$. (See [6] or [7] for more details.)

In order to construct the representation we coerce the root elements into the rational field and then pass them to the Magma function Representation.

```
>    ρ := Representation(G, ChangeUniverse(e11, Q), ChangeUniverse(f11, Q))
>              where Q is MatrixRing(Rationals(), 8);
```

The image of G under the homomorphism ρ is

```
>    imG := sub< GL(8, F) | [ ρ(x) : x in Generators(G) ] >;
```

If $x \in \mathbb{O}(F)$ is not a scalar, the subalgebra $F[x]$ is isomorphic to one of $F \oplus F$, the dual numbers $F[X]/(X^2)$ over F, or a quadratic extension of F (in this case $\mathbb{F}_{25}$), according to whether the quadratic equation

$$X^2 - \mathrm{Tr}(x)X + q(x) = 0$$

has 2, 1 or 0 roots in F. The alternative laws imply that $\mathbb{O}(F)$ is a left $F[x]$-module.

If $F[x]$ is a field, the subgroup of $G_2(F)$ fixing x is the special unitary group SU$(3, F)$. Furthermore, if the characteristic of F is not 2, then the stabilizer of $\{x, -x\}$ is a maximal subgroup of $G_2(F)$ and contains the stabilizer of x as a subgroup of index 2. We use Magma to verify these assertions in the case of $\mathbb{F}_5$.

```
>    O_5 := SplitOctonions(F);
>    X_2 := { {W ! x, −W ! x} : x in O_5 | x² eq 2 }
>        where W is VectorSpace(F, 8);
>    _, P_2, _ := OrbitAction(imG, X_2); P_2;
```

    Permutation group P2 acting on a set of cardinality 7750
    Order = 2^6 * 3^3 * 5^6 * 7 * 31

```
>    IsPrimitive(P_2);
```

    true

```
>    CompositionFactors(Stabilizer(P_2, 1));
```

    G
    | Cyclic(2)
    *
    | 2A(2, 5)   = U(3, 5)
    *
    | Cyclic(3)
    1

Note that the centre of $SU(3,5)$ has order 3 and the quotient of $SU(3,5)$ modulo its centre is the simple group $U(3,5)$.

There is a construction of the split octonions that directly brings out the rôle of the group $SU(3, F)$. Begin with a vector space E of dimension 3 over the field K, where K is a quadratic extension of F with an automorphism σ of order 2 such that F is the fixed field of σ. If γ is a non-degenerate, isotropic, σ-hermitian form on E, there is a bi-additive skew vector product $u \times v$ on E such that

$$u \times u = 0,$$
$$(au) \times v = u \times (av) = \sigma(a)(u \times v),$$
$$\gamma(u, u \times v) = 0,$$
$$\gamma(v_1 \times v_2, u_1 \times u_2) = \gamma(u_1, v_1)\gamma(u_2, v_2) - \gamma(u_1, v_2)\gamma(u_2, v_1), \text{ and}$$
$$(u \times v) \times w = \gamma(u, w)v - \gamma(v, w)u.$$

The unitary group $SU(3, F)$ of linear transformations of determinant 1 that preserve γ can also be described as the group of linear transformations T of E such that $uT \times vT = (u \times v)T$.

Consider the space $C := K \oplus E$ as a vector space of dimension 8 over F and define a product and quadratic form on C by

$$(a, u)(b, v) = (ab - \gamma(u, v), av + \sigma(b)u + u \times v), \quad \text{and}$$
$$Q((a, v)) = \sigma(a)a + \gamma(v, v).$$

A direct calculation shows that the conjugate of $\xi = (a, u)$ is $\bar{\xi} = (\sigma(a), -u)$ and that $Q(\xi\eta) = Q(\xi)Q(\eta)$. Therefore C is the split composition algebra

$\mathbb{O}(F)$ over F. If x generates K, then the minimal polynomial of x is irreducible and the automorphisms T that fix x act on the orthogonal complement E of $K = F \oplus Fx$. Furthermore we see that a linear transformation T of E extends to an automorphism of C if and only if T preserves the vector product and the form γ; i.e., if and only if $T \in \mathrm{SU}(3, F)$. The field automorphism σ extends to an automorphism of C that normalizes $\mathrm{SU}(3, F)$ and interchanges x and $\sigma(x)$.

The next case to consider is when $F[x] = F \oplus F$. Then the subgroup fixing x is $\mathrm{SL}(3, F)$ and it is a subgroup of index 2 in the stabilizer of $\{x, -x\}$.

```
>     X₁ := { {W ! x, -W ! x} : x in O₅ | x² eq 1 and
>                   x ne 1 and x ne -1} where W is VectorSpace(F,8);
>     P₁ := OrbitImage(imG, X₁); P₁;

      Permutation group P1 acting on a set of cardinality 7875

>  CompositionFactors(Stabilizer(P₁, 1));

          G
          |  Cyclic(2)
          *
          |  A(2, 5)                    = L(3, 5)
          1
```

There is a construction of the split octonions that clearly exhibits $\mathrm{SL}(3, F)$ as a subgroup of its automorphism group; it uses Zorn's "vector matrices". In order to describe this construction we let $U := F^3$ be a vector space of dimension 3 over F and let (u, v) denote the standard inner product; namely if $u = (u_1, u_2, u_3)$ and $v = (v_1, v_2, v_3)$, then $(u, v) = u_1 v_1 + u_2 v_2 + u_3 v_3$. There is a bilinear vector product $u \times v$ on U such that

$$u \times u = 0,$$
$$(u, u \times v) = 0,$$
$$(u_1 \times u_2, v_1 \times v_2) = (u_1, v_1)(u_2, v_2) - (u_1, v_2)(u_2, v_1), \text{ and}$$
$$(u \times v) \times w = (u, w)v - (v, w)u.$$

The *vector matrices* are symbols of the form $\begin{pmatrix} a & u \\ v & b \end{pmatrix}$ where $a, b \in F$ and $u, v \in U$. Addition is defined componentwise and the product of two vector matrices is

$$\begin{pmatrix} a_1 & u_1 \\ v_1 & b_1 \end{pmatrix} \begin{pmatrix} a_2 & u_2 \\ v_2 & b_2 \end{pmatrix} = \begin{pmatrix} a_1 a_2 + (u_1, v_2) & a_1 u_2 + b_2 u_1 + v_1 \times v_2 \\ a_2 v_1 + b_1 v_2 + u_1 \times u_2 & b_1 b_2 + (v_1, u_2) \end{pmatrix}.$$

The quadratic form is

$$Q \begin{pmatrix} a & u \\ v & b \end{pmatrix} = ab - (u, v).$$

and a direct calculation shows that $Q(AB) = Q(A)Q(B)$. It follows that the algebra $\mathcal{Z}$ of all vector matrices is the split octonions $\mathbb{O}(F)$.

If T is a linear transformation of U of determinant 1, then there is a uniquely determined linear transformation T^* such that

$$(uT, vT^*) = (u, v), \quad \text{and}$$
$$uT \times vT = (u \times v)T^*$$

for all $u, v \in U$. Thus T extends to an automorphism of $\hat{T}$ of $\mathcal{Z}$ such that

$$\begin{pmatrix} a & u \\ v & b \end{pmatrix} \hat{T} = \begin{pmatrix} a & uT \\ vT^* & b \end{pmatrix}.$$

It follows that the elements $\hat{T}$ form a subgroup L of $\mathrm{Aut}(\mathcal{Z})$ isomorphic to $\mathrm{SL}(3, F)$. Furthermore, the map that sends $\begin{pmatrix} a & u \\ v & b \end{pmatrix}$ to $\begin{pmatrix} b & v \\ u & a \end{pmatrix}$ is an automorphism τ that normalizes L. Its effect on L is to interchange $\hat{T}$ and $\hat{T}^*$. If $x = \begin{pmatrix} 1 & 0 \\ 0 & -1 \end{pmatrix}$, then $x^2 = 1$ and $\tau(x) = -x$. The elements of L fix x and it is clear that $F[x]$ is isomorphic to $F \oplus F$.

Finally, consider the case that $F[x]$ is the ring of dual numbers $F[X]/(X^2)$.

```
>      X₀ := { {a*W ! x: a in F | a ne 0} : x in O₅ | x² eq 0 and x ne 0 }
>        where W is VectorSpace(F,8);
>      P₀ := OrbitImage(imG, X₀); P₀;
           Permutation group PO acting on a set of cardinality 3906

>      CompositionFactors(Stabilizer(P₀,1));
                G
                |   Cyclic(2)
                *
                |   Alternating(5)
                *
                |   Cyclic(2)
                *
                |   Cyclic(2)
                *
                |   Cyclic(5)
                *
                |   Cyclic(5)
                *
                |   Cyclic(5)
                *
                |   Cyclic(5)
                *
                |   Cyclic(5)
                1
```

The subgroup of $G_2(F)$ fixing the one-dimensional subspace Fx for $x \in X_0$ is an extension of a nilpotent group of order 5^5 (and nilpotency class 3) by the central product of a cyclic group of order 4 and $GL(2, 5)$. This is a parabolic subgroup of $G_2(F)$.

References

1. J. F. Adams, *Spin(8), Triality, F_4 and all that*, pp. 435–445 in: S. W. Hawking, M. Roček (eds.), *Superspace and Supergravity*, Proceedings of the Nuffield Workshop, Cambridge. June 16 – July 12, 1980, Cambridge University Press, 1981.
2. John C. Baez, *The octonions*, Bull. Amer. Math. Soc. **39** (2001), 145–205.
3. Wieb Bosma, John Cannon, Catherine Playoust, *The Magma algebra system I: The user language*, J. Symbolic Comput. **24** (1997), 235–265.
   See also the Magma home page at http://magma.maths.usyd.edu.au/magma/.
4. E. Cartan, *Le principe de dualité et la théorie des groupes simples et semisimple*, Bull. Sci. Math. (2) **49** (1925), 361–374.
5. J. Cannon, B. Souvignier, *Structure Constant Algebras*, Chapter 69, pp. 2089–2097 in: John Cannon, Wieb Bosma (eds.), *Handbook of Magma Functions*, Version 2.11, Volume **6**, Sydney, 2004.
6. Roger W. Carter, *Simple groups of Lie type*, Pure and Applied Mathematics, Vol. **28**, London-New York-Sydney: John Wiley & Sons, 1972,
7. Arjeh M. Cohen, Scott H. Murray, D. E. Taylor, *Computing in groups of Lie type*, Math. Comp. **73** (2004), 1477–1498.
8. J. R. Faulkner, J. C. Ferrar, *Exceptional Lie algebras and related algebraic and geometric structures*, Bull. London Math. Soc. **9** (1977), 1–35.
9. James E. Humphreys, *Introduction to Lie Algebras and Representation Theory*, Graduate Texts in Mathematics, vol. **9**, New York Heidelberg Berlin:Springer-Verlag, 1972.
10. Adolf Hurwitz, *Über die Composition der quadratischen Formen von beliebig vielen Variablen*, Nachrichten Ges. Wiss. Göttingen (1898), 309–316.
11. Nathan Jacobson, *Composition algebras and their automorphisms*, Rend. Circ. Mat. Palermo **7** (1958), 55–80.
12. Nathan Jacobson, *Basic Algebra. I*, second ed., New York: W. H. Freeman and Company, 1985.
13. S. Murray, W. de Graaf, D. E. Taylor, *Lie Theory*, Chapters 82–91, pp. 2359–2553 in: John Cannon, Wieb Bosma (eds.), *Handbook of Magma Functions*, Version 2.11, Volume **6**, Sydney, 2004.
14. M. Rost, *On the dimension of a composition algebra*, Documenta Mathematica **1** (1996), 209–214.
15. L. J. Rylands, D. E. Taylor, *Constructions for octonion and exceptional Jordan algebras*, Designs, Codes and Cryptography **21** (2000), 191–203.
16. T. A. Springer, F. D. Veldkamp, *Octonions, Jordan Algebras and Exceptional Groups*, Berlin Heidelberg New York: Springer-Verlag, 2000.
17. F. van der Blij, *History of the octaves*, Simon Stevin **34** (1960/1961), 106–125.
18. F. van der Blij, T. A. Springer, *The arithmetics of octaves and of the group G_2*, Nederl. Akad. Wetensch. Proc. Ser. A **62** = Indag. Math. **21** (1959), 406–418.
19. F. van der Blij, T. A. Springer, *Octaves and triality*, Nieuw Arch. Wisk. (3) **8** (1960), 158–169.

Support varieties for modules

*Jon F. Carlson**

Department of Mathematics
University of Georgia
Athens GA, USA
jfc@math.uga.edu

Summary. The support variety of a module over a group algebra is an affine variety that encodes many of the homological properties of the module. Although the definition of the support variety is given in terms of the cohomology ring of the group, it can be computed directly from the actions of the elementary abelian subgroups. We present a Magma implementation that determines the support variety of a module in the case that the characteristic of the base field is 2. The construction is illustrated with examples that verify a theorem on the support varieties of truncations of the syzygy modules of the identity module of a group algebra.

1 Introduction

Support varieties for modules were introduced more than twenty years ago to describe modules over group algebras. In that context, the varieties are defined in terms of the maximal ideal spectrum of the group cohomology ring and they measure many of the homological properties of modules. For example, the support variety of a module detects whether the module is projective and under some circumstances can predict whether one module has extension with another module. The notions of support varieties have been extended to modules over restricted p-Lie algebras and infinitesimal group schemes for algebraic groups. In general, they can be applied to any finite group scheme. See, for example [9] or [10] for some details.

In the setting of modular group algebras of finite groups, there are two theorems that make the support variety an especially effective tool. First, the support variety of a module is the union of images under maps induced by the restriction of the support varieties of the module over the elementary abelian subgroups of the group. Consequently, in some sense, the support variety of a module can be computed by looking at the support varieties of the restriction of the module to elementary abelian subgroups. The second theorem says that over an elementary abelian p-group, the support variety of a module is

---
*This work was partly supported by a grant from NSF.

isomorphic (as a set) to a rank variety, a variety computed by looking at the ranks of certain matrices coming from the actions of the group elements on the module. The rank variety is what we actually compute.

In this paper we demonstrate a method for the computation of the support variety of a module over the group ring of an elementary abelian p-group. The method consists mainly of finding a collection of subalgebras of the group algebra of a very special type, that fail to act freely on the module. Each of these yields one or more points on the variety which is a Zariski closed subset of k^n for some field k. The variety is now determined by finding a collection of polynomials which are satisfied by all of the points. We end the paper with a couple of examples of modules obtained from cohomological calculations of extraspecial groups. The examples illustrate a theorem that was proved by the author and Dave Benson using very complicated spectral sequence methods.

At this point we should emphasize a few things. First of all, our methods involve finding points on a variety by a random process. It is always possible that the points will not be distributed sufficiently to give a correct answer. That is, this is a Monte Carlo algorithm and there is always the possibility that the computation does not return the correct answer. Secondly, we are not actually computing the support variety, but only the components of maximal dimension. Again we emphasize that the method is to collect random points on the variety. In the event that the variety has components of unequal dimensions, then the probability is that all of the selected points will lie on the components of maximal dimension and the method will not detect the smaller components.

Throughout the paper, G denotes a finite group, usually an elementary abelian p-group. Let k be a field of prime characteristic p. Although the theory of support varieties applies equally to finitely generated kG-modules without regard to the prime, the computational algorithms are easier in the case that $p = 2$. For that reason, our demonstration will assume that the characteristic p is even. The same general methods can be employed in the odd characteristic case, but there are several additional complications. All of the algorithms are implemented in Magma [4] and the displayed output* is from Magma sessions.

The paper is organized as follows. In the next section, we outline a few facts about projectivity of modules over local algebras. In Section 3 we introduce the varieties. We present some discussion of the cohomological support varieties, but this is mostly for motivational purposes. For understanding the computation, the most important parts are the definition of the rank variety and its properties. An outline of the basic algorithm is included in Section 4. In addition, we present the functions that find the points on the variety. This is the most complicated part of the algorithm. Section 5 discusses the analysis of the data and the functions that find the polynomial generators of the ideal of the variety. An application is developed in Section 6.

---

*See the Preface to this volume for style conventions regarding Magma code; code appearing in this book is available at http://magma.maths.usyd.edu.au/magma/.

2 Notes on projectivity

Our algorithms for support varieties depend very much on knowing when a module is projective. In the case that G is a p-group and k is a field of characteristic p there is an easy computational criterion for projectivity. A more general discussion of the algorithm is given in [5] in this volume. It is based on the idea that kG is a split local ring. Hence projective kG-modules are free. If M is a finitely generated kG-module, the algorithm checks that the dimension of $M/\operatorname{Rad} M$ is $(1/|G|)\operatorname{Dim} M$. The function that we write is the same as the one in [5].

```
IsProjective := function(M, B, n);
    V := VectorSpace(BaseRing(B[1]), Dimension(M));
    B := Basis(M);
    S := &+[ sub<V | RowSpace(x)> : x in B ];
    if (Dimension(M) − Dimension(S))*n eq Dimension(M) then
        return true;
    end if;
    return false;
end function;
```

Here M is a kG-module, B is a collection of nilpotent elements of kG that generate a subalgebra. More precisely, the function assumes that the elements of the sequence B are the action matrices of generators of the subalgebra on the space M. The final argument n is the dimension of the subalgebra generated by B.

A particular case of the above algorithm is written directly into the routines for computing support varieties. We present it as a lemma.

Lemma 2.1. *Suppose that $G = \langle x \rangle$ is a cyclic group of order 2. Assume that k has characteristic 2, and let M be a finitely generated kG-module on which x acts by the k-matrix A. Then M is projective as a kG-module if and only if the rank of $A - I$ is half the dimension of M, where I is the identity matrix.*

Group algebras are self-injective rings. That is, projective kG-modules are injective and *vice versa*. Consequently, if Q is a projective submodule of a finitely generated kG-module M, then $M \cong Q \oplus M/Q$. In the sections that follow we need a function that factors out any projective summands of a module. So the end result should be a module that has no projective submodules. The following function does exactly that. It is a Monte Carlo algorithm in that it depends on the fact that a randomly chosen sequence of elements should include an element that generates a projective submodule, if a projective submodule exists. It would be possible to write a deterministic algorithm for such an exercise, but what we have is faster and works very well in practice.

The algorithm begins by restricting the module M to the group G. This allows the function to be applied when G is only a subgroup of the group of M. Then we choose a sequence of random elements from M and test each element to see whether it generates a projective submodule of M. For this it is only necessary to check whether the dimension of the submodule generated by the element is equal to the order of G. If the submodule is projective then we factor it out and repeat the process. After a string of five failures to produce a projective submodule, we exit the loop and return the module that we have computed to that point.

```
LeftProjectiveStrip := function(M, G);
    MM := Restriction(M, G);
    repeat
        flag := true;
        for i := 1 to 5 do
            a := Random(MM);
            SM := sub<MM | a>;
            if Dimension(SM) eq #G then
                MM := quo<MM | SM>;
                flag := false;
                break;
            end if;
        end for;
    until flag;
    return MM;
end function;
```

3 Support varieties and rank varieties

In this section we present a brief overview of the theory of support varieties for group algebras. The material on cohomological varieties is mainly intended to be motivational. For the computational methods that we develop, the important material in this section is the discussion of the rank variety. See any of the books [1], [7] or [8] for references.

Suppose that G is a finite group and that k is an algebraically closed field of characteristic p. The assumption of algebraic closure is important for the geometry. In practice, a sufficiently large field extension will suffice. The cohomology ring $H^*(G, k)$ is a finitely generated graded-commutative k-algebra. As such, it has a maximal ideal spectrum $V_G(k)$ which is a homogeneous affine variety. An example is the situation in which $G \cong (\mathbb{Z}/2\mathbb{Z})^n$ is an elementary abelian 2-group and $p = 2$. Then $H^*(G, k) \cong k[X_1, \ldots, X_n]$, a polynomial ring in n variables. For p odd, the quotient of the cohomology ring of an elementary abelian group of order p^n by its radical is a polynomial ring in n variables. Thus the maximal ideal spectrum of $H^*(G, k)$ is the affine variety $V_G(k) = k^n$, when G is an elementary abelian group of order p^n.

When M is a finitely generated kG-module, then the cohomology ring $\operatorname{Ext}^*_{kG}(M, M)$ is a finitely generated module over $H^*(G, k)$. So let $J(M)$ denote the annihilator of $\text{Ext}^*_{kG}(M, M)$ in $\text{H}^*(G, k)$, and let $V_G(M) = V_G(J(M))$, the closed set of all maximal ideals that contain $J(M)$. Then $V_G(M)$ is a homogeneous subvariety of $V_G(k)$ that carries much information about the module M. One property is that M is projective if and only if $V_G(M) = \{0\}$. Some more properties are listed in the paper [5] in this volume.

If H is a subgroup of G then the restriction map $\text{H}^*(G, k) \longrightarrow \text{H}^*(H, k)$ induces maps on varieties $\text{res}^*_{G,H} : V_H(k) \longrightarrow V_G(k)$ and $\text{res}^*_{G,H} : V_H(M) \longrightarrow V_G(M)$. It is important for many purposes to know that

$$V_G(M) \cong \bigcup_{E \in \mathcal{E}\mathcal{A}} \text{res}^*_{G,E}(V_E(M))$$

where the union is over the set $\mathcal{E}\mathcal{A}$ of elementary abelian p-subgroups of G (see Theorem 9.4.2 of [7]). This result was first proved independently by Alperin and Evens and by Avrunin. It is a generalization of Quillen's Dimension Theorem (see Theorem 8.6.4 of [7]). In some sense, it reduces the computational problem of finding the support variety over G to that of finding the support variety over all elementary abelian subgroups E.

So suppose that $G = \langle x_1, \dots, x_n \rangle \cong (\mathbb{Z}/p)^n$ is an elementary abelian p-group. We define another variety for a kG-module M. For $\alpha = (\alpha_1, \dots, \alpha_n) \in k^n$ let $u_\alpha = \sum_{i=1}^n \alpha_i(x_i - 1)$. Note that u_α is a unit in the group algebra kG, and because the characteristic of the field is p, $u_\alpha^p = 1$. We call the subgroup $U_\alpha = \langle u_\alpha \rangle$ of $kG^\times$ (the group of units of kG) generated by u_α, a cyclic shifted subgroup. A shifted subgroup is subgroup of the group of units of kG generated by $u_\alpha, u_\beta, \dots$ where $\alpha, \beta, \dots$ are linearly independent in k^n. If W is a shifted subgroup, then kG is a free kW-module where the action is induced by the embedding of kW into kG.

Definition 3.1 *Suppose that M is a finitely generated kG-module. The rank variety of M is the set*

$$V^r_G(M) = \{\alpha | M{\downarrow}_{U_\alpha} \text{ is not free}\} \cup \{0\}.$$

Here $M{\downarrow}_{U_\alpha}$ is the restriction of M to U_α. The origin $\{0\}$ must be included to insure that we have a homogeneous variety. Indeed, it is a theorem that $V^r_G(M)$, the collection of all cyclic shifted subgroups that fail to act freely on M, is a subvariety of k^n (a Zariski closed subset). Moreover, we have the following theorem proved by Avrunin and Scott.

Theorem 3.2 *(See Theorem 9.5.5 of [7]) If G is an elementary abelian p-group and k is an algebraically closed field of characteristic p, then*

$$V^r_G(M) \cong V_G(M).$$

If $p > 2$, then the connection between the varieties is not really an isomorphism but only an isogeny. That is, the map between the varieties involves a twist by the Frobenius homomorphism. It is injective and surjective as a map of sets, but it does not have an inverse that is a polynomial map.

The proof of the theorem is reasonably straightforward. But perhaps it is best illustrated by the following example. The example also points toward our algorithm.

Assume that $G = \langle x_1, x_2, x_3 \rangle$ is an elementary abelian group of order 8 and that $p = 2$. Fix three elements β, γ and θ in k. Then define M to be the kG module of dimension 4 on which (for some choice of basis) the generators of G act by the matrices

$$
\rho_{x_1} = \begin{pmatrix} 1 & 0 & 1 & 0 \\ 0 & 1 & 0 & 1 \\ 0 & 0 & 1 & 0 \\ 0 & 0 & 0 & 1 \end{pmatrix}, \quad
\rho_{x_2} = \begin{pmatrix} 1 & 0 & \beta & 0 \\ 0 & 1 & 0 & \gamma \\ 0 & 0 & 1 & 0 \\ 0 & 0 & 0 & 1 \end{pmatrix}, \quad
\rho_{x_3} = \begin{pmatrix} 1 & 0 & 0 & \theta \\ 0 & 1 & 1 & 0 \\ 0 & 0 & 1 & 0 \\ 0 & 0 & 0 & 1 \end{pmatrix}
$$

Then for $\alpha = (\alpha_1, \alpha_2, \alpha_3) \in k^3$ we have that

$$
\rho_{u_\alpha} = \begin{pmatrix} 1 & 0 & \alpha_1 + \alpha_2\beta & \alpha_3\theta \\ 0 & 1 & \alpha_3 & \alpha_1 + \alpha_2\gamma \\ 0 & 0 & 1 & 0 \\ 0 & 0 & 0 & 1 \end{pmatrix}
$$

It is easy to check that the matrices of the action commute and that the square of any ρ_{x_i} is the identity. So this really is a kG-module. Now by Lemma 2.1 and the definition, $\alpha \in V_G^r(M)$ if and only if the rank of $u_\alpha - \mathrm{Id}$ is less than 2, or the determinant of the upper right corner of ρ_{u_α} is zero. That is, $V_G^r(M)$ is the set of all solutions (in $(\alpha_1, \alpha_2, \alpha_3)$) of the equation

$$
(\alpha_1 + \beta\alpha_2)(\alpha_1 + \gamma\alpha_2) - \theta\alpha_3^2 = 0.
$$

Thus the equation that determines the variety is the determinant of a matrix in the variables $\alpha_1, \alpha_2, \alpha_3$. This is the situation in general.

It should be noted that the rank variety is independent of the choice of the generators $x_1, \ldots, x_n$ of the group. In particular, for any $\alpha \in k^n$ and any $\omega \in \mathrm{Rad}^2(kG)$, the unit u_α acts freely on M if and only if the unit $u_\alpha + \omega$ acts freely on M.

4 Finding points on the variety

In this section, we outline our algorithm for finding the variety $V_G^r(M)$ of a finitely generated kG-module, in the case that G is an elementary abelian 2-group. It was stated in the introduction, and we emphasize it again, that we are actually only going to find the components of maximal dimension in $V_G^r(M)$. Also the algorithm involves making random choices and could fail if one were very unlucky. For the algorithm, we assume that the field k is finite. Algebraic closure is necessary for the application of Theorem 3.2: however in practice, we approximate algebraic closure by taking large extensions of the field k to compute the variety.

The principle on which the algorithm is based, is Bezout's theorem. This is a theorem about the intersection of algebraic varieties. It tells us, for example that quadratic curve and a line in 2-space intersect (generically) in two points. Of course, its statement is about varieties in projective space. However, it can be usefully interpreted in the context of homogeneous varieties in the affine space k^n. The reader who is interested in the technicalities is referred to any of the standard texts on algebraic geometry. Another good reference is the paper [3]. For our purposes, we need the following two statements. Assume that V is a homogeneous variety in k^n.

B_1 Suppose that $W = \langle \alpha(1), \ldots, \alpha(t) \rangle$ is a linear subspace in k^n such that $W \cap V = \{0\}$. Then the dimension of V is at most $n - t$.

B_2 Let the dimension of V be s. Suppose that for every linear subspace W such that W intersects V transversely and W has dimension $n - s + 1$, the intersection $W \cap V$ consists of at most d lines. Then the degree of V is at most d.

Statement B_2 is really the classical definition of the degree of the variety. What we use is the fact that the degree of the variety is larger than the maximum of the degrees of the polynomials that define the variety.

The first two steps in our algorithm are the following. We are interested in finding points on $V = V_G^r(M)$.

$\mathfrak{A}_1$ Beginning with $r = 1$, we test to see if there is a linear subspace W of dimension r in k^n such that $W \cap V = \{0\}$. This is done by choosing a basis $\alpha(1), \ldots, \alpha(r)$ for W, forming the shifted subgroup $U = \langle u_{\alpha(1)}, \ldots, u_{\alpha(r)} \rangle$ and testing to see if M_U, the restriction of M to U is projective. The function described in Section 2 is used for the last step. If such a W with $W \cap V = \{0\}$ is found, then replace r by $r + 1$ and test again. When after a reasonable number of tests, we have not found such a space W, then we know with high probability that the dimension of V is $n - r + 1$.

$\mathfrak{A}_2$ For a large number of subspaces W of dimension r as found above, record the intersections. From this information, we will extract the collection of points on the variety V.

The method of finding the points in $\mathfrak{A}_2$ requires some explanation. The point is that when $W \cap V$ has dimension 1, then it is easy to find $\alpha \in W$ such that α is in the rank variety. As an illustration, consider the example at the end of the last section. Suppose that the basis for W, $\alpha(1)$, $\alpha(2)$ is chosen so that $\alpha(1) = (1, 0, 0)$. Then $u_{\alpha(1)} = x_1$. Then we can rearrange the basis for M so the matrix of $u_{\alpha(2)}$ has the form

$$\begin{pmatrix} I_2 & A \\ 0 & I_2 \end{pmatrix}$$

where I_2 is the 2×2 identity matrix, and A is some 2×2 matrix. Then $\gamma\alpha(1) - \alpha(2)$ is in $V = V_G^r(M)$ for any eigenvalue γ of A.

The method we use for larger groups is a generalization of the above. Assume that $H = \langle u_{\alpha(1)}, \ldots, u_{\alpha(r-1)} \rangle$ and that M is free as a kH-module. Then let $L = M/\operatorname{Rad}^2(kH)M$. We can find a basis for L so that the actions of the generators of H on L are given by

$$u_{\alpha(1)} \leftrightarrow \begin{pmatrix} I & I & 0 & \ldots & 0 \\ & I & 0 & & 0 \\ & & I & & 0 \\ & & & & \\ & & & & I \end{pmatrix}, u_{\alpha(2)} \leftrightarrow \begin{pmatrix} I & 0 & I & \ldots & 0 \\ & I & 0 & & 0 \\ & & I & & 0 \\ & & & & \\ & & & & I \end{pmatrix}, \ldots, u_{\alpha(r-1)} \leftrightarrow \begin{pmatrix} I & 0 & 0 & \ldots & I \\ & I & 0 & & 0 \\ & & I & & 0 \\ & & & & \\ & & & & I \end{pmatrix}$$

Then the matrix of $u_{\alpha(r)}$ has the form

$$u_{\alpha(r)} \leftrightarrow \begin{pmatrix} I & A_1 & A_2 & \ldots & A_{r-1} \\ & I & 0 & & 0 \\ & & I & & 0 \\ & & & & \\ & & & & I \end{pmatrix}$$

where the I is the $t \times t$ identity matrix for $t = \operatorname{Dim}(M/\operatorname{Rad}(kH)M) = 1/|H| \operatorname{Dim}(M)$. Because the square of the matrix of $u_{\alpha(r)}$ on M is the identity it can be easily shown that the matrices $A_1, A_2, \ldots, A_{r-1}$ commute with each other. Then we have that $\alpha = \gamma_1 \alpha(1) + \cdots + \gamma_{r-1} \alpha(r-1) - \alpha(r)$ is in V whenever $\gamma_1, \ldots, \gamma_{r-1}$ is a sequence of simultaneous eigenvalues for $A_1, \ldots, A_{r-1}$. This follows from the fact that $u_\alpha - 1$ multiplies some generator of M into $\operatorname{Rad}^2(kH)M$. In practice we find all eigenvalues of each A_i and sort out those that are simultaneous eigenvalues by trial and error.

We now present the Magma code for producing points on the variety. The first function computes only the dimension of the variety and raw data for points on the variety. Here G is the elementary abelian 2-group. It may be only a subgroup of the group of M. The integer n is the number of points on the variety that we want to compute. This would depend on the size of the module M which would in turn affect the degree of the variety.

```
DimensionOfVariety := function(M, G, num)
    φ := Representation(M);
    n := Ngens(G);
    dim := Dimension(M);
    ff := ext< CoefficientRing(M) | 6 >;
```

The field *ff* is our approximation to an algebraically closed field. The guess is that an extension of degree 6 should be sufficient to generate enough points on the variety V to make the determination.

```
    NM := ExtendField(M, ff);
    V := VectorSpace(ff, n);
    MA := MatrixAlgebra(ff, dim);
    newmatlist := [MA ! Representation(NM)(G.i)−Id(MA): i in[1..n]];
```

This is the list of actions of the elements $(x_i - 1)$.

```
for r := 1 to n do
    g := AbelianGroup(GrpPerm, [2: i in [1..r]]);
    RRL := [ ];
    flag := true;
    for j := 1 to num do
        a := [Random(V): i in [1..r]];
```

We are selecting $\alpha(1), \ldots, \alpha(r)$ in k^n. Next we get the action of $u_{\alpha(i)}$ for each i and the action of $U = \langle u_{\alpha(i)}, \ldots, u_{\alpha(r)} \rangle$ on M.

```
        mmmlist := [ Id(MA) + &+[ a[i][j]*newmatlist:
                           j in [1..n] ] : i in [1..r] ];
        mmm := GModule(g, mmmlist);
        str := LeftProjectiveStrip(mmm, g);
```

Projective summands tell us nothing about the variety, and so they can be factored out.

```
        if Dimension(str) eq 0 and r lt n then
            flag := false;
            break j;
        end if;
```

The above conditional statement checks to see if the restriction of M to U is projective. If so, then we replace r by $r + 1$ and start again.

```
        if Dimension(str) eq 0 and r eq n then
            return g, [<a, str>], newmatlist;
        end if;
```

When this conditional is true, then the module M is projective over G, and we know that the variety is $V_G^r(M) = \{0\}$. If it is not true, then we know that M restricted to U is not projective, and we can store the data that we have collected.

```
        RRL := Append(RRL, <a, str>);
        if #RRL eq 8 and r eq 1 then
            return g, RRL, newmatlist;
        end if;
```

If the restriction of M to a shifted cyclic subgroup is never free then the dimension of V is equal to n. In this case it must be the case that $V = k^n$. It should be that eight tests will confirm this fact.

```
    end for;
    if flag then
        return g, RRL, newmatlist;
    end if;
end for;
end function;
```

At this point we have completed step $\mathfrak{A}_1$ of our algorithm. The next function extracts the matrices $A_1, \ldots, A_{r-1}$. It is a matter of factoring out $\mathrm{Rad}^2(kH)M$ and then making the appropriate base change.

```
StandardForm := function(M);
  r := Ngens(Group(M));
  id := MatrixAlgebra(BaseRing(M), Dimension(M)) ! 1;
  LL := [ ActionGenerator(M, i) − id : i in [1..r] ];
  W := sub<M | &cat[ &cat[ Basis(RowSpace(LL[i]*LL[j])) :
                      j in [1..i] ]: i in [1..r−1] ]>;
  MM := quo<M | W>;
  idd := MatrixAlgebra(BaseRing(MM), Dimension(MM)) ! 1;
  Rad := sub<MM | &cat[ Basis(RowSpace(
                  ActionGenerator(MM, i) − idd)): i in [1..r−1]]>;
  Gen := [ MM ! ExtendBasis(Rad, MM)[i]:
                  i in [Dimension(Rad)+1..Dimension(MM)] ];
  Bas := Gen cat &cat[ [ x*ActionGenerator(MM, i) − x :
                  x in Gen]: i in [1..r−1] ];
  b := Matrix(Dimension(MM), Dimension(MM), [ Vector(x): x in Bas]);
  NL := [ b*ActionGenerator(MM, i)*b⁻¹: i in [1..r] ];
  return NL;
end function;
```

The next function uses the output of the **DimensionOfVariety** function to compute the points on the variety.

```
PointsOnVariety := function(M);
  G := Group(M);
```

We are assuming that the elementary abelian group G is the group of M.

```
  PL := [ ];
  BL := [ ];
  Npoints := 2^Ngens(G)*(Dimension(M) div #G +1) + 20;
  GG, vl, ACT := DimensionOfVariety(M, G, Npoints);
  r := Ngens(GG);
  if r eq 1 then
```

This is the case in which $V = V_G(k)$.

```
    return [ ], 0, 0;
  end if;
  ff := CoefficientRing(vl[1][2]);
  if Dimension(vl[1][2]) eq 0 then
```

This is the case in which M is projective.

```
    return [0], Ngens(G), 0;
  end if;
  mini := Minimum([ Dimension(x[2]) : x in vl ]);
```

We assume that the nonprojective part of of the restriction to $\langle u_{\alpha(1)}, \ldots, u_{\alpha(r)} \rangle$ is as small as possible. We should be able to get more than enough points in this case.

$w := \text{mini } \textbf{div } (2^{(r-1)})$;

Here w is the dimension of the blocks A_i. Now we extract the blocks and compute their minimal polynomials.

```
for x in vl do
   mm := x[2];
   if Dimension(mm) eq mini then
      bool := IsProjective(mm, [ ActionGenerator(mm, i)
             − MatrixAlgebra(ff, mini) ! 1 : i in [1..r−1] ], 2^(r−1));
      if bool then
         SF := StandardForm(mm);
         blocks := [ Submatrix(SF[r], 1, w*i+1, w, w) : i in [1..r−1]];
         minpolys := [ MinimalPolynomial(y) : y in blocks ];
         Append(∼PL, <x[1], minpolys>);
         Append(∼BL, blocks);
      end if;
   end if;
end for;
```

To get the eigenvalues of the blocks $A_1, \ldots, A_{r-1}$ we must extend the field and split the minimal polynomials.

```
degrees := [ Degree( Factorization(PL[i][2][j])[k][1] ) : k in
         [1..#Factorization(PL[i][2][j])] ], j in [1..#PL[i][2]], i in [1..#PL]];
xd := LCM(degrees);
if xd eq 1 then
```

In this case no field extension is necessary to split the minimal polynomials.

```
RL := [ < PL[i][1], [Roots(PL[i][2][j]) :
          j in [1..#PL[i][2]]]>: i in [1..#PL]];
NPL := PL;
fff := ff;
else
```

Otherwise we take the field extension fff and coerce everything into the appropriate algebra over the field extension.

```
NPL:= [ ];
RL := [ ];
fff, γ := ext<ff | xd>;
VV := VectorSpace(fff, Ngens(G));
Q := PolynomialRing(fff);
for i := 1 to #PL do
   npl := <[ VV ! a : a in PL[i][1]], [Q ! f:f in PL[i][2]] >;
```

```
            Append(~NPL, npl);
            rl := <npl[1], [Roots(y): y in npl[2]]>;
            Append(~RL, rl);
        end for;
    end if;
```

Next we collect the roots of the minimal polynomials; that is, the eigenvalues
of the matrices A_i.

```
    VL₁ := [ ];
    for w in RL do
        rt := w[2];
        rl := [[rt[j][k][1]:k in [1..#rt[j]]]: j in [1..#rt]];
        if #rl eq 1 then
            RL₁ := [[x, 1] : x in rl[1]];
        else
            RL₁ := [[x] : x in rl[1]];
            for i := 2 to #rl do
                RL₁ := [ Append(x, y) : x in RL₁, y in rl[i] ];
            end for;
            RL₁ := [ Append(x, 1):x in RL₁ ];
        end if;
```

Each of the sequences RL_1 is a list of possible points on the variety. Each entry
is an n-tuple $[\alpha_1, \ldots, \alpha_{r-1}]$ where each α_i is an eigenvalue of the matrix A_i.
Now we want to interpret each as a cyclic shifted cyclic subgroup and see if
it gives a point on the variety. Note RL_2 is a list of elements in k^n that are
possible variety points. VL_1 is the entire collection and VL is the subset of the
list consisting of the points on V.

```
        RL₂ := [ &+[w[1][j]*RL₁[k][j] : j in [1..r]] : k in [1..#RL₁] ];
        VL₁ := VL₁ cat RL₂;
    end for;
    VL := [ ]; m := Nrows(ACT[1]);
    if xd gt 1 then
        NACT := [ MatrixAlgebra(fff, m) ! x : x in ACT ];
    else
        NACT := ACT;
    end if;
    for w in VL₁ do
        MAT := &+[w[i]*NACT[i]: i in [1..Ngens(G)]];
        if 2*Rank(MAT) lt m then
            Append(~VL, w);
        end if;
    end for;
    return VL, r−1, w;
end function;
```

The function returns the list VL of points on the variety as well as $r - 1$ (the codimension of V in k^n) and w which is an upper bound on the degree of the variety.

5 Computing the variety from a set of points

At this point we have computed a collection of points $\beta_1, \ldots, \beta_m$ on the variety V that should be sufficiently large to determine V. We should be clear what we mean by this. Perhaps an example illustrates it best.

Suppose that $n = 3$ and that V is the zero set of the set $\{YX, ZX\}$ of polynomials in $k[Z, Y, X]$. Then V has two components, the plane $X = 0$ and the line $Z = 0 = Y$. The dimension of V is two and we have found points on V by intersecting V with randomly chosen planes $\gamma_1 Z + \gamma_2 Y + \gamma_3 X = 0$. In all probability, none of the random planes will include the line $Z = 0 = Y$ and consequently all of the points will have the form $\beta_i = (\beta_{i1}, \beta_{i2}, 0)$.

The next step in the algorithm is the following.

$\mathfrak{A}_3$ For $i = 1, \ldots, d$, find in $F[X_1, \ldots, X_n]$ all homogeneous polynomials of degree i that are satisfied by all of the points that we have computed. The method is to list all monomials $f_1, \ldots, f_t$ of degree i and then to find all of the solutions in $\mu_1, \ldots, \mu_t$ of the equations $\sum_{i=1}^{t} \mu_i f_i(\beta_j) = 0$ for all j.

In the example, the algorithm would find the polynomial $X = 0$ in degree 1, if no point on the line $Z = 0 = Y$ were in the list of points. If a nonzero point $(0, 0, \beta_{i3})$ were in the list then only the polynomials ZX and ZY would be returned in degree 2.

Of course, it is important that the number of points be much larger than the maximum degree d of homogeneous polynomials that we compute. Otherwise, incorrect results are guaranteed. In the example, if m, the number of computed points, is less than d, then the polynomial $\prod_{i=1}^{m}(\beta_{i1}Y - \beta_{i2}Z)$ is satisfied by all of the points. But this polynomial is not in the ideal of the variety.

Now we present the function that computes the homogeneous polynomials. Here P is the polynomial ring, *pts* is the collection of points on the variety and *deg* is the degree in which we are searching for homogeneous polynomials.

```
VarietyPolynomials := function(P, pts, deg)
    V := Parent(pts[1]);
    F := CoefficientField(V);
    Mons := MonomialsOfDegree(P, deg);
    M := RMatrixSpace(F, #Mons, #pts);
    φ := [hom< P → F | ElementToSequence(pts[i]) >:
                                i in [1..#pts]];
    A := M ! &cat[ [ φ[i] (Mons[j]): i in [1..#pts], j in [1..#Mons] ] ];
    _, Null := Solution(A, Zero(VectorSpace(F, #pts)));
```

```
        return [ &+[ Basis(Null)[j][i] * Mons[i] :
                  i in [1.. #Mons] ]: j in [1.. Dimension(Null)] ];
    end function;
```

Finally, we have the function that computes the support variety.

```
    SupportVariety := function(M);
      G := Group(M);
      MM := LeftProjectiveStrip(M, G);
      if Dimension(MM) eq 0 then
```

In this case M is a projective module.

```
      P := PolynomialRing(CoefficientRing(M), Ngens(G));
      return [P.i: i in [1.. Rank(P)]], P;
    end if;
    pts, codim, deg := PointsOnVariety(MM);
    if #pts eq 0 then
```

In this case the variety is all of k^n.

```
      P := PolynomialRing(CoefficientRing(M), Ngens(G));
      return [ ], P;
    end if;
    s := Ngens(G);
    ff := BaseRing(pts[1]);
    P := PolynomialRing(ff, s);
    POLS := &cat[ VarietyPolynomials(P, pts, t) : t in [1.. deg] ];
```

At this point, we have a set of generators for the ideal of the variety. We want a minimal set of generators. To get the minimal set we take the Gröbner basis for the ideal and discard redundant elements of that basis.

```
    I := ideal< P | POLS >;
    Groebner(I);
    B := Basis(I);
    dd := [ TotalDegree(B[j]):j in [1..#B] ];
    flag := true;
    MinGens := [ ];
    for i := 1 to #Basis(I) do
      a, b := Minimum(dd);
      if flag then
        II := ideal< P | MinGens >;
      end if;
      if B[b] notin II then
        Append( ~MinGens, B[b] );
        flag := true;
      else
        flag := false;
      end if;
```

```
    Remove(~B, b);
    Remove(~dd, b);
  end for;
```

Finally, to make the output easy to read, we assign symbols to the variables
of the polynomial ring.

```
    variables := ["z","y","x","w","v","u","t","s"];
    AssignNames(~P, [ variables[i]: i in [1.. Rank(P)] ]);
    return MinGens, P;
  end function;
```

6 Varieties of truncated syzygy modules

We end the chapter with some examples that illustrate a theorem of the author
and Dave Benson. The theorem concerns groups that are similar in presentation to extraspecial groups. It was used in the original paper [2] to prove
a result about the periods of periodic modules. Most recently, the theorem
has been crucial in the characterization of endotrivial modules (see CT[1] for
background). A p-group is extraspecial if its center, commutator subgroup
and Frattini subgroup coincide and are cyclic of order p. Examples, and the
ones that we consider, are central products of dihedral groups. Suppose that
$H_1, \ldots, H_s$ are dihedral groups of order 8. Let z_i be the unique nonidentity
element in the center of H_i. Then the group

$$E_s = H_1 \times H_2 \times \cdots \times H_s/\langle z_i z_j | 1 \le i < j \le s\rangle$$

is an extraspecial group of order 2^{2s+1}. Let $H = \langle z\rangle$ denote the center of E_s
which has order 2.

Let k denote the trivial kG-module of dimension 1. If F_0 is a projective
cover of k, then let $\Omega(k)$ denote the kernel of the surjection $F \longrightarrow k$. Induc-
tively, let $\Omega^t(k)$ denote the kernel of the surjection $F_{t-1} \longrightarrow \Omega^{t-1}(k)$ where
F_{t-1} is a projective cover of $\Omega^{t-1}(k)$.

In the case of E_s, the theorem [2] says the following. Let $m = 2^t$ for
$1 \le t \le s$. Let $M = (z-1)\Omega^m(k)$. Then the module M can be regarded
as a module over $G = E_s/H$ which is an elementary abelian group of order
2^{2s}. Then the variety of M as a kG-module is the variety of the kernel J_t of
the partial inflation up to the $(m+1)^{st}$ page of the Lyndon–Hochschild–Serre
spectral sequence

$$E_2^{r,s} = \mathrm{H}^r(E_s/H, \mathrm{H}^s(H,k)) \Rightarrow \mathrm{H}^{r+s}(E_s,k).$$

That is, the cohomology ring $\mathrm{H}^*(G,k)$ inflates injectively onto the bottom
row of the E_2 page of the spectral sequence and has a homomorphic image
into subsequent pages. The variety of M is the variety of the kernel of the
homomorphism into the $(m+1)^{st}$ page.

Without getting technical, we should just note that the spectral sequence is very well understood due to work of Quillen [11]. The cohomology ring of G is $H^*(G,k) \cong k[X_1, \ldots, X_{2s}]$. The partial inflation at the E_3 page has ideal generated by $f_2 = X_1X_2 + X_3X_4 + \cdots + X_{2s-1}X_{2s}$. At the E_5 page it is generated by f_2 and

$$f_3 = \sum_{i=1}^{s} X_{2i-1}^2 X_{2i} - X_{2i-1}X_{2i}^2.$$

At the E_9 it is generated by f_2, f_3 and

$$f_5 = \sum_{i=1}^{s} X_{2i-1}^4 X_{2i} - X_{2i-1}X_{2i}^4.$$

This is what we verify in the exercise.

The first thing that we need is a function for the computation of the syzygy modules $\Omega^j(k)$. The following works, though there are better functions that can be used in the context of basic algebras. Here M is the module whose syzygy $\Omega(M)$ we want to find. The argument FM is the free module isomorphic to kG.

```
SyzygyModule := function(M, FM);
    r := Dimension(M)−Dimension(JacobsonRadical(M));
    F, _, prj := DirectSum([FM: i in [1..r]]);
    Hom := GHom(FM, M);
    flag := true;
    while flag do
        If := [Random(Hom): i in [1..r]];
        f := &+[ MapToMatrix(prj[i])*If[i] : i in [1..r] ];
        if Rank(f) eq Dimension(M) then
            flag := false;
        end if;
    end while;
    OM := Kernel(f);
    return OM;
end function;
```

Next we compute the syzygy modules $\Omega^i(k)$ for $i = 1, \ldots, 4$. Here $E = E_2$ is an extraspecial group of order 32.

```
>    E := ExtraSpecialGroup(2, 2);
>    S := TrivialModule(E, GF(2));
>    kE := PermutationModule(E, sub<E | >, GF(2));
>    O_1 := SyzygyModule(S, kE);
>    O_2 := SyzygyModule(O_1, kE);
>    O_3 := SyzygyModule(O_2, kE);
>    O_4 := SyzygyModule(O_3, kE);
```

We first look at the truncation of $\Omega^2(k)$.

```
>    Z := ActionGenerator(O_2, 5)−Representation(O_2)(E ! 1);
>    N := sub<O_2 | RowSpace(Z)>;
```

So $N \cong (z - 1)\Omega^2(k)$, but as a kE_2-module. We must make it into a kG-module.

```
>    G := AbelianGroup(GrpPerm, [2,2,2,2]);
>    L_2 := GModule(G, [ ActionGenerator(N, i) : i in [1..4] ]);
>    var_2 := SupportVariety(L_2);
>    var_2;

         [ z*y + x*w ]
```

This is what we expect. Now consider the truncation of $\Omega^4(k)$.

```
>    Z_4 := ActionGenerator(O_4, 5)−Representation(O_4)(E ! 1);
>    N_4 := sub<O_4 | RowSpace(Z_4)>;
>    L_4 := GModule(G, [ActionGenerator(N_4, i): i in [1..4]]);
>    time var_4 := SupportVariety(L_4);

         Time: 109.710

>    var_4;

         [
              z*y + x*w,
              z*x*w + y*x*w + x^2*w + x*w^2
         ]
```

Notice that the second polynomial is $(z + y)f_2 + f_3$. So this too is what we expect. The theorem is verified in these cases.

References

1. D. J. Benson, *Representations and Cohomology* II: *Cohomology of Groups and Modules*, Cambridge Studies in Advanced Mathematics **31**, Cambridge: Cambridge University Press, 1991.
2. D. J. Benson, J. F. Carlson, *Periodic modules with large period*, Quart. J. Math. **43** (1992), 283–296.
3. E. Boda, W. Vogel, *On system of parameters, local intersection multiplicity and Bezout's theorem*, Proc. Amer. Math. Soc. **75** (1980), 1–7.
4. Wieb Bosma, John Cannon, Catherine Playoust, *The Magma algebra system I: The user language*, J. Symbolic Comput. **24** (1997), 235–265.
   See also the Magma home page at http://magma.maths.usyd.edu.au/magma/.
5. J. F. Carlson, *When is projectivity detected on subalgebras?*, pp. 205–220 in this volume.
6. J. F. Carlson, J. Thévenaz, *Torsion endo-trivial modules*, Algebras and Rep. Theory **3** (2000), 303–335.
7. J. F. Carlson, L. Townsley, L. Valero-Elizondo, M. Zhang, *The cohomology rings of finite groups*, Dordrecht: Kluwer Academic Publishers, 2003.

8. L. Evens, *The Cohomology of Groups*, New York: Oxford University Press, 1991.
9. E. M. Friedlander, B. Parshall, *Support varieties for restricted Lie algebras*, Invent. Math. **86** (1986), 553–562.
10. E. Friedlander, J. Pevtsova, *Representation-theoretic support spaces for finite group schemes*, preprint.
11. D. Quillen, *The mod 2 cohomology of extra-special 2-groups and the spinor groups*, Math. Ann. **194** (1971), 197–212.

When is projectivity detected on subalgebras?

*Jon F. Carlson**

Department of Mathematics
University of Georgia
Athens GA, USA
jfc@math.uga.edu

Summary. A well known theorem of Chouinard states that if G is a finite group and k is a field of characteristic $p > 0$, then a kG-module is projective precisely when its restrictions to all elementary abelian p-subgroups of G are projective. We investigate some similar situations in which the restrictions to subalgebras detect the projectivity of a module over an algebra. The examples played a crucial role in the classification of torsion endotrivial modules by the author and Jacques Thévenaz.

1 Introduction

Modular representation theory of finite groups is founded on the principle that many aspects of the representations are determined locally, *i.e.* at the level of the p-subgroups, or at worst, at the level of the normalizers of the p-subgroups. A perfect example of this is given by the theorem of Chouinard which says that if M is a kG-module for G a finite group and k a field of characteristic $p > 0$, then M is projective if and only if the restriction of M to every elementary abelian p-subgroup of G is projective. In other words, the subalgebras of kG generated by the elements of the elementary abelian p-subgroups of G detect the projectivity of kG-modules on restriction. The question that we investigate in this paper is, when does this happen for other algebras? That is, given an algebra A, does there exist a collection of subalgebras $\{A_i\}$ such that an A-module M is projective if and only if it is projective on restriction to every A_i.

Our particular interest is in quotient algebras of the group algebras of certain p-groups. We will produce by some experimentation, examples where there exist families of subalgebras that detect projectivity. In contrast, we present another similar example where the same sort of family of subalgebras does not detect the projectivity. These examples were important in the formation of a strategy to solve the 25 year old problem of classifying the torsion endotrivial modules for a finite p-group [4, 5].

---

*This work was partly supported by a grant from NSF.

In general, determining whether a module over an algebra is projective can be a difficult problem. However, if the algebra A over the field k is split local, *i.e.* if the radical of A has codimension 1 in A, then there is a straightforward criterion for projectivity that we can easily program into the computer. In the first section we show how this is done. This function for determining projectivity is useful in the examples.

Chouinard's theorem is actually an application to modules of the methods of Quillen's Dimension Theorem [9] (see [3] for an overview of the applications to modules). The ideas depend in a fundamental way on results concerning group cohomology. In particular, a module M over the group algebra of a p-group is projective if and only if the cohomology $\mathrm{H}^n(G, M) = \{0\}$ for all $n > 0$. Similar statements can be made concerning any split local self-injective algebra. So in some sense, it is the case that a collection of subalgebras detects projectivity of modules if and only if that collection also detects the cohomology $\mathrm{H}^*(G, k) \cong \mathrm{Ext}^*_{kG}(k, k)$ except for nilpotent elements. It is this fact, that the cohomology ring is detected by the subalgebras, that we actually demonstrate in the examples. We also show how the failure of the subalgebras to detect cohomology leads to the construction of an example of a nonprojective module that is projective on restriction to the subalgebras. We give further explanation of the connection in Section 3 of the paper.

All of the computational functions that we use are written in the Magma language and run on the Magma platform [2]. For the purposes of the calculation, a basic algebra over a field k is a finite dimensional algebra whose simple modules all have dimension one. In the literature such an algebra would be called a split basic algebra. The type AlgBas is a type of algebra that is optimized for homological applications such as the construction of complexes and projective resolutions and computations of homology and cohomology.

Note that to be consistent with operations in Magma, our modules are right modules and maps are written on the right. In particular, if $f : U \longrightarrow V$ and $g : V \longrightarrow W$ are maps, then the composition is denoted $f \circ g$.

2 Criterion for projectivity

Suppose that k is a field and that A is a split local algebra over k. Then A has a unique maximal ideal which is the radical $\mathrm{Rad}\, A$ of A and $A/\mathrm{Rad}\, A \cong k$. It is well known that a module over such a local ring is projective if and only if it is free. Consequently, a module over A is projective if and only if it is a direct sum of copies of A as a right module over itself. Suppose that we are given a projective A-module M. Then $M \cong \sum_{i=1}^{n} m_i A$ for some positive integer n and some generators $m_1, \ldots, m_n$ of M. Therefore $\mathrm{Dim}\, M = n \,\mathrm{Dim}\, A$, and moreover $M/\mathrm{Rad}\, M \cong \sum_{i=1}^{n} k$ has dimension n. In fact, the converse of this statement is also true.

Lemma 2.1 *Suppose that A is a split local finite dimensional algebra over a field k. Let M be a finitely generated A-module. Then M is projective if and only if $\mathrm{Dim}(M/\mathrm{Rad}\,M) \cdot \mathrm{Dim}\,A = \mathrm{Dim}\,M$.*

Proof. Let $n = \mathrm{Dim}(M/\mathrm{Rad}\,M)$ and suppose that $m_1, \ldots, m_n$ is a set of generators of M. Then $(m_1)\phi, \ldots, (m_n)\phi$ is a basis for $M/\mathrm{Rad}\,M$, where $\phi : M \longrightarrow M/\mathrm{Rad}\,M$ is the natural quotient. Let F be a free A-module with A-basis $f_1, \ldots, f_n$ and define $\theta : F \longrightarrow M$ by $(f_i)\theta = m_i$. From the construction we have that $(F)\theta + \mathrm{Rad}(M) = M$ and, by Nakayama's Lemma, $(F)\theta = M$. Hence θ is surjective. If we assume the condition on the dimension of M then θ must also be injective, since $\mathrm{Dim}\,M = n \cdot \mathrm{Dim}\,A = \mathrm{Dim}\,F$. Hence θ is an isomorphism in that case.

This condition is easily written into Magma code* as a test for the projectivity of a module over a split basic algebra. Our setup consists of some finite dimensional module M over a k-algebra A which is split local. The sequence L is a list of actions of generators of a subalgebra with identity, which we may call B. Our interest is in whether M is projective as a B-module. We assume that the elements in L are all nilpotent, which is the same as saying that they are all in the radical of A. Hence, the radical of B is generated by the elements of L. As a vector subspace of M, the radical of M_B, the restriction of M to B, is spanned by the collection of all vectors $m \cdot \ell$, for m in a k-basis for M and for $\ell \in L$. So the subspace S in the function given below is the B-radical of M in the space V which we may consider to be the underlying vector space of M. Finally, the input n is the dimension of B. As this dimension may be difficult to compute in some cases, we simply assume that it is known and include it in the input. Such will be the situation in the examples to follow. The function checks that the codimension of S in M, when multiplied by n, is the same as the dimension of M. By the lemma, the module is projective if and only if the answer is yes.

```
IsProjective := function(M, L, n);
    V := VectorSpace( BaseRing(L[1]), Dimension(M) );
    S := &+[ sub< V | RowSpace(x) > : x in L ];
    if (Dimension(M) − Dimension(S))*n eq Dimension(M) then
        return true;
    end if;
    return false;
end function;
```

Note here that elements of L are actually the matrices of the action of the generators of subalgebra. Perhaps we should also note that we have not tried to compute the submodule of M generated by $m \cdot b$ for $m \in M$ and $b \in L$. The problem is that this would give us an A-submodule which would likely be much larger than the B-radical of M.

---

*See the Preface to this volume for style conventions regarding Magma code; code appearing in this book is available at http://magma.maths.usyd.edu.au/magma/.

3 Basic algebras and homological algebra on the computer

The type AlgBas in Magma is the category of split basic algebras. The type is optimized for the computation of resolutions, cohomology, and general homological algebra. We assume that our algebras are finite dimensional over a field k. So, for example, an injective module is the dual of a projective module over the opposite algebra. For this and other reasons, we can concentrate on the projective modules and on homomorphisms from projective modules. To construct projective resolutions and for many other applications, it is necessary to have the capability to lift homomorphisms to projective modules. Namely, given maps α and φ as in the diagram

$$
\begin{array}{ccc}
 & & P \\
 & & \downarrow \varphi \\
L & \xrightarrow{\ \alpha\ } & M & \longrightarrow 0
\end{array}
$$

with P a projective module and with the row exact, we must find the homomorphism $\theta : P \longrightarrow L$ satisfying $\theta\alpha = \varphi$.

A basic algebra in Magma has the solution to this problem built in to the data structure. Specifically, a structure of type AlgBas consists of a sequence of indecomposable projective modules together with a path tree that solves the homomorphism lifting problem for each indecomposable projective module. A projective module is a sequence of matrices, one for each generator of the algebra. The generators must be chosen so that the first few are idempotent and the remainder are nonidempotent generators that are in the radical of the algebra and hence are nilpotent. The identity element of the algebra is the sum of the idempotents, and the underlying vector space of the algebra is the direct sum of the vector spaces of the projective modules. At the heart of all of the operations are linear algebra manipulations.

An indecomposable projective module P is generated by an idempotent e. Consequently, any homomorphism from P to L (as in the diagram) is completely determined by the image of the element e. We set the first basis element of P to be e, so that the first row of the matrix of θ is $(e)\theta$. The path tree for the projective module P tells us how we get the other rows of the matrix. The path tree consists of a sequence of pairs of numbers. If the ith pair in the sequence is $\langle j, k \rangle$, then it indicates that the ith basis element in the chosen basis for P is obtained by multiplying the jth basis element by the kth generator for A. Thus the ith row of the matrix for θ is obtained by multiplying the jth row by the matrix for the kth generator of A on M. By this means, the matrix for θ is constructed very rapidly from the vector $(e)\theta$ in M.

There are a number of ways of generating basic algebras as is demonstrated in the examples below. One method is to simply input the projective modules as sequences of matrices for the generators of the algebra and also input the path trees. The group algebra of a p-group over a field of characteristic p is naturally a basic algebra, and that algebra can be created by the computer simply by supplying the group and the field. There is also a method of creating basic algebras from generators and relations using Steve Linton's vector enumerator. In the near future, this method is to be replaced by a function whose input is a noncommutative polynomial ring. The relations will be analyzed using noncommutative Gröbner basis techniques.

4 Support varieties for modules over group algebras

One of the first things that we learn in a course on homological algebra is that a module M over an algebra A is projective if and only if $\mathrm{Ext}_A^1(M, L) = 0$ for every A-module L. This is the basic relationship between projectivity of modules and cohomology. For group algebras, there is a more specific connection in the form of support varieties. In this section, we outline the definitions and some of the properties of support varieties. One of the most important results is the generalization of Quillen's Theorem [9] which connects the support variety of a module with the support varieties of the restrictions of the module to the elementary abelian p-subgroups. We refer the reader to any of several texts for group cohomology such as [1, 6, 8].

Assume that G is a finite group and that k is a field of characteristic $p > 0$. The cohomology ring $\mathrm{H}^*(G, k)$ is a finitely generated graded-commutative k-algebra. As such, its maximal ideal spectrum is a homogeneous affine variety which we denote $V_G(k)$. For a finitely generated kG-module M, the cohomology ring $\mathrm{Ext}_{kG}^*(M, M)$ is a finitely generated module over $\mathrm{H}^*(G, k)$. Letting $J(M)$ denote the annihilator in $\mathrm{H}^*(G, k)$ of $\mathrm{Ext}_{kG}^*(M, M)$, we define the support variety of M to be the closed subset $V_G(M) = V_G(J(M)) \subseteq V_G(k)$ consisting of all maximal ideals that contain $J(M)$. The support variety measures the homological properties of the module. A few of the properties are outlined in the following results.

Theorem 4.1 *Suppose that L, M and N are finitely generated kG-modules. Then the following are true.*

1. *M is projective if and only if $V_G(M) = \{0\}$.*
2. *(Quillen, Alperin–Evens/Avrunin)*

$$V_G(M) = \bigcup_{E \in \mathcal{E}\mathcal{A}} \mathrm{res}_{G,E}^*(V_E(M))$$

   where $\mathcal{E}\mathcal{A}$ is the collection of maximal elementary abelian subgroups of G.
3. *$V_G(M \otimes_k N) = V_G(M) \cap V_G(N)$.*
4. *If $0 \longrightarrow L \longrightarrow M \longrightarrow N \longrightarrow 0$ is a short exact sequence, then $V_G(M) \subseteq V_G(L) \cup V_G(N)$.*

The first statement is evident from the fact that kG is a self-injective algebra. That is, projective kG-modules are injective and *vice versa*. So any module that has finite projective dimension is in fact projective. The second statement is Quillen's Dimension Theorem in the case that $M = k$. The generalization to general finitely generated modules was proved independently by Jon Alperin and Leonard Evens and by George Avrunin. The third result depends on the fact that kG is a Hopf algebra. That is, we have a coalgebra structure $kG \longrightarrow kG \otimes_k kG$ that sends an element $g \in G$ to $g \otimes g$. This makes the tensor product (over k) of modules M and N into a kG-module by letting $g \in G$ act by $(m \otimes n)g = mg \otimes ng$ for $m \in M$ and $n \in N$. The last item can be derived from the long exact sequence on cohomology.

The theorem of Chouinard is a consequence. The key point is that by part (2) of the theorem, $V_G(M) = \{0\}$ if and only if $V_E(M) = \{0\}$ for all elementary abelian subgroups E.

Corollary 4.2 *A kG-module M is projective if and only if its restriction to every elementary abelian subgroup is projective.*

There is a method for realizing closed subsets of $V_G(k)$ as varieties of modules. It works as follows. Suppose that $\zeta \in \mathrm{H}^n(G, k)$ for some n. If (P_*, ε) is a minimal projective resolution of k, then ζ is represented by a cocycle $\zeta' : P_n \longrightarrow k$ as in the diagram

$$\cdots \longrightarrow P_{n+1} \xrightarrow{\partial_{n+1}} P_n \xrightarrow{\partial_n} \cdots \longrightarrow P_0 \xrightarrow{\varepsilon} k \longrightarrow 0$$
$$\downarrow \zeta'$$
$$k$$

The fact that ζ' is a cocycle means that $\partial_{n+1}\zeta' = 0$, and $(P_{n+1})\partial_{n+1}$ is in the kernel of ζ'. Hence, ζ' induces a homomorphism $\hat{\zeta} : \Omega^n(k) \longrightarrow k$, where $\Omega^n(k) \cong P_n/(P_{n+1})\partial_{n+1}$ is isomorphic to the kernel of ∂_n. Then we have an exact sequence

$$0 \longrightarrow L_\zeta \longrightarrow \Omega^n(k) \xrightarrow{\hat{\zeta}} k \longrightarrow 0,$$

where L_ζ denotes the kernel of $\hat{\zeta}$.

Proposition 4.3 *The support variety of the module L_ζ is $V_G(L_\zeta) = V_G(\zeta)$, the closed subvariety of $V_G(k)$ consisting of all maximal ideals that contain ζ.*

This idea will be used in the examples. Perhaps the reader who is unfamiliar with support varieties can get some insight from the following example. Suppose that ζ is a regular element (a nondivisor of zero) in $\mathrm{H}^*(G, k)$. Then the long exact sequence on cohomology tells us that

$$\cdots \to \mathrm{H}^m(G, L_\zeta) \longrightarrow \mathrm{H}^m(G, \Omega^n(k)) \xrightarrow{\zeta} \mathrm{H}^m(G, k) \longrightarrow \mathrm{H}^{m+1}(G, L_\zeta) \to \cdots$$

is exact. But $\mathrm{H}^m(G, \Omega^n(k)) \cong \mathrm{H}^{m-n}(G, k)$ for $m > n$, and multiplication is injective in that case. So for $m > n$, we have that

$$\mathrm{H}^{m+1}(G, L_\zeta) \cong \mathrm{H}^m(G, k)/(\mathrm{H}^{m-n}(G, k)\zeta)$$

which is annihilated by ζ.

5 Some notes on cohomology and computations

In the examples we consider finite dimensional algebras and finitely generated modules. In such a situation, every module has a minimal projective resolution. If A is a finite dimensional algebra and M is a finitely generated A-module, then a projective resolution (P_*, ε) given as

$$\ldots \longrightarrow P_n \xrightarrow{\partial_n} P_{n-1} \xrightarrow{\partial_{n-1}} \ldots \longrightarrow P_1 \xrightarrow{\partial_1} P_0 \xrightarrow{\varepsilon} M \longrightarrow 0$$

is minimal provided for every n, we have that $\partial_n(P_n) \subseteq \operatorname{Rad} P_{n-1}$. A minimal projective resolution injects into any other resolution and is also a surjective image of any other resolution. We will make use of the following aspect of minimal resolutions.

Lemma 5.1 *Suppose that (P_*, ε) is a minimal projective resolution of an A-module M. If S is a simple module, then any cohomology class $\zeta \in \operatorname{Ext}_A^n(M, S)$ is uniquely represented by a cocycle $\hat{\zeta} : P_n \longrightarrow S$.*

Proof. Because S is simple, any map $\theta : N \longrightarrow S$ from any A-module N must have $\operatorname{Rad} N$ contained in the kernel of θ. Hence the induced map $\partial_n^* : \operatorname{Hom}_A(P_{n-1}, S) \longrightarrow \operatorname{Hom}_A(P_n, S)$ is the zero map. It follows that all of the coboundaries are zero and every cocycle represents a distinct cohomology class.

In the case of a group algebra, the product of cohomology elements can be defined using the Hopf algebra structure. This is equivalent to the Yoneda splice operation which holds for any algebra. That is, if $\zeta \in \operatorname{Ext}_A^m(L, M)$ and $\theta \in \operatorname{Ext}_A^n(M, N)$, then each is represented by an exact sequence. The product is the splice of the sequences at the module M (see [6] for details on products). For computational purposes, the cohomology product is given in a more convenient form. Any element $\zeta \in \operatorname{Ext}_A^m(L, M)$ is represented by a cocycle $\hat{\zeta} : P_m \longrightarrow M$, where P_* is a projective resolution of L. But if Q_* is a projective resolution of M, then we can lift $\hat{\zeta}$ to a chain map $\zeta_* : P_* \longrightarrow Q_*$ such the following diagram commutes.

$$
\begin{array}{ccccccccc}
\ldots \longrightarrow & P_{m+1} & \longrightarrow & P_m & \longrightarrow & P_{m-1} & \longrightarrow \ldots \longrightarrow P_0 & \longrightarrow L \xrightarrow{\varepsilon} 0 \\
& \zeta_1 \downarrow & & \zeta_0 \downarrow & & \searrow \hat{\zeta} & & \\
\ldots \longrightarrow & Q_1 & \longrightarrow & Q_0 & \xrightarrow{\varepsilon} & M & \longrightarrow 0 &
\end{array}
$$

Any two such chain maps are chain homotopic. Similarly, if we have a chain map $\{\zeta_*\}$ of degree $-m$ as above then the composition $\zeta_0 \circ \varepsilon$ is a cocycle and represents a cohomology element. That is, the cohomology $\mathrm{Ext}_A^m(L, M)$ can be defined as the classes of all chain maps from P_* to Q_* modulo chain homotopy. Most importantly, if $\theta \in \mathrm{Ext}_A^n(M, N)$ is represented by a chain map $\theta_* : Q_* \longrightarrow R_*$, where R_* is a projective resolution of N then the cup product of ζ and θ is represented by the composition of the chain maps. Composition of chain maps is an efficient method for computing cohomology products.

If (P_*, ε) is a minimal projective resolution of an A-module M, then the image of the boundary map $(P_n)\partial_n$ is denoted $\Omega^n(M)$. Of course, it is also isomorphic to the kernel of the boundary map $\partial_{n-1} : P_{n-1} \longrightarrow P_{n-2}$. As noted before, any cocycle $\hat{\zeta} : P_n \longrightarrow N$ induces a homomorphism $\tilde{\zeta} : \Omega^n(M) \longrightarrow N$ which also represents the cohomology class ζ of $\hat{\zeta}$ in $\mathrm{Ext}_A^n(M, N)$. Indeed, two maps $\mu, \mu' : \Omega^n(M) \longrightarrow N$ represent the same cohomology class if and only if they differ by a homomorphism that factors through a projective module. Hence analysis of $\mathrm{Hom}_A(\Omega^n(M), N)$ is another way of studying the cohomology $\mathrm{Ext}_A^n(M, N)$. The following lemma will be useful to us.

Lemma 5.2 *Suppose that A is a split local algebra and that $B \subseteq A$ is a subalgebra generated by a single nilpotent element b (and the identity element). Suppose that A is free as a B-module. Then a cohomology element $\zeta \in \mathrm{Ext}_A^{2n}(k, k)$ restricts to nonzero element of $\mathrm{Ext}_B^{2n}(k, k)$ if and only if the kernel L_ζ of the corresponding homomorphism $\tilde{\zeta} : \Omega^n(k) \longrightarrow k$ is projective as a B-module.*

Proof. By hypothesis, the algebra B must have the form $B \cong k[X]/(X^t)$ where t is the least positive integer such that $b^t = 0$. Then the minimal B-projective resolution of k has the form

$$\cdots \xrightarrow{b^{t-1}} Y_3 \xrightarrow{b} Y_2 \xrightarrow{b^{t-1}} Y_1 \xrightarrow{b} Y_0 \longrightarrow k \longrightarrow 0$$

where for every i, $Y_i \cong B$ as B-modules and the boundary maps alternate as multiplication by b or b^{t-1}. In particular, $\Omega^{2n}(k_B) \cong k_B$ for all n.

Suppose that P_* is an A-projective resolution of k. The fact that A is projective (hence free) as a B-module means that every P_j is projective as a B-module, and hence P_* becomes a B-projective resolution of k on restriction. To compute the actual restriction map, we must find a B-module chain map $\mu : Y_* \longrightarrow P_*$. The chain map induces a B-module homomorphism $\hat{\mu} : \Omega^n(k_B) \longrightarrow (\Omega^n(k))_B$. It follows that $(\Omega^n(k))_B \cong k \oplus Q$ where Q is a projective B-module and the summand k, isomorphic to the trivial module, is $(\Omega^n(k_B))\hat{\mu}$. We have an exact sequence

$$0 \longrightarrow L_\zeta \longrightarrow \Omega^n(k) \xrightarrow{\tilde{\zeta}} k \longrightarrow 0.$$

The restriction of ζ to B is zero if and only if $(\Omega^n(k_B))\hat{\mu}\tilde{\zeta}$ is zero. But this fails to be zero precisely when the sequence splits on restriction to B.

One last lemma that we will use is the following.

Lemma 5.3 *Suppose that A is a split local algebra and that (P_*, ε) is a minimal projective resolution of k. Then a chain map $\mu_* : P_* \longrightarrow P_*$ of degree $-n$ represents a nonzero cohomology class if and only if the rank of $\mu_0 : P_n \longrightarrow P_0$ is $\operatorname{Dim} A$.*

Proof. The point is that $P_0 \cong A$ as an A-module. Consequently, the rank of μ_0 is less than $\operatorname{Dim} A$ if and only if the image of μ_0 is contained in the radical of A, in which case, $\mu_0 \circ \varepsilon = 0$.

6 An algebra whose projective modules are detected on proper subalgebras

We want to consider certain algebras which are quotients of group algebras and investigate the question as to whether certain specific subalgebras detect projectivity. The subalgebras are the quotients of the subgroup algebras of the elementary abelian subgroups. The groups of interest are extraspecial p-groups, groups whose centers, commutator subgroups and Frattini subgroups coincide and are cyclic of order p. Firstly, we create the group by generators and relations.

```
>    F<x, y, z> := FreeGroup(3);
>    R := [ x³=1, y³=z, z³=1, zˣ=z, zʸ=z, yˣ=y*z ];
>    G := quo< GrpPC : F | R >;
```

A check shows that G is a PC group of order 27 with three generators and relations $G.2^3 = G.3, G.2^{G.1} = G.2 * G.3$. So G is an extraspecial group of exponent 9. Next we create the basic algebra. As mentioned before, the group algebra of a p-group is a basic algebra.

```
>    A := BasicAlgebra(G, GF(3));
```

The basic algebra A has dimension 27, one projective module and 4 generators. The first generator is the unique idempotent. The other three generators are $X = x - 1$, $Y = y - 1$ and $Z = z - 1$. The algebra that we want to construct is $B = A/(Z^2)$ where (Z^2) is the ideal generated by the square of the central element Z. Note that G has a unique elementary abelian subgroup $\langle x, z \rangle$. The corresponding subgroup algebra is generated by X and Z. Let E be the image of the subgroup algebra under the quotient map $A \longrightarrow B$.

Question 6.1 *Does the subalgebra E detect projectivity of B-modules?*

Our first objective is to create the algebra B on the computer. Then we study the question by investigating the cohomology ring of B. We obtain the algebra of B by taking a quotient of the matrix algebra of A, modifying the path tree of the projective module and then feeding this into our basic algebra machinery.

```
>    Z := Generators(A)[4];
>    P := ProjectiveModule(A, 1);
>    M := quo< P | [ Basis(P)[i] * Z² : i in [1..#Basis(P)] ] >;
```

Note that P is the unique projective module of A and is isomorphic to A as a right A-module. Thus M is the quotient of P by the ideal generated by Z^2. So the algebra B that we want is the action of A on M. Next we get the desired path tree. The path tree of P is PathTree(A, 1) which is the sequence

```
>    PathTree(A, 1);

      [ <1, 1>, <1, 2>, <2, 2>, <1, 3>, <2, 3>, <3, 3>, <4, 3>,
        <5, 3>, <6, 3>, <1, 4>, <2, 4>, <3, 4>, <4, 4>, <5, 4>,
        <6, 4>, <7, 4>, <8, 4>, <9, 4>, <10, 4>, <11, 4>, <12, 4>,
        <13, 4>, <14, 4>, <15, 4>, <16, 4>, <17, 4>, <18, 4> ]
```

Notice from the path tree that the last nine elements are multiples of Z^2. So we need only truncate the tree.

```
>    PT := [ PathTree(A, 1)[i]: i in [1 .. 18] ];
```

We now form the basic algebra.

```
>    B := BasicAlgebra( [ <Action(M), PT> ] );
```

We are interested in the cohomology of the unique simple B-module.

```
>    S := SimpleModule(B, 1);
>    P := CompactProjectiveResolution(S, 20);
```

This last function computes a projective resolution in a compact form out to 20 steps. We can check the size of the resolution with the following function.

```
>    SimpleHomologyDimensions(S);

      [ 20, 19, 18, 17, 16, 15, 14, 13, 12, 11, 10, 9,
        8, 7, 6, 5, 4, 3, 2, 1 ]
```

The sequence that is returned is a list of the ranks (as free B-modules) of the terms of the projective resolution of S in degrees 19 down to degree 0. We can see that the terms grow at the same rate as the homogeneous pieces of a polynomial ring in two variables. This is an indication that the kernel of the restriction map from B to E is likely to be a nilpotent ideal. The reason is that $E \cong k[X, Z]/(X^3, Z^2) \cong k[X]/(X^3) \otimes_k k[Z]/(Z^2)$. It is not difficult to show that the projective resolution for k as an E-module grows at the same rate as above. Indeed, we could use exactly the same methods to compute SimpleHomologyDimensions(T) where T is the unique simple E-module. The answer would come out the same as what was returned above.

   Another indication of this fact comes from the computation of the ring structure of $\mathrm{Ext}^*_B(k, k)$. To get the generators we compute:

```
>    cg := CohomologyRingGenerators(P);
>    DegreesOfCohomologyGenerators(cg);

      [ 1, 1, 2, 2 ]
```

So we see that there are two generators in degree 1 and two in degree 2. We expect the two generators in degree 1 to anticommute and to be nilpotent, and the two generators in degree 2 to be central and to generate a polynomial subring. The expectations are based on the model of group cohomology and require some proof. Here is some verification. First we have to transform the projective resolution into an actual complex of B-modules. The first function uses the compact data that we computed earlier. Then we transform the cohomology generators into chain maps on the projective resolution.

```
>    PP := ProjectiveResolution(P);
>    PP;
```

```
        Chain complex with terms of degree 20 down to 0
        Dimensions of terms: 378 360 342 324 306 288 270 252 234
        216 198 180 162 144 126 108 90 72 54 36 18
```

```
>    g₁ := CohomologyGeneratorToChainMap(PP, cg, 1);
```

Similarly, we let g_2, g_3, and g_4 be the chain maps of the second, third and fourth generators.

```
>    Rank( ModuleMap(g₁*g₁, 2) );
```

```
        15
```

```
>    Rank( ModuleMap(g₁*g₂+g₂*g₁, 2) );
```

```
        15
```

```
>    Rank( ModuleMap(g₂*g₂, 2) );
```

```
        17
```

```
>    Rank( ModuleMap(g₃*g₄+(−1)*g₄*g₃, 4) );
```

```
        0
```

The function ModuleMap returns the homomorphism of the given chain map in the given degree. Note that the ranks that we computed are all less than the dimension of A which is 18. It then follows from Lemma 5.3 that $\zeta_1^2 = 0 = \zeta_1\zeta_2 + \zeta_2\zeta_1 = \zeta_2^2 = \zeta_3\zeta_4 - \zeta_4\zeta_3$ where ζ_i is the cohomology class of the chain map g_i.

All of this points to the following theorem.

Theorem 6.2 *The kernel of the restriction map* $\mathrm{Ext}_B^*(k, k) \longrightarrow \mathrm{Ext}_E^*(k, k)$ *is a nilpotent ideal. Hence, projectivity of B-modules is detected on restriction to E.*

What we have given is only a strong indication and does not constitute a proof. A complete proof requires spectral sequences and can be found in [4].

7 An example in which projectivity is not detected on subalgebras

In this section we want to investigate the truncated group algebra of the other extraspecial group of order 27. By contrast with the previous example, the projectivity of modules is not detected on the subalgebras that are trunca-tions of the group algebras of the elementary abelian subgroups. We see how this result is indicated by the cohomology and then use the cohomology to construct a module which is free on restriction to the subalgebras but which is not projective.

The example that we are about to present was something of a disappoint-ment when we first discovered it. Jacques Thévenaz and the author had used the theorem of the last section to treat some important cases in the classifi-cation of torsion endotrivial modules [4]. The example of this section showed that other methods were going to be necessary to complete the classification.

The group of interest is the extraspecial group of order 27 and exponent 3. It can be given by generators and relations as follows.

$$G = \langle x, y, z \mid x^3 = y^3 = z^3 = 1, \ xy = yxz, \ xz = zx, \ yz = zy \rangle.$$

As before we let $X = x - 1$, $Y = y - 1$ and $Z = z - 1$. Let $A = kG$ and let $B = A/(Z^2)$. The group G has four maximal elementary abelian subgroups $U_1 = \langle x, z \rangle$, $U_2 = \langle xy, z \rangle$, $U_3 = \langle xy^2, z \rangle$ and $U_4 = \langle y, z \rangle$. Note that $(xy - 1) = (x-1) + (y-1) + (x-1)(y-1) = X + Y + XY$. The corresponding subalgebras are

$$W_1 = \langle X, Z \rangle, \qquad\qquad\qquad W_2 = \langle X + Y + XY, Z \rangle$$
$$W_3 = \langle X - Y + Y^2 - XY + XY^2, Z \rangle, \qquad W_4 = \langle Y, Z \rangle.$$

We set $E_i = W_i/(Z^2)$.

Question 7.1 *Do the subalgebras E_1, E_2, E_3 and E_4 detect projectivity of B-modules?*

We begin by constructing the algebras. This time we will use generators and relations to construct B. This function depends on noncommutative Gröbner basis methods whose implementations in Magma are still under development. At the time of this writing, the version of the BasicAlgebra function that we use is not available in Magma. However it will be included in a release in the near future.

```
>    gf := GF(3);
>    F<X,Y,Z> := FPA(gf,3);
>    Rel := [ X^3, Y^3, Z^2, X*Z − Z*X, Y*Z − Z*Y,
>                  (1+X)*(1+Y) − (1+Y)*(1+X)*(1+Z)];
>    B := BasicAlgebra(F,Rel);
```

An FPA is a finitely presented algebra, a polynomial ring in noncommuting variables. The sequence *Rel* is the relations. With this input, the function assumes that the algebra is a split local algebra and, in particular, that it has only one idempotent. Now we compute the projective resolution and cohomology as before.

```
>    S := SimpleModule(B, 1);
>    P := CompactProjectiveResolution(S, 12);
>    cg := CohomologyRingGenerators(P);
>    SimpleHomologyDimensions(S);

        [ 57, 51, 40, 35, 26, 22, 15, 12, 7, 5, 2, 1 ]}

>    DegreesOfCohomologyGenerators(cg);

        [ 1, 1, 2, 2, 2, 2, 2 ]
```

The indication is that this cohomology ring is growing at the same rate as a polynomial ring in three (not two) variables. Note that there are five new generators in degree 2 and that P_2 is the direct sum of five copies of B. It must be the case that all products of elements of degree 1 are zero. The problem here is that the subalgebras on which we would like to detect projectivity have cohomology rings that grow at a quadratic, not cubic rate. So we suspect that the subalgebras do not detect projectivity of B-modules. To prove this fact, we build a B-module M with the property that M is projective on restriction to every E_i, but M is not projective. The process is a bit tricky. The point is that we want to take a tensor product of modules using something like the Tensor Product Theorem 4.1(3). However, B is not a Hopf algebra and there is no tensor product operation. So what we want to do is to inflate appropriate B-modules to the group ring kG, and take the tensor product there. We explain this process as we go along.

First we need a module that is projective on restriction to the subalgebra $\hat{Z} = k[Z]/(Z^2)$. There should be a generator for $\mathrm{Ext}_B^2(k, k)$ that restricts nontrivially to $\hat{Z}$, so we need only consider the kernel of such an element on the second syzygy module $\Omega^2(k)$. Here we use the results 4.3 and 5.2 as models. It is probably possible find such an element by deterministic methods, but perhaps much faster to do so by random methods.

```
>    OO := SyzygyModule(S, 2);
>    HH := AHom(OO, S);
>    flag := true;
>    while flag do
>      θ := Random(HH);
>      N := Kernel(θ);
>      if IsProjective( N, [ActionGenerator(N, 4)], 2 ) then
>        flag := false;
>      end if;
>    end while;
```

Remember that the element Z acts on N as the fourth action generator. We can check independently that the module N is free on restriction to $\hat{Z}$.

```
>    Rank( ActionGenerator(N,4) );
        9
```

```
>    Dimension(N);
        18
```

So, N, the kernel of θ, is indeed free on restriction to $\hat{Z}$. Next we desire a B-module on which each $\overline{E}_i = E_i/(Z)$ acts freely. Note that $\overline{E}_1 = k[X]/(X^3)$, and $\overline{E}_2 = k[T]/(T^3)$ where $T = xy-1 = X+Y+XY$. A similar thing happens for $\overline{E}_3$ and $\overline{E}_4$. In particular, we are looking for a module over $B/(Z) \cong k\overline{G}$ where $\overline{G} \cong G/\langle z \rangle$ is an elementary abelian group of order 9. We could take the kernel of a cohomology element of degree two, but we are likely to have better success if we go to a higher degree.

Firstly we form the algebra $k\overline{G}$, and take the cohomology. Once again we make the construction using finitely presented algebras.

```
>    F1<e, a, b> := FPA(gf,3);
>    Rel1 := [ aP, bP, a*b − b*a ];
>    C := BasicAlgebra(F1,Rel1);
```

Next we go to the 8th syzygy of the trivial module and take the kernel of a random homomorphism.

```
>    T := SimpleModule(C,1);
>    O8 := SyzygyModule(T,8);
>    HH1 := AHom(O8,T);
```

We run a loop to find the desired module.

```
>    flag := true;
>    while flag do
>        μ := Random(HH1);
>        N1 := Kernel(μ);
>        XX := ActionGenerator(N1,2);
>        YY := ActionGenerator(N1,3);
>        LL := [ XX, XX+YY+XX*YY, XX−YY+YY2−XX*YY+XX*YY2, YY];
>        if forall(t){IsProjective(N1,x,3):x in LL} then
>            flag := false;
>        end if;
>    end while;
```

Here, N_1 is a module over the algebra C. We want to inflate it to a module over the extraspecial group. Note that we could construct the group algebra as a basic algebra, but it is easy enough in this case to use the GModule capabilities in Magma.

```
>    G := ExtraSpecialGroup(3, 1);
```

```
>    Mat₁ := MatrixAlgebra(gf, Dimension(N₁));
>    L₂ := GModule(G, [ActionGenerator(N₁,i) + Mat₁ ! 1 : i in [2..3]]
>                      cat [Mat₁ ! 1]);
```

Likewise we should inflate the module N. Then we can take the tensor product as G-modules.

```
>    Mat := MatrixAlgebra(gf, Dimension(N));
>    L₁ := GModule(G, [ActionGenerator(N,i) + Mat ! 1 : i in [2..4]]);
>    L := TensorProduct(L₁, L₂);
```

The point of the tensor product is that $Z = (z - 1)$ annihilates L_2 and Z^2 annihilates L_1. Therefore, it is not difficult to check that Z^2 must annihilate $L_1 \otimes L_2$, and $L_1 \otimes L_2$ is the inflation of a B-module. Finally, we want to check that L is projective on restriction to E_1, E_2, E_3 and E_4.

```
>    ll := Dimension(L);
>    a := ActionGenerator(L,1) − MatrixAlgebra(gf, ll) ! 1;
>    b := ActionGenerator(L,2) − MatrixAlgebra(gf, ll) ! 1;
>    c := ActionGenerator(L,3) − MatrixAlgebra(gf, ll) ! 1;
>    [ IsProjective(L, [a,c], 6), IsProjective(L, [a+b+a*b,c], 6),
>      IsProjective(L, [a−b+b²−a*b+a*b²,c], 6), IsProjective(L, [b,c], 6)];

        [ true, true, true, true ]
```

We emphasize that the function IsProjective(L, $[a,c]$, 6) tests the projectivity of L on restriction to E_1, because $kE_1 \subseteq kG$ is generated by the operators a and c. Similarly, the others test for projectivity on restriction to E_2, E_3 and E_4.

Finally, we need to be certain that L is not projective on all of B. Note that the action of B is given by the matrices a, b and c. Recall that the dimension of B is 18.

```
>    IsProjective(l, [a,b,c], 18);

        false
```

So we have demonstrated the following.

Proposition 7.2 *There exists a nonprojective B-module L such that L is free on restriction to E_i for $1 \leq i \leq 4$.*

Finally, we remark that both the example of this section and the last were run on the computer for $p = 5$ and the extraspecial groups of order 125. The same results were obtained in both cases.

References

1. D. J. Benson, *Representations and Cohomology* II: *Cohomology of Groups and Modules*, Cambridge Studies in Advanced Mathematics **31**, Cambridge: Cambridge University Press, 1991.

2. Wieb Bosma, John Cannon, Catherine Playoust, *The Magma algebra system I: The user language*, J. Symbolic Comput. **24** (1997), 235–265.
   See also the Magma home page at http://magma.maths.usyd.edu.au/magma/.
3. J. F. Carlson, *Cohomology and induction from elementary abelian subgroups*, Quarterly J. Math. **51** (2000), 169–181.
4. J. F. Carlson, J. Thévenaz, *Torsion endo-trivial modules*, Algebras and Rep. Theory **3** (2000), 303–335.
5. J. F. Carlson, J. Thévenaz, *The classification of torsion endotrivial modules*, Ann. of Math. **162** (2005), 823–883.
6. J. F. Carlson, L. Townsley, L. Valero-Elizondo, M. Zhang, *The Cohomology Rings of Finite Groups*, Dordrecht: Kluwer Academic Publishers, 2003.
7. L. Chouinard, *Projectivity and relative projectivity over group algebras*, J. Pure Appl. Algebra **7** (1976), 278–302.
8. L. Evens, *The Cohomology of Groups*, New York: Oxford University Press, 1991.
9. D. Quillen, *The spectrum of an equivariant cohomology ring*, I, II, Ann. Math. **94** (1971), 549–602.

Cohomology and group extensions in Magma

Derek F. Holt

Mathematics Institute
University of Warwick
Coventry, CV4 7AL, UK
dfh@maths.warwick.ac.uk

Summary. We describe the theory and implementation of some new and more flexible Magma functions for computing cohomology groups of finite groups, and their application to the computation of group extensions.

1 Introduction

In this paper, we shall describe the theory, implementation and application of some new functions in the Magma language [1] for computing the cohomology groups $H^1(G, M)$ and $H^2(G, M)$ of finite groups G acting on finite-dimensional modules M. We shall assume that the reader is familiar with basic definitions and results relating to group cohomology; for example, the treatment in Chapter 7 of [13] would provide enough background knowledge.

These implementations have been designed with flexibility of use in mind. The group G may be a permutation, matrix, or a finite soluble group defined by a power conjugate (PC) presentation. The module M may be defined over various commutative rings with identity, including finite fields and the integers, or M may be an arbitrary finitely generated abelian group with specified action of G. Cocycles corresponding to the elements of $H^1(G, M)$ and $H^2(G, M)$ can be computed explicitly, and user-supplied cocycles can be identified.

We shall also describe some successful applications of these functions. These include one or two new results, such as the verification of the conjecture that $H^2(\Omega^-(12, 2), M)$ has dimension 1 over $\mathbb{F}_2$, where M is the natural 12-dimensional module for $\Omega^-(12, 2)$.

Other, more general applications, involve using additional machinery from Magma, some of which has itself been developed only recently. It is well-known that $H^2(G, M)$ corresponds to equivalence classes of extensions of M by G in which the module action is induced by conjugation in the extension. But group-theorists are typically more interested in classification up to group isomorphism than up to equivalence of extension, and extensions may be isomorphic as groups without being equivalent as extensions. To classify extensions up to group isomorphism, we need to introduce the actions of the automorphism group of G and of the module M on $\mathrm{H}^2(G, M)$.

Even more generally, some applications require a complete list of extensions, up to group isomorphism, of a specific finite abelian or even soluble group A by the finite group G. Solving this problem may involve finding all modules for G over a given prime field up to a specified dimension in addition to the cohomological computations.

1.1 Other implementations

The author's implementations of functions for computing Schur Multipliers and covering groups of finite groups given as permutation groups, and the first and second cohomology groups $\mathrm{H}^1(G, M)$, $\mathrm{H}^2(G, M)$ of finite groups G acting on finite dimensional modules M over a prime field have been available in CAYLEY and then in Magma for about 15 years. They are described in the papers [7, 8, 9], and the Magma functions, including *p*Multiplicator, *p*Cover, CohomologicalDimension, Extension, are documented in the chapter on group cohomology in the Magma Handbook [11].

Despite their age, they still perform reasonably efficiently, particularly when G and M are both small. Their principal disadvantage is their relative inflexibility: it is not easy to use the results computed as components of other computations.

Other implementations generally apply to soluble groups G defined by a PC-presentation, but the methods involved are very similar to ours in many places. Such implementations are described for finite soluble groups in [5] and for general polycyclic groups in [6].

2 Computing cohomology groups

In this section, we give a brief description of our method for computing $\mathrm{H}^i(G, M)$ ($i \leq 2$) for a finite group G acting on a finitely generated module M. The method for $i = 2$ when G is a permutation or matrix group has not been published elsewhere, so we shall include more details for that case.

The algorithms require a finite presentation $\langle X \mid R \rangle$ of the group G, where X is a finite generating set and R is a finite set of defining relators of G. If G is soluble and defined by means of a PC-presentation (see the chapter on finite soluble groups in the Magma Handbook [4]), then the given PC-presentation is used for this purpose. Otherwise, if the input group G is a permutation or a matrix group, then the presentation is computed automatically by Magma, using one of the functions FPGroup and FPGroupStrong. In any case, let $|X| = r$, $|R| = s$ and $X = \{g_1, \ldots, g_r\}$.

The module M may be input to Magma as a KG-module, where K is a finite field or the integers $\mathbb{Z}$. It is also possible for M to be an arbitrary finitely generated abelian group defined by its abelian invariants $[e_1, e_2, \ldots, e_d]$, together with a list of r $d \times d$ matrices over $\mathbb{Z}$, which define the action of the generators of G on M. In any case, let $[\mu_1, \mu_2, \ldots, \mu_r]$ be this list of $d \times d$ matrices over either a finite field or the integers, which define the action of the generators of G on M.

2.1 Computing $\mathrm{H}^0(G, M)$ and $\mathrm{H}^1(G, M)$

The group $\mathrm{H}^0(G, M)$ can be identified with the additive group of vectors in M that are fixed by all $g \in G$. This is easily computed as follows. For $v \in M$, regarded as a $1 \times d$ row vector over K, we have $v \in \mathrm{H}^0(G, M)$ if and only if $v\mu_j = v$ for $1 \leq j \leq r$. Hence $\mathrm{H}^0(G, M)$ is just the nullspace of the $d \times dr$ matrix C_0, where C_0 is the horizontal join of the matrices $\mu_j - I_d$ for $1 \leq j \leq r$, which can be calculated immediately in Magma.

It should be noted, however, that here, and also in computations of $\mathrm{H}^i(G, M)$ for $i = 1, 2$, there is an extra complication in the case when M is given as an abelian group with specified invariants $[e_1, e_2, \ldots, e_d]$, which arises from the fact that the entries of the i-th column of the matrices μ_j need to be regarded as integers modulo e_i. This problem is solved by introducing dr extra rows into C_0, each of which contains a single non-zero entry: for $1 \leq i \leq d$ and $1 \leq j \leq r$, there is an extra row which contains e_i in the column corresponding to the i-th column of μ_j. We omit the details, and we have not included this modification in the code below.

The method for computing $\mathrm{H}^1(G, M)$ is described in detail in Section 5 of [3]. This approach was probably first proposed by Zassenhaus in [15]. The algorithm for finite soluble groups defined by a PC-presentation described in [5] uses this method. We have $\mathrm{H}^1(G, M) = \mathrm{Z}^1(G, M)/\mathrm{B}^1(G, M)$, where $\mathrm{Z}^1(G, M)$ and $\mathrm{B}^1(G, M)$ are respectively the groups of 1-cocycles an 1-coboundaries of G on M. Here is an outline of how we compute $\mathrm{Z}^1(G, M)$.

A 1-cocycle is a crossed homomorphism $\chi : G \to M$, and is specified uniquely by the images $\chi(x_j)$ of the generators of G in M. We shall represent r-tuples of module elements $(m_1, \ldots, m_r)$ $(m_j \in M)$ by vectors $v \in K^{rd}$, where, for $1 \leq j \leq r$, the components of v in positions $d(j-1)+1, \ldots, dj$ define m_j. In particular, we shall represent $\chi \in \mathrm{Z}^1(G, M)$ by the corresponding vector in K^{rd} for $(\chi(x_1), \ldots, \chi(x_r))$.

For an arbitrary r-tuple $(m_1, \ldots, m_r)$, we need to test whether or not there exists $\chi \in \mathrm{Z}^1(G, M)$ with $\chi(x_j) = m_j$ for $1 \leq j \leq r$. Such a χ satisfies $\chi(x_j^{-1}) = -\chi(x_j)^{x_j^{-1}}$ and, for a general element $w = x_{k_1}^{\varepsilon_1} \cdots x_{k_l}^{\varepsilon_l} \in G$ with each $\varepsilon_i = \pm 1$, we have

$$\chi(w) = \sum_{i=1}^{l} \varepsilon_i \chi(x_{k_i})^{g_i}, \tag{$\dagger$}$$

where $g_i = x_{k_{i+1}}^{\varepsilon_{i+1}} \cdots x_{k_l}^{\varepsilon_l}$ or $x_{k_i}^{\varepsilon_i} \cdots x_{k_l}^{\varepsilon_l}$ when $\varepsilon_i = 1$ or -1, respectively. It can be shown that $\chi \in \mathrm{Z}^1(G, M)$ if and only if the above expression evaluates to 0 for each $w \in R$.

Let $m_j = (z_{j1}, \ldots, z_{jd}) \in K^d$ for $1 \leq j \leq r$. Then, for a word w, the condition $\chi(w) = 0$ reduces to a system of d equations in the rd unknowns

$$z_{11}, \ldots, z_{1d}, z_{21}, \ldots, z_{2d}, \ldots, z_{r1}, \ldots, z_{rd}.$$

To see this, note that, in the expression (†) for $\chi(w)$, $\chi(x_{k_i})^{g_i}$ is the vector $m_{k_i} \cdot \mu_i$, of which the i_2-th component (for $1 \leq i_2 \leq d$) is $\sum_{i_1=1}^{d} z_{k_i i_1} (\mu_i)_{i_1 i_2}$. So, since there are s relators, each giving rise to d equations, the K-module $Z^1(G, M)$ is given by the solution set of a system of sd equations in rd unknowns; that is, by the nullspace of a certain $rd \times sd$ matrix C_1 over K.

It is straightforward to see that $B^1(G, M)$ is isomorphic to the rowspace of the matrix C_0 defined above, and so we can compute $H^1(G, M)$ as Nullspace(C_1)/Rowspace(C_0),

Here is the Magma code* for the computation of C_0 and C_1. The matrix C_1 is constructed in **C1Matrix** as an $r \times s$ matrix of $d \times d$ blocks. The call of the procedure **add_block** has the effect of adding or subtracting the $d \times d$ matrix **mat** to the block of C_1 having its top left entry in position (*sg,si*).

```
COMatrix := function(M)
    local I;
    I := IdentityMatrix(BaseRing(M), Dimension(M));
    return HorizontalJoin(
        [ActionGenerator(M, k) − I : k in [1..Nagens(M)]] );
end function;


add_block := procedure(∼M, i, j, M₂)
    local rows, cols, block;
    rows := NumberOfRows(M₂);
    cols := NumberOfColumns(M₂);
    block := ExtractBlock(M, i, j, rows, cols);
    InsertBlock(∼M, block+M₂, i, j);
end procedure;


C1Matrix := function(M)
    local K, d, G, RG, r, s, rel, w, C₁, mat, g, si, sg;
    K := BaseRing(M);
    d := Dimension(M);
    G := Group(M); // must be of type GrpFP
    RG := Relations(G);
    r := #Generators(G);
    s := #RG;
    C₁ := RMatrixSpace(K, r*d, s*d) ! 0;
```

---

*See the Preface to this volume for style conventions regarding Magma code; code appearing in this book is available at http://magma.maths.usyd.edu.au/magma/.

```
// Now fill in the entries of C₁
for i in [1..s] do
    // The i-th relation gives rise to the d columns of C₁
    // from (i−1)*d+1 to i*d. First turn it into a relator.
    rel := RG[i];  w := LHS(rel)*RHS(rel)⁻¹;
    si := (i−1)*d + 1;
    mat := IdentityMatrix(K, d);
    // We will scan the relator w from right to left.
    // mat is the matrix of the action of the current suffix of w.
    for g in Reverse(Eltseq(w)) do
        // The rows of C₁ from (g−1)*d+1 to g*d correspond to
        // generator g, and will be changed by this generator.
        sg := (Abs(g)−1)*d + 1;
        if g lt 0 then
            mat := ActionGenerator(M, −g)⁻¹ * mat;
        end if;
        add_block(∼C₁, sg, si, Sign(g)*mat);
        if g gt 0 then
            mat := ActionGenerator(M, g) * mat;
        end if;
    end for;
end for;
return C₁;
end function;
```

The groups $H^0(G, M)$ and $H^1(G, M)$ can then be calculated as follows:

```
>    H0 := Nullspace(C0);
>    H1 := quo< Nullspace(C1) | RowSpace(C0) >;
```

In this quotient, Nullspace(C_1) and Rowspace(C_0) are isomorphic to the groups $Z^1(G, M)$ and $B^1(G, M)$ of 1-cocycles and 1-coboundaries, respectively. Let v be an element of Nullspace(C_1) corresponding to $\nu \in Z^1(G, M)$. Then v is a $1 \times dr$ row vector, which is equal to the horizontal join of the $1 \times d$ row vectors $\nu(g_1), \ldots, \nu(g_r)$. Since a 1-cocycle ν is determined by its action on the generators of G, this enables the map $\nu : G \to M$ to be computed. It is also possible to express a given 1-cocycle input by the user as a linear sum of the generating rows of Nullspace(C_1).

2.2 Computing $H^2(G, M)$

Elements of $H^2(G, M)$ correspond to equivalence classes of extensions E of M by G in which the module action is induced by conjugation in the extension. Recall that two extensions of M by G are said to be *equivalent* if there is an isomorphism between them which induces the identity map on both M and on G.

Given such an extension E, choose a transversal T of M in E, and let $x_1, \ldots, x_r \in T$ map onto the generators $g_1, \ldots, g_r$ of G. If we substitute x_i for each g_i in one of the s defining relators $\rho \in R$ and evaluate the resulting word in E, then we get an element $m_\rho \in M$. By the standard theory of group presentations [12, Chapter 10], we can construct a group presentation of E by including group relations of M, relations defining the conjugation action of E on M, and relations which specify the elements of $m_\rho \in M$ defined by the relators $\rho \in R$ evaluated in E. So E is defined up to equivalence of extensions of M by G by the s elements $m_\rho \in M$. We can combine all of the m_ρ to form a single $1 \times ds$ row vector representing E. Let Z_2 be the set of $1 \times ds$ row vectors that arise in this way from extensions E of M by G.

With the transversal T, we have an associated 2-cocycle $\nu = \nu_T \in Z^2(G, M)$ defined by $x_g x_h = x_{gh} \nu(g, h)$ for $g, h \in G$, where $x_g \in E$ maps onto $g \in G$. Group multiplication in E is determined by ν and hence so is the $1 \times ds$ row vector described above. Indeed, this vector is determined uniquely by the choice of the r elements $x_i \in T$. So we have a surjective map $\psi : Z^2(G, M) \to Z_2$, which is easily seen to be a homomorphism of additive groups. In fact, it is a homomorphism of K-modules when M is a KG-module.

Let B_2 denote Rowspace(C_1), and let $m_1, \ldots, m_d$ be the generators of M as a K-module (when M is given as a KG-module) or as an abelian group (when M is given as an abelian group with prescribed action of G). Suppose we change our choice of the transversal T in E such that, for some for $1 \leq i \leq r$ and $1 \leq j \leq d$, x_i is replaced by $x_i m_j$, and x_k is unchanged for $k \neq i$. Then it is straightforward to check that $\psi(\nu_T)$ is replaced by $\psi(\nu_T) + v$, where v is row $(i-1)d + j$ of C_1. Furthermore, any change to the transversal elements $x_1, \ldots, x_r$ is a composite of changes of this type. So ψ induces $\overline{\psi} : H^2(G, M) = Z^2(G, M)/B^2(G, M) \to Z_2/B_2$, and the fact that the extension E is determined up to equivalence by $\psi(\nu)$ implies that $\overline{\psi}$ is an isomorphism.

Our aim is to compute $H^2(G, M)$ as Z_2/B_2, but to do this we need a method of computing Z_2. By analogy with the calculations for $H^1(G, M)$, we might hope that $Z_2 = $ Nullspace(C_2), where C_2 is some suitable matrix with ds rows.

In the case in which G is soluble and $\langle X \mid R \rangle$ is a PC-presentation, this all works out nicely. We have $s = r(r+1)/2$, and C_2 is a $ds \times dt$ matrix, where $t = r(r-1)(r-2)/6 + r(r-1) + r$. (We remark that columns of C_2 correspond to the consistency conditions for the resulting presentation of E.) This algorithm for the soluble case is described in detail in [5] and for possibly infinite polycyclic groups in [6], so we shall not give any further details here.

When G is given initially as a permutation or matrix group, then we can do something similar, but unfortunately it appears to be necessary to construct and calculate the nullspace of a much larger matrix C_2' in this situation. Since this method has not yet been published, we shall now give a brief description of it in the case when G is a permutation group. The method for matrix groups is very similar.

It depends heavily on the theory and machinery of bases and strong generating sets (BSGS), introduced originally by Sims in [14], and we assume that the reader has some familiarity with this topic. We start by computing a fixed base $[\beta_1, \ldots, \beta_b]$ of G and associated strong generating set $S = \{g_1, \ldots, g_r\}$. For $1 \leq k \leq b+1$, let $G^{(k)}$ be the k-th basic stabilizer $G_{\beta_1, \ldots, \beta_{k-1}}$, let $S^{(k)} = S \cap G^{(k)}$, and let $\Delta^{(k)}$ be the k-th basic orbit $\beta_k^{G^{(k)}}$. Then, by the definition of a BSGS, we have $G^{(b+1)} = 1$ and $\langle S^{(k)} \rangle = G^{(k)}$ for $1 \leq k \leq b+1$, where $S^{(b+1)}$ is empty. (We assume that $1 \notin S$.)

In what follows, by a word *over* a set X, we shall mean a word $w_1 \cdots w_t$ such that $w_i \in X$ or $w_i \in X^{-1}$ for each i. For $1 \leq k \leq b$ and $\gamma \in \Delta^{(k)}$, let $u_{k,\gamma}$ be a word over $S^{(k)}$ with $\beta_k^{u_{k,\gamma}} = \gamma$. These words can easily be computed from the data structures associated with the BSGS; furthermore, each element of G can be written as $u_{b,\gamma_b} \cdots u_{1,\gamma_1}$ for uniquely determined $\gamma_k \in \Delta^{(k)}$, and this expression is also readily computable from g. We call this word the *normal form* for g.

Fix a k with $1 \leq k \leq b$. We shall assume that $\{u_{k,\gamma} \mid \gamma \in \Delta^{(k)}\}$ is prefix closed and hence forms a *Schreier system* of right coset representatives of $G^{(k+1)}$ in $G^{(k)}$. For each $\gamma \in \Delta^{(k)}$ and $g \in S^{(k)}$, we have an equation in $G^{(k)}$ of the form $u_{k,\gamma}g = w_{k,\gamma,g}u_{k,\gamma^g}$, where $w_{k,\gamma,g}$ is a word in normal form in $G^{(k+1)}$. For exactly $|\Delta^{(k)}| - 1$ of these equations, one for each $\gamma^g \in \Delta^{(k)} \setminus \{\beta_k\}$, $w_{k,\gamma,g}$ is the empty word and the free reductions of the $u_{k,\gamma}g$ and u_{k,γ^g} are equal. We call these equations the *definitions* of the associated $\gamma^g \in \Delta^{(k)} \setminus \{\beta_k\}$. The remaining $|\Delta^{(k)}|(|S^{(k)}| - 1) + 1$ equations define the *Schreier generators* $w_{k,\gamma,g}$ of $G^{(k+1)}$.

For a presentation $\langle S \mid R \rangle$ of G on the strong generating set S, and $1 \leq k \leq b$, let $R^{(i)}$ be the subset of R consisting of those relators which are words over $S^{(k)}$. Then we call $\langle S \mid R \rangle$ a *strong presentation* of G if $G^{(k)} = \langle S^{(k)} \mid R^{(k)} \rangle$ for $1 \leq k \leq b$. Such a presentation can be computed by the Magma function **FPGroupStrong**, and we shall assume that this has been done.

Now let E be an extension of M which induces the prescribed action of G on M by conjugation. For each $g_i \in S$, choose an inverse image x_i of g_i in E, let $X = \{x_i \mid g_i \in S\}$, and let $X^{(k)}$ be the subset of X mapping onto $S^{(k)}$ for $1 \leq k \leq b$. For a word w over S, let $\hat{w}$ be the word over X obtained by substituting x_i for each $g_i \in S$. Then, for a relator $\rho \in R$, we have $\hat{\rho} = m_\rho \in M$. As we remarked above, the extension E, and hence also the element of $\mathrm{H}^2(G, M)$ defined by E, is uniquely determined by the list of elements $m_\rho \in M$. Unfortunately, a knowledge of the m_ρ alone does not easily allow us to calculate within E.

The normal form words $u_{b,\gamma_b} \cdots u_{1,\gamma_1}$ described above for elements of G enable us to define a transversal T of M in E consisting of the elements of E defined by the corresponding words $\hat{u}_{b,\gamma_b} \cdots \hat{u}_{1,\gamma_1}$. Since the free reductions of the left and right hand sides of the definitions $u_{k,\gamma}g = u_{k,\gamma^g}$ of γ^g are equal, we also have $\hat{u}_{k,\gamma}\hat{g} = \hat{u}_{k,\gamma^g}$ in this case. However, the equations

$u_{k,\gamma}g = w_{k,\gamma,g}u_{k,\gamma^g}$ which are not definitions, correspond to equations in E of the form $\hat{u}_{k,\gamma}\hat{g} = \hat{w}_{k,\gamma,g}\hat{u}_{k,\gamma^g}m_{k,\gamma,g}$, where $m_{k,\gamma,g} \in M$.

A knowledge of all of the elements $m_{k,\gamma,g}$ enables us easily to put an arbitrary element of E given as a word over generators of M and X into a normal form word $\hat{u}_{b,\gamma_b}\cdots\hat{u}_{1,\gamma_1}m$ with $m \in M$. We proceed as follows. Elements of m will always be moved to the right end of the word by using the prescribed action of G on M. Let $v = v_1$ be a word over X. Then, since $\hat{u}_{1,\beta_1}$ is the empty word, we have $v = \hat{u}_{1,\beta_1}v$ and, by using the equations $\hat{u}_{1,\gamma}\hat{g} = \hat{w}_{1,\gamma,g}\hat{u}_{1,\gamma^g}m_{1,\gamma,g}$ and their inverses $\hat{u}_{1,\gamma^g}\hat{g}^{-1} = \hat{w}_{1,\gamma,g}^{-1}\hat{u}_{1,\gamma}m_{1,\gamma,g}^{-g^{-1}}$, we can move the $\hat{u}$ term to the right of the word and end up with $v = v_2\hat{u}_{1,\gamma_1}m_1$ for some $\gamma_1 \in \Delta^{(1)}$, $m_1 \in M$, and a word v_2 over $X^{(2)}$. Now we just repeat this process in $G^{(2)}$ using v_2 in place of v_1 to get $v = v_3\hat{u}_{2,\gamma_2}\hat{u}_{1,\gamma_1}m_2$, and so on.

In order to compute $\mathrm{H}^2(G,M)$, we consider a general extension E of M by G inducing the prescribed action, and treat the $|R| = s$ elements m_ρ ($\rho \in R$) and the $s' := \sum_{k=1}^{b}(|\Delta^{(k)}|(|S^{(k)}|-1)+1)$ elements $m_{k,\gamma,g}$ as variable elements of M. Each of these variables will, of course, give rise to d variable integers or field elements. In the algorithm, we set up a matrix C_2', which corresponds to a system of homogeneous equations over these variables. So C_2' has $d(s+s')$ rows. The solutions to these equations — that is, the elements of the nullspace Z_2' of C_2' — will then define the required extensions E. Let Z_2 be the space of $1 \times ds$ row vectors defined by taking the first ds components of the vectors in Z_2'. Then the vectors in Z_2 specify the values of m_ρ in the associated extensions and, as we explained above, we can calculate $\mathrm{H}^2(G,M)$ as Z_2/B_2.

It remains to explain where the equations come from! We do the following for each k with $1 \le k \le b$. For each relator $\rho \in R^{(k)}$ and each $\gamma \in \Delta^{(k)}$, we have

$$\hat{u}_{k,\gamma}\hat{\rho}\hat{u}_{k,\gamma}^{-1} = \hat{u}_{k,\gamma}m_\rho\hat{u}_{k,\gamma}^{-1}.$$

We use the procedure described above to put the left and right hand sides of this equation into normal form in E, where we treat the m_ρ and $m_{k,\gamma,g}$ as variables in M. Each such equation reduces to an equation in M and gives rise to d linear equations, one for each component of M. In addition, if $k < b$, then for each $g \in S^{(k+1)}$ we can put the word $\hat{g}$ into normal form in two different ways, treating it either as a word over $S^{(k+1)}$ or as a word over $S^{(k)}$. This results in further equations in M. So the total number of equations, which is the number of columns of C_2', is dt, where $t = \sum_{k=1}^{b}(|\Delta^{(k)}||R^{(k)}| + |S^{(k+1)}|)$.

We shall not attempt to write down a formal proof that a solution of this system of equations really does define an extension E of M by G. The method is based on the Reidemeister–Schreier algorithm for computing a presentation of a subgroup of finite index in a finitely presented group; see for example [12, Section 9.1]. The process outlined in the previous paragraph produces the relations of such a presentation for $G^{(k+1)}$ in $G^{(k)}$ for $1 \le k \le b$. The property that m_ρ and $m_{k,\gamma,g}$ satisfy these equations is equivalent to these relations holding in the group E defined by these values of m_ρ and $m_{k,\gamma,g}$, which is equivalent to E being a genuine extension of M by G. (Values of m_ρ and $m_{k,\gamma,g}$ that did not satisfy these equations would result in the group E collapsing and having order smaller than $|M|\,|G|$.)

In the Magma implementation, C_2' is discarded once its nullspace has been computed, since it is no longer required, but Z_2', Z_2, and $\mathrm{H}^2(G,M) = Z_2/B_2$ are all stored. The group $\mathrm{H}^2(G,M)$ is returned as a module over the base-ring K of M, or as an abelian group if M is defined by means of its invariants as an abelian group. For an element of $\mathrm{H}^2(G,M)$ supplied by the user, an inverse image $v \in Z_2$ can be computed, as can a corresponding inverse image $v' \in Z_2'$.

From v we can immediately write down a group presentation of the corresponding extension E. Furthermore, we can use v' to compute the associated 2-cocycle $\nu : G \times G \to M$ as follows. If u_g is the normal form word over S for $g \in G$, then we have chosen $\hat{u}_g$ as the corresponding element of the transversal T, and so by definition of ν we have $\hat{u}_g \hat{u}_h = \hat{u}_{gh} \nu(g,h)$ for all $g,h \in G$. Hence $\nu(g,h)$ can be found by computing the normal form word in E for $\hat{u}_{gh}^{-1} \hat{u}_g \hat{u}_h$. Conversely, if we are given a 2-cocycle as a function $G \times G \to M$, then we can use it on the words $\hat{\rho}$ ($\rho \in R$) to evaluate the module elements m_ρ in the corresponding extension E, and hence to compute the associated $v \in Z_2$ and its image in $\mathrm{H}^2(G,M)$.

2.3 Some examples

Here is a sample computation.

```
>    G := SL(2,5);
```

A permutation representation makes it easier to construct modules.

```
>    G := CosetImage(G, Sylow(G,5));
>    PM := PermutationModule(G, GF(2));
>    CP := Constituents(PM);
>    CP;

        [
              GModule of dimension 1 over GF(2),
              GModule of dimension 4 over GF(2)
        ]
```

```
>    M := CP[2];
```

Since $Z(\mathrm{PSL}(2,5))$ has trivial action on this module, M comes from a 4-dimensional module for $\mathrm{PSL}(2,5) \cong A_5$. In fact A_5 has two 4-dimensional irreducible modules over $\mathbb{F}_2$, and this is the one which is not absolutely irreducible and arises from the natural module for $\mathrm{SL}(2,4) \cong A_5$.

```
>    μ1 := ActionGenerator(M,1);
>    μ2 := ActionGenerator(M,2);
```

```
>    μ₁; μ₂;
          [1 1 1 1]
          [0 0 1 0]
          [0 1 0 0]
          [0 0 0 1]

          [1 1 0 1]
          [1 0 1 0]
          [0 0 1 1]
          [0 0 1 0]
```

Before embarking upon a computation of cohomology groups, we call a function CohomologyModule, which constructs an object that is used as the first argument for all of the cohomology functions. This object is used to store known information about the group G and the module M, such as the BSGS and strong presentation of G, and also to store the data computed by the cohomology functions themselves, such as the matrices C_0 and C_1 and the spaces Z_2, Z'_2, B_2 described earlier.

```
>    C := CohomologyModule(G, M);
>    H₀ := CohomologyGroup(C, 0);
>    H₁ := CohomologyGroup(C, 1);
>    H₂ := CohomologyGroup(C, 2);
>    H₀; H₁; H₂;
          Full Vector space of degree 0 over GF(2)
          Full Vector space of degree 2 over GF(2)
          Full Vector space of degree 2 over GF(2)

>    ν := OneCocycle(C, H₁.1);
>    ν(G.1); ν(G.2); ν(G.1*G.2);
          (0 0 0 1)
          (0 0 0 0)
          (0 0 1 0)
```

Note that $\nu(G.1*G.2) = \nu(G.1)*\mu_1 + \nu(G.1)$.

```
>    IdentifyOneCocycle(C, ν);
          (1 0)
```

We now construct a non-split extension of M by G.

```
>    E := Extension(C, H₂.1); #E;
          1920

>    Category(E);
          GrpFP

>    Ngens(E);
          7
```

The last four generators generate the module. We use this property to get a faithful permutation representation.

```
>   P := CosetImage(E, sub< E | E.5, E.6, E.7 >);
>   #P;
        1920

>   ChiefFactors(P);
            G
            |   Alternating(5)
            *
            |   Cyclic(2)
            *
            |   Cyclic(2) (4 copies)
            1
```

We can verify that the extension is non-split.

```
>   N := MinimalNormalSubgroups(P);
>   #N;

        1

>   #N[1];
        16

>   Complements(P, N[1]);
        [ ]
```

Finally, we demonstrate how to calculate the restriction of a 2-cocycle to a subgroup of G.

```
>   S := Sylow(G, 2);
>   CS := CohomologyModule(S, Restriction(M, S));
>   H2S := CohomologyGroup(CS, 2);
>   H2S;
        Full Vector space of degree 4 over GF(2)

>   IdentifyTwoCocycle(CS, TwoCocycle(C, H_2.1));
        (1 1 0 1)

>   IdentifyTwoCocycle(CS, TwoCocycle(C, H_2.2));
        (1 0 1 0)

>   #Extension(CS, $1);
        128
```

To give the reader an idea of the sizes of the matrix C'_2, here are a few examples. These dimensions depend on a number of random choices within Magma, such as a base and strong generating set for G, so they can vary slightly from one run to another.

- $G = A_5 : 33d \times 54d$;
- $G = A_8 : 193d \times 586d$;
- $G = M_{24} : 1057d \times 5305d$.

2.4 $\mathbf{H^2(G, M)}$ as a subgroup of $\mathbf{H^2(P, M)}$

It is clear from the above examples that the method for $\mathrm{H}^2(G, M)$ is limited to moderately small groups. Fortunately, if M is a KG-module for a finite field K of characteristic p, and we only require the dimension of $\mathrm{H}^2(G, M)$ over K rather than the ability to compute explicitly with extensions and 2-cocycles, then there is an alternative approach, which is applicable to much larger groups. This is based on the same idea as was used in the author's earlier implementations described in [7] and [9].

Let $P \in \mathrm{Syl}_p(G)$. Then $\mathrm{H}^2(G, M)$ is isomorphic to a certain subgroup of $\mathrm{H}^2(P, M)$. This is the subgroup consisting of those elements which are *stable* with respect to each $g \in G$, where $\tau \in \mathrm{H}^2(P, M)$ is defined to be stable with respect to $g \in G$ if the restriction of τ to $\mathrm{H}^2(P \cap P^g, M)$ is the same as the image of τ under the composite of the conjugation map $\mathrm{H}^2(P, M) \to \mathrm{H}^2(P^g, M)$ and the restriction map $\mathrm{H}^2(P^g, M) \to \mathrm{H}^2(P \cap P^g, M)$.

Since P is soluble, we can calculate a PC-presentation for P, and then $\mathrm{H}^2(P, M)$ can be computed relatively easily, using the algorithm that was mentioned earlier and which is described in [5]. For a given $g \in G$, we can also compute $\mathrm{H}^2(P^g, M)$ and $\mathrm{H}^2(P \cap P^g, M)$, and the relevant conjugation and restriction maps are straightforward to evaluate, so we can calculate the subgroup of elements of $\mathrm{H}^2(P, M)$ which are stable with respect to g. We do not want to have to do this calculation for every $g \in G$, and fortunately the theory allows us to reduce the number of g which we need to consider.

More precisely, we use standard Magma functions to find a chain of subgroups $P = H_0 < H_1 < \ldots H_m = G$ between H and G. We have written a function to find such subgroups H_i, which works by experimentally trying a variety of possibilities; for example, an H_i might be chosen as the centralizer in G of some central element of P, the normalizer in G of some normal subgroup of P, or perhaps the subgroup of G that fixes all of the orbits of P or permutes the orbits of P among themselves. For the best performance, we try to choose these subgroups such that the indices $|H_{i+1} : H_i|$ are as small as possible, while avoiding expensive computations. Then for $i = 0, 1, \ldots m-1$, we calculate a set D_i of double coset representatives of H_i in H_{i+1} and carry out the stability test only for $g \in D_i$. See [7] for more details.

We shall now give a brief description of some successful computations using the method just outlined. The first of these had already been done using the older functions mentioned in Subsection 1.1, but with great effort. As far as the author is aware, the results of the second and third of these calculations are new.

The group $\Omega^-(8,3)$. Here we take the simple orthogonal group $\Omega^-(8,3)$ acting on its natural module over the field of order 3. As can be seen from the output below, the chain of subgroups chosen had length $m = 5$ (using the notation above) with $|P| = |H_0| = 3^{12}$, $|H_1 : H_0| = 8$, $|H_2 : H_1| = 10$, $|H_3 : H_2| = 112$, $|H_4 : H_3| = 2$, $|H_5 : H_4| = 1066$.

In the reproduction of the calculation below, we have kept only part of the verbose output! The line 'Setting up 2912 equations in 624 unknowns' means that the matrix C_2 for which the nullspace was calculated in the computation of $H^2(P, M)$ had size 624×2912. The 120 seconds taken by CohomologyModule were consumed by the calculation of the BSGS related information and strong presentation of G.

The intermediate dimensions $H^2(P, M)$ and $H^2(H, M)$ are the dimensions of $H^2(H_i, M)$ for $i = 0, 1, 2, 3, 4, 5$. The stability test was carried out for a total of 13 elements g. For each such g, the computation of $H^2(P \cap P^g, M)$ involved finding the nullspace of a matrix which was generally a little smaller than that for P itself.

```
>   G := OmegaMinus(8,3); M := GModule(G);
>   time C := CohomologyModule(G, M);

          Time: 120.540

>   time dim := CohomologicalDimension(C, 2);
          Computing H^2(P,M): |P| = 3^12.
          Setting up 2912 equations in 624 unknowns
          Dimension of H^2(P,M) = 18

          Next subgroup - index 8
          Dimension of H^2(H,M) = 5

          Next subgroup - index 10
          Dimension of H^2(H,M) = 4

          Next subgroup - index 112
          Dimension of H^2(H,M) = 4

          Next subgroup - index 2
          Dimension of H^2(H,M) = 4

          Next subgroup - index 1066
          Dimension of H^2(H,M) = 2
          Time: 470.010

>   dim;
          2
```

It turns out to be a little quicker to do the computation using a permutation representation rather than a matrix representation of G. We omit the details.

The group $\Omega^-(12, 2)$. The next example is $\Omega^-(12, 2)$ on its natural module. Until recently, this was the only unknown case in the list of dimensions of $H^2(G, M)$ for the finite classical groups acting on their natural modules. As had been widely expected, the answer turns out to be dimension 1, so there is a unique non-split extension of the module by the group.

This time the calculation ran out of space when using the matrix representation of G, but worked with a permutation representation. The stability test was carried out for a total of 29 elements g.

```
>    G := OmegaMinus(12, 2); M := GModule(G);
>    // Use a permutation representation of G.
>    V := VectorSpace(G);
>    P := OrbitImage(G, V.1);
>    Degree(P);

        2015

>    mats := [ ActionGenerator(M, i) : i in [1 .. Ngens(G)] ];
>    M := GModule(P, mats);
>    time C := CohomologyModule(P, M);

        Time: 1596.930

>    time dim := CohomologicalDimension(C, 2);

        Computing H^2(P,M): |P| = 2^30.
        Setting up 59520 equations in 5580 unknowns
        Dimension of H^2(P,M) = 28

        Finding subgroup chain
        Indices (from bottom up): [ 3, 135, 119, 495, 2015 ]

        Next subgroup - index 3
        Dimension of H^2(H,M) = 16

        Next subgroup - index 135
        Dimension of H^2(H,M) = 5

        Next subgroup - index 119
        Dimension of H^2(H,M) = 1

        Next subgroup - index 495
        Dimension of H^2(H,M) = 1

        Next subgroup - index 2015
        Dimension of H^2(H,M) = 1
        Time: 7009.210
```

The group $\Omega^+(10,2)$. Our final example is the group $\Omega^+(10,2)$ acting on a certain irreducible module of dimension 16. A non-split extension of this module by $\Omega^+(10,2)$ occurs as a quotient of the maximal subgroup 2^{10+16} . $\Omega^+(10,2)$ of the Monster simple group, and this computation shows that this non-split extension is unique. The stability test was carried out for a total of 39 elements g.

The module is easily constructed as a constituent of a permutation module of the group.

```
>    G := POmegaPlus(10,2);
>    Degree(G);
        527

>    M := PermutationModule(G, GF(2));
>    I := Constituents(M);
>    I;
        [
            GModule of dimension 1 over GF(2),
            GModule of dimension 10 over GF(2),
            GModule of dimension 16 over GF(2),
            GModule of dimension 16 over GF(2),
            GModule of dimension 44 over GF(2),
            GModule of dimension 100 over GF(2),
            GModule of dimension 164 over GF(2)
        ]

>    time C := CohomologyModule(G, I[3]);
        Time: 94.180

>    time dim := CohomologicalDimension(C,2);
        Computing H^2(P,M): |P| = 2^20.
        Setting up 24640 equations in 3360 unknowns
        Dimension of H^2(P,M) = 21
        Finding subgroup chain
        Indices (from bottom up): [ 315, 135, 527 ]

        Next subgroup - index 315
        Dimension of H^2(H,M) = 3

        Next subgroup - index 135
        Dimension of H^2(H,M) = 2

        Next subgroup - index 527
        Dimension of H^2(H,M) = 1
        Time: 2187.410
```

3 Finding group extensions

In this section, we discuss how we can apply the cohomological methods just described to the problem of finding group extensions.

3.1 Generalities

The most general form of the problem to be addressed is: given two finite groups N and G, find all extensions of N by G up to group isomorphism. In other words, find the isomorphism classes of groups E having a normal subgroup N with $E/N \cong G$.

There is already a potential ambiguity in this problem. Let E_1 and E_2 be groups having normal subgroups N_1 and N_2 both isomorphic to N and with quotients E_1/N_1 and E_2/N_2 both isomorphic to G. It is possible for there to be an isomorphism $\phi : E_1 \to E_2$ but no such isomorphism that maps N_1 to N_2. An example of this phenomenon is the group

$$E_1 = E_2 = \langle\, x, y, z \mid x^{29} = y^{29} = z^7 = 1,\, xy = yx,\, x^z = x^7,\, y^z = y^{16} \,\rangle$$

of order 5887, where N_1 and N_2 are the normal subgroups $\langle x \rangle$ and $\langle y \rangle$ of order 29.

Since our aim is to classify the isomorphism classes of extensions of N by G, in this situation we prefer to regard E_1 and E_2 as being distinct as extensions of N by G. In other words, we regard extensions E_1 of $N_1 \cong N$ by G and E_2 of $N_2 \cong N$ by G as being isomorphic extensions of N by G if there is an isomorphism from E_1 to E_2 that maps N_1 to N_2. Notice that this is not the same as the extensions being equivalent, because two extensions of N by G are equivalent only if there is an isomorphism between them which induces the identity on both N and G. So equivalent extensions are isomorphic, but the converse is not necessarily true.

We are not yet in a position to offer a solution to this problem for arbitrary N. Provided that N is soluble, however, by finding a series of characteristic subgroups of N with elementary abelian layers, we can reduce to the case in which N is elementary abelian. With our current rather naive implementation, this reduction is rather costly, particularly when the elementary abelian layers have large numbers of generators, or when there are more than two or three such layers. It could undoubtedly be improved with further work, but we shall say no more about that here, and assume now that N is an elementary abelian group of order p^d for a prime p.

As we saw earlier, in an extension E of an elementary abelian p-group N by G, the conjugation action on N within E makes N into a KG-module, where K is the field of order p. For a given KG-module of dimension d over K, there is a natural bijection between $\mathrm{H}^2(G, N)$ and the equivalence classes of extensions of N by G that induce that module action by conjugation.

So the problem of finding all extensions of N by G splits into two parts. First we find representatives of the isomorphism classes of KG-modules of dimension d, and then we calculate their second cohomology groups $H^2(G, N)$ and use the results to classify extensions up to isomorphism. We shall discuss these two parts of the problem in the following two subsections.

3.2 Finding the modules

This problem also splits into two parts! The first is to find the irreducible KG-modules. There is now a standard Magma function to do this, which we shall discuss very briefly here. For finite fields K, there are very efficient 'Meataxe' based methods for testing KG-modules for irreducibility, absolute irreducibility, finding their composition series, and testing two such modules for isomorphism. Using these tools, we can find all irreducible modules by starting with any faithful module, such as a permutation module, and then repeatedly finding the composition series of this module, and then of the tensor products of the already known irreducible modules. The number of inequivalent absolutely irreducibles is equal to the number of conjugacy classes of p-regular elements, and we can use this fact to check whether or not we have found them all.

For larger groups, it is worthwhile to compute the Brauer characters for the prime p of the irreducible modules that have been found so far, since this enables us to decide in advance which tensor products will yield new irreducibles. Further details will be published in [2].

Now a KG-module M of dimension d is either irreducible, or it has an irreducible submodule N of dimension $c < d$. In the second case, M is a module extension of N by M/N. So, we can find all KG-modules up to a given dimension d, provided that we can solve the extension problem for KG-modules; that is, given two KG-modules N and L, find all KG-modules (up to isomorphism) with submodule isomorphic to N and quotient isomorphic to L. (Of course, with this approach, we will get each module several times, once for each of its irreducible submodules, but we can use our KG-module isomorphism function to eliminate repetitions.)

The extension problem can be solved as follows. We can make the K-space $\mathrm{Hom}_K(L, N)$ of K-linear maps from L to N into a KG-module, by defining $\phi^g(l) = \phi(l^{g^{-1}})^g$ for $\phi \in \mathrm{Hom}_K(L, N)$ and $g \in G$. There is then a natural isomorphism between $H^1(G, \mathrm{Hom}_K(L, N))$ and the isomorphism classes of KG-module extensions of N by L. To be explicit, let $\nu : G \to \mathrm{Hom}_K(L, N)$ be a 1-cocycle representing an element of $H^1(G, \mathrm{Hom}_K(L, N))$. Then we can construct a corresponding KG-module extension of N by L as follows. First let $M = L \oplus N$ be the direct sum of the K-spaces L and N. Then we make M into a KG-module by defining $(l, n)^g = (l^g, n^g + \nu(g)(l^g))$ for $l \in L$, $n \in N$ and $g \in G$.

It is a routine calculation to check that this makes M into a KG-module, and that every KG-module extension of N by L is isomorphic to a module defined in this way for some $\nu \in Z^1(G, \mathrm{Hom}_K(L, N))$. We can use the methods described in Section 2 to calculate $H^1(G, \mathrm{Hom}_K(L, N))$ and the representative cocycles ν, and we can then use these to construct the corresponding KG-modules as just described.

3.3 Classifying the group extensions

After finding representatives of the isomorphism classes of the KG-modules of the required dimension d, we can compute $\mathrm{H}^2(G, N)$ for each such module N, and then construct the corresponding group extensions of N by G as described in Section 2. This will give us all of the required extensions, but in general many of these will be isomorphic as group extensions. Of course, we could simply test each pair of them for isomorphism as groups, but this would be very slow in many cases, particularly when the group $\mathrm{H}^2(G, N)$ is large.

In fact the classification of group extensions of N by G into isomorphism classes can be carried out much more efficiently by making use of the automorphism group $\mathrm{Aut}(G)$ of G, so we first compute this. For any $\alpha \in \mathrm{Aut}(G)$ and KG-module M, we can define another KG-module M^α with the same underlying K-space by defining m^{g^α} in M^α to be equal to m^g in M. This defines an action of $\mathrm{Aut}(G)$ on the set of isomorphism classes of KG-modules of dimension d.

If two KG-modules N_1 and N_2 lie in the same orbit under this action, where $N_1 \cong_K N_2 \cong_K N$, then the extensions of N by G that induce the module actions corresponding to N_1 and N_2 are easily seen to be isomorphic as group extensions. Indeed, we get from one to the other simply by renaming the elements of the extensions using the group automorphism α with $N_1^\alpha \cong_{KG} N_2$. Conversely, if the group extensions inducing N_1 and N_2 are isomorphic, then we have $N_1^\alpha \cong_{KG} N_2$ for some $\alpha \in \mathrm{Aut}(G)$. Hence we just need to compute $\mathrm{H}^2(G, N)$ for KG-modules N lying in a set of orbit representatives of $\mathrm{Aut}(G)$ on the KG-modules of dimension d.

Now we can concentrate on a fixed KG-module N. Let $A \leq \mathrm{Aut}(G)$ be the stabilizer of N in the action defined above, and let E be the group of KG-isomorphisms of N. For $\alpha \in A$, let $\phi_\alpha : N \to N$ be a module isomorphism from N^α to N, and let $\psi \in E$. Then $\phi_\alpha\psi$ is also a module isomorphism from N^α to N, and any such isomorphism is of the form $\phi_\alpha\psi$ for some $\psi \in E$. If $\nu \in \mathrm{Z}^2(G, N)$ is a 2-cocycle, then we can define a 2-cocycle $\nu^{(\alpha,\phi_\alpha\psi)}$ by

$$\nu^{(\alpha,\phi_\alpha\psi)}(g^\alpha, h^\alpha) = \nu(g, h)^{\phi_\alpha\psi}$$

for all $g, h \in G$. It is not hard to prove that the extensions of N by G defined by 2-cocycles ν and μ are isomorphic if and only if $\mu = \nu^{(\alpha,\phi_\alpha\psi)}$ for some $\alpha \in A$ and $\psi \in E$.

As we saw in Subsection 2.2, the data computed by our algorithm for $\mathrm{H}^2(G, N)$ allows us to compute and identify 2-cocycles. It is therefore straightforward to compute the extension defined by $\nu^{(\alpha,\phi_\alpha\psi)}$ from that defined by ν. It is not difficult to calculate A and E, but it turns out that the automorphism group B of the semidirect product $G \ltimes N$ consists precisely of the pairs

$(\alpha, \phi_\alpha \psi)$ with the obvious action on the elements of $G \ltimes N$. So we prefer to compute B, and then we can calculate the induced action of B on $\mathrm{H}^2(G, M)$. The extensions corresponding to the orbit representatives of this action are then representatives of the isomorphism classes of extensions of N by G.

Here are two easy examples, with verbose output. In our first example, we find all extensions of an elementary abelian group of order 81 by A_4.

```
>   time E := ExtensionsOfElementaryAbelianGroup(3, 4, Alt(4));
        2 irreducible modules
        5 modules of dimension 4
        5 modules of dimension 4 under automorphism action
          Next module
            Dimension of cohomology group = 1
              2 extensions
          Next module
            Dimension of cohomology group = 4
              2 extensions
          Next module
            Dimension of cohomology group = 3
              3 extensions
          Next module
            Dimension of cohomology group = 1
              2 extensions
          Next module
            Dimension of cohomology group = 2
              2 extensions
        Time: 1.960

>   #E;
        11

>   E := ExtensionsOfElementaryAbelianGroup(2, 6, Alt(5));
        3 irreducible modules
        10 modules of dimension 6
        9 modules of dimension 6 under automorphism action
          Next module
            Dimension of cohomology group = 2
              2 extensions
          Next module
            Dimension of cohomology group = 2
              2 extensions
          Next module
            Dimension of cohomology group = 2
              3 extensions
          Next module
            Dimension of cohomology group = 2
              2 extensions
          Next module
```

```
              Dimension of cohomology group = 2
                 3 extensions
           Next module
              Dimension of cohomology group = 1
                 2 extensions
           Next module
              Dimension of cohomology group = 1
                 2 extensions
           Next module
              Dimension of cohomology group = 2
                 2 extensions
           Next module
              Dimension of cohomology group = 6
                 2 extensions
         Time: 4.340
```

> #E;

```
         20
```

As our final example, we calculate the extensions of a nonabelian soluble group Q_8 (the quaternion group of order 8) by a group G, which is achieved by repeated application of the elementary abelian extension calculation. The verbose output here is not easily comprehensible, so we have omitted most of it!

> Q_8 := **sub**< Sym(8) | (1,2,3,4)(5,6,7,8), (1,8,3,6)(2,7,4,5) >;
> **time** E := ExtensionsOfSolubleGroup(Q_8, CyclicGroup(6));

```
         LEVEL: 1
           Layer size [ <2, 2>]
             2 irreducible modules
             3 modules of dimension 2
             3 modules of dimension 2 under automorphism action
               Next module
                 Dimension of cohomology group = 0
               Next module
                 Dimension of cohomology group = 2
               Next module
                 Dimension of cohomology group = 0
         4 EXTENSIONS
         LEVEL: 2
           Layer size [ <2, 1>]
         6 EXTENSIONS
         Time: 6.420
```

> #E;

```
         6
```

References

1. Wieb Bosma, John Cannon, Catherine Playoust, *The Magma algebra system I: The user language*, J. Symbolic Comput. **24** (1997), 235–265.
   See also the Magma home page at http://magma.maths.usyd.edu.au/magma/.
2. J. J. Cannon, D. F. Holt, *Finding the irreducible modules of a finite group over a finite field*, to appear.
3. J. J. Cannon, B. Cox, D. F. Holt, *Computing the subgroups of a permutation group*, J. Symb. Comput. **31** (2001), 149–161.
4. J. Cannon, M. Slattery, *Finite Soluble Groups*, Chapter 20, pp. 519–586 in: John Cannon, Wieb Bosma (eds.), *Handbook of Magma Functions*, Version 2.11, Volume **4**, Sydney, 2004.
5. F. Celler, J. Neubüser, C. R. B. Wright, *Some remarks on the computation of complements and normalizers in soluble groups*, Acta Applicandae Mathematicae **21** (1990), 57–76.
6. Bettina Eick, *Algorithms for Polycyclic Groups*. Habilitationsschrift, University of Kassel, 2000.
7. D. F. Holt, *The calculation of the Schur multiplier of a permutation group*, pp. 307–319 in: M. Atkinson (ed.), *Computational Group Theory*, Academic Press 1984.
8. D. F. Holt, *A computer program for the calculation of the covering group of a finite group*, J. Pure and Applied Algebra **35** (1985), 287–295.
9. D. F. Holt, *The mechanical computation of first and second cohomology groups*, J. Symb. Comput. **1** (1985), 351–361.
10. D. F. Holt, S. Rees, *Testing modules for irreducibility*, J. Austral. Math. Soc. Ser. A **57** (1994), 1–16.
11. D. F. Holt, *Cohomology and Extensions*, Chapter 25, pp. 647–664 in: John Cannon, Wieb Bosma (eds.), *Handbook of Magma Functions*, Version 2.11, Volume **4**, Sydney, 2004.
12. D. L. Johnson, *Presentations of Groups*, London Math. Soc. Student Texts **15**, Cambridge: Cambridge University Press, 1990.
13. Joseph J. Rotman, *An Introduction to the Theory of Groups*, Berlin: Springer-Verlag, 4th edition, 1994.
14. Charles C. Sims, *Computational methods in the study of permutation groups*, pp. 169–183 in: *Computational problems in abstract algebra*, Oxford: Pergamon Press, 1970,
15. H. Zassenhaus, *Über einen Algorithmus zur Bestimmung der Raumgruppen*, Comment. Math. Helvet. **21** (1948), 117–141.

Computing the primitive permutation groups of degree less than 1000

Colva M. Roney-Dougal and William R. Unger

School of Mathematics and Statistics
The University of Sydney
Sydney, Australia
colva@mcs.st-and.ac.uk
billu@maths.usyd.edu.au

Summary. In this chapter we describe how Magma was used to complete the classification of the primitive groups of degree less than 1000.

1 Some background

The classification of primitive groups of small degree has a long and rich history. The first major work in this area was a paper published by Jordan in 1872 [12] which enumerates the primitive groups of degrees 4 to 17. In 1874, Jordan correctly stated that every transitive group of degree 19 is either alternating, symmetric, or of affine type. In a series of papers at the turn of the century [15, 16, 17, 18, 19, 20, 21], Miller corrected the determination of the primitive groups of degrees 12 to 17, and by 1913 the classification had been extended to degree 20 [3, 14]. Many other people worked on this classification. The reader is referred to [24] for more details

In the 1960s, Sims used new computational techniques to redetermine this list for degree $d \leq 20$ [25], and by 1970 he had extended these methods to classify the primitive groups of degree less than 50. This latter classification was made available in 1977 as the PRMGPS database in V3.5 of CAYLEY [4], and was later turned into databases in both GAP [9] and Magma [2].

Dramatic progress was made in 1988, when Dixon and Mortimer [6] used the Classification of Finite Simple Groups and the O'Nan–Scott Theorem to classify the primitive groups with insoluble socles of degree less than 1000. This list was incorporated into GAP by Theißen, along with the primitive groups of degree less than 256 having soluble socles [24, 26]. However, several groups were missing from both of these classifications.

Thus, since 1987 the major open problem has been to classify the primitive groups with soluble socles of degree less than 1000. That is, to determine the primitive affine subgroups of $\mathrm{AGL}(n, p)$, for prime p and $p^n < 1000$. This chapter describes how Magma was used to perform this classification.

Finding the primitive affine subgroups of $\mathrm{AGL}(n, p)$ is equivalent to finding the irreducible subgroups of the matrix group $\mathrm{GL}(n, p)$. There are two main problems here. One is to find all such subgroups, the second is to reduce the subgroups found to one representative for each conjugacy class. Describing our solution to these two problems is the main part of this chapter.

The first problem is solved using Aschbacher's Theorem (see below), plus a heavy reliance on new Magma algorithms for computing maximal subgroups, as described in [5]. Our subgroup algorithm is described in § 2. The second problem required new techniques for finding conjugating elements in matrix groups, again based on Aschbacher's Theorem, as described in [23]. We describe our conjugacy finding methods in § 3.

Using the methods described in this chapter, we successfully classified the primitive groups of affine type of degree less than 1000.

2 Mathematical preliminaries

In this section we recall some basic mathematical definitions and results, before describing how we calculate the input to our subgroup algorithm.

We start begin by recalling some elementary permutation group definitions. The reader is referred to [7] for general background. A permutation group G acting on a set Ω is *primitive* if G is transitive and preserves no proper nontrivial equivalence relation on Ω. The *socle* of a finite group is the product of its minimal normal subgroups.

The O'Nan–Scott Theorem, as given in [7], states that that a finite primitive permutation group G must belong to one of 5 classes. One of these classes is the class of primitive groups with soluble socle. These are precisely the groups of affine type.

Let $V := \mathbb{F}_p^{(n)}$. An *affine transformation* of V is a map $t_{a,w} : V \to V$ where $a \in \mathrm{GL}(V)$, $w \in V$ and $t_{a,w}(v) := va + w$. The group of all affine transformations of V forms the *affine general linear group*, denoted $\mathrm{AGL}(n, p)$ or $\mathrm{AGL}(V)$. We consider $\mathrm{AGL}(V)$ as a permutation group, acting on the p^n vectors of V. The socle of $\mathrm{AGL}(V)$ may be identified with $(V, +)$, and the point stabiliser with $\mathrm{GL}(V)$. A subgroup $G \leq \mathrm{AGL}(V)$ is called a *group of affine type* if $V \trianglelefteq G$. It can be shown that G is primitive if and only if its point stabiliser is an irreducible subgroup of $\mathrm{GL}(V)$. Therefore to classify the primitive groups with soluble socles of degree less than 1000, we must classify the conjugacy classes of irreducible subgroups of $\mathrm{GL}(V)$ for $|V| < 1000$.

We will require the following elementary properties of linear groups. Let q be a prime power and let $G \leq \mathrm{GL}(n, q)$. Then G is *irreducible* if G does not stabilise any proper nontrivial subspace of $\mathbb{F}_q^{(n)}$. We say that G is *absolutely irreducible* if the image of G under the natural embedding into $\mathrm{GL}(n, \mathbb{F})$ is irreducible for all field extensions $\mathbb{F}$ of $\mathbb{F}_q$. The matrix group G is *imprimitive* if G is irreducible and preserves a direct sum decomposition $V = V_1 \oplus \cdots \oplus V_t$. If there exists a divisor s of n such that G can be embedded in $\Gamma L(n/s, q^s)$ then G is *semilinear*. The class of semilinear groups includes all groups that are not absolutely irreducible.

What follows is a simplified version of Aschbacher's theorem on subgroups of linear groups: the full version of the theorem describes classes of subgroups of all of the classical groups, but we give only the description for $GL(n, q)$. See [1] for a full statement of the theorem.

Theorem 2.1 (Aschbacher's theorem) *Let $G \leq GL(n, q)$ be given, let $q = p^e$, let $V := \mathbb{F}_q^n$ and let $Z := Z(GL(n, q))$. Then at least one of the following holds:*

1. *G is reducible.*
2. *G is imprimitive.*
3. *G is semilinear.*
4. *G preserves a tensor product decomposition $V = V_1 \otimes V_2$.*
5. *A conjugate of G can be embedded in $GL(n, q_0)Z$ for some q_0 dividing q.*
6. *G normalises an extraspecial r-group for some prime r dividing $q - 1$, or a 2-group of symplectic type.*
7. *G preserves a tensor induced decomposition $V = V_1 \otimes \cdots \otimes V_t$.*
8. *$G \leq N_{GL(n,q)}(C)$ for some classical group C.*
9. *For some nonabelian simple group T, the group $G/(G \cap Z)$ is almost simple with socle T. The quasisimple normal subgroup $(G \cap Z).T$ acts absolutely irreducibly and does not preserve any nondegenerate form.* □

Groups lying in class i of this classification are called $\mathcal{C}_i$ *groups*. Groups lying in $\cup_{i=1}^{8} \mathcal{C}_i$ are called *geometric groups*. Let $G \leq GL(n, q)$ be irreducible. We define G to be *potentially maximal* if either G is a $\mathcal{C}_9$ group, or G is geometric and is not a proper subgroup of a geometric group. The idea is that if there is any possibility that a group might be maximal, then we describe it as potentially maximal.

We write $G.H$ to denote an extension of a group G by a group H. For a split extension we write $G : H$, and for a central product we write $G \circ H$. When naming groups, the natural number n is the cyclic group of order n, and the symbol p^{1+2k} denotes an extraspecial p-group.

While we follow the notation of [13] for the classical groups, for the sake of clarity we will briefly review some notation. Let $\Gamma L(n, p^e)$ be the full semilinear group over $\mathbb{F}_{p^e}^{(n)}$, so that $\Gamma L(n, p^e) = GL(n, p^e) : e$ with the cyclic group acting on the general linear group by field automorphisms. For odd q, let $SL^{\pm}(n, q)$ be the subgroup of $GL(n, q)$ consisting of all matrices of determinant ± 1. Let $O^\epsilon(n, q)$, where ϵ is $+$, $-$ or omitted, be the largest subgroup of $GL(n, q)$ which preserves a nondegenerate quadratic form of type ϵ: these are the groups returned by the Magma functions $GO(n, q)$, $GOPlus(n, q)$ and $GOMinus(n, q)$.

We will require two lists of subgroups of $G := GL(n, p)$. The first list, $\mathcal{M}_1$, consists of the potentially maximal subgroups of G. The second list,

$\mathcal{M}_2$, consists of all groups H such that $H \leq K$ for some $K \in \mathcal{M}_1 \cup \mathcal{M}_2$, but $H \not\leq L$ for any $L \in \mathcal{M}_1 \cup \mathcal{M}_2$ whose subgroups may be computed using built in Magma functions. The idea is to place enough groups in $\mathcal{M}_2$ to ensure that we can find all irreducible subgroups of the groups in $\mathcal{M}_1$. Using Magma V2.9, one can compute the subgroups of a permutation group G whenever the order of each nonabelian simple composition factor of G is less than 1.6×10^7. The lists $\mathcal{M}_1$ and $\mathcal{M}_2$ are calculated using theoretical methods (see [22] for details).

The subgroups machinery of Magma V2.9 is sufficiently powerful that it can immediately compute all primitive subgroups of AGL(n, p) for all (n, p) with $p^n < 1000$ other than $\{(n, 2) : n \in \{6, 7, 8, 9\}\} \cup \{(5, 3), (6, 3), (4, 5)\}$. We note that, for Magma V2.11, the only group that remains on this list is AGL$(6, 3)$.

3 Determining conjugacy

We wish to determine the irreducible subgroups of GL(n, p). However, we only want to determine these groups up to conjugacy in the general linear group, as conjugate subgroups of GL(n, p) will yield permutation isomorphic subgroups of AGL(n, p). Unfortunately, subgroup conjugacy is in general a difficult problem, and standard subgroup conjugacy algorithms, based on backtrack search, can take a long time to reach a conclusion.

If two groups $H, K \leq G$ are close to being maximal subgroups of G, then it is usually comparatively quick to determine their conjugacy. However, if H and K are small and G is large then this can be a very time-consuming process.

In this section we describe some techniques that were used to speed up the determination of conjugacy of subgroups of GL(n, p). We write $H \sim_G K$ to denote that H is conjugate to K under G.

3.1 Proving that groups are not conjugate

The first stage in determining whether or not $H \sim_{\mathrm{GL}} K$ is to compute various group-theoretic invariants of H and K, and to examine H and K's permutation actions on $V := \mathbb{F}_p^{(n)}$. If any of these differ then we have a proof that $H \not\sim_{\mathrm{GL}} K$.

The following function[*] is used as a preliminary test to try to prove that H and K are not conjugate under GL(n, q), where H and K are represented as permutation groups. It returns two booleans: the first is true if and only if the function has been able to determine whether or not the groups are conjugate. If the first value is true then the second is true if the two groups are identical,

---

[*]See the Preface to this volume for style conventions regarding Magma code; code appearing in this book is available at http://magma.maths.usyd.edu.au/magma/.

and false if the function has proved them not conjugate. If the first value is false then the second value remains undefined.

```
NotConjTest := function(H, K)
    if #H ne #K or Transitivity(H) ne Transitivity(K) then
        return true, false;
    end if;
    if H eq K then
        return true, true;
    end if;
```

We now check that the orbit lengths of the k-point stabiliser are equal, where H and K are k-transitive. The function BasicStabiliser(L, $i+1$) takes as input a permutation group L for which a base and strong generating set are known, and an integer i that is less than the size of the base. It returns the subgroup of L that stabilises the first i points of the base. The function OrbitRepresentatives(L) takes as input a group L and returns a sequence of tuples $<l, r>$, one for each orbit, where l is the orbit length and r is a representative from the orbit.

```
    k := Transitivity(K);
    Hs := {* t[1]:t in OrbitRepresentatives(BasicStabilizer(H, k+1)) *};
    Ks := {* t[1]:t in OrbitRepresentatives(BasicStabilizer(K, k+1)) *};
    if Hs ne Ks then
        return true, false;
    end if;
```

If one group is soluble and the other is not then the groups are not conjugate.

```
    h_solv := IsSolvable(H);
    k_solv := IsSolvable(K);
    if not h_solv eq k_solv then
        return true, false;
    end if;
```

Most of our groups are soluble, so we pay particular attention to this case. First we check that H and K have the same number of conjugacy classes.

```
    if h_solv then
        h_classes := Classes(H); k_classes := Classes(K);
        if not #h_classes eq #k_classes then
            return true, false;
        end if;
```

Next we check that the classes can be matched up, in terms of having the same size and containing elements with the same cycle structure. $h\_classes[i][2]$ is the size of the i-th conjugacy class of H, and $h\_classes[i][3]$ is a representative element of that class.

```
        classlength := #h_classes;
        for i in [1 .. classlength] do
```

```
if not exists(t){ cl : cl in k_classes | h_classes[i][2] eq cl[2] and
    CycleStructure(h_classes[i][3]) eq CycleStructure(cl[3])} then
    return true, false;
  end if;
end for;
end if;
```

If we reach this part of the code, then we have not been able to prove non-conjugacy of H and K, nor have we found a conjugating element. Therefore, we return false. Since this function must return two values, we place a marker _ where the second return value would appear.

```
    return false, _;
  end function;
```

The function NotConjTest will be an invaluable cheap test for establishing that groups are not conjugate.

3.2 Proving that groups are conjugate

If NotConjTest(H, K) returns false, _ then we strongly suspect that $H \sim_{\mathrm{GL}} K$. To prove this, we need to find a conjugating element. The principal observation is that if two groups H and K are both subgroups of some maximal subgroup C of $\mathrm{GL}(n,p)$, and $H \sim_{\mathrm{GL}} K$, then it is very probable that $H \sim_C K$. See [23] for more details.

We use the functions for *matrix groups of large degree* to identify a type of Aschbacher structure that is preserved by both H and K. We then construct a group Overgroup which is a maximal subgroup of $\mathrm{GL}(n,p)$ that preserves a structure of this type, and conjugate H and K so that they are subgroups of Overgroup. We then use standard Magma functions to look inside Overgroup for a conjugating element. This algorithm is a *one-sided Monte Carlo* algorithm – if it finds a conjugating element then this is guaranteed to be correct, however there is a small probability that it may decide that two groups are not conjugate when in fact they are. For a full analysis, see [23].

For the irreducible matrix groups of degree less than 1000, the only cases for which the time taken to determine conjugacy proves to be a real obstacle are $\mathrm{GL}(4,5)$ and $\mathrm{GL}(6,3)$. For both of these, most of the irreducible subgroups are either semilinear or imprimitive (or both). We therefore only require functions to deal with these two cases.

Semilinear groups

This subsection describes our code which can be used as a fast method of searching for a conjugating element inside $\mathrm{GL}(n,p)$ in the semilinear case. It is a simplified version of the algorithm described in [23].

If a group H is not absolutely irreducible then some conjugate of H can be embedded in $\mathrm{GL}(n/s,p^s)$ for some s: in particular, the class of semilinear groups contains all groups which are not absolutely irreducible. However, a group may be absolutely irreducible and still semilinear.

Let $H \leq \mathrm{GL}(n,p)$ be semilinear. Then for some s dividing n, the group H embeds into $\Gamma\mathrm{L}(n/s, p^s)$, and so some $N \trianglelefteq H$ embeds into $\mathrm{GL}(n/s, p^s)$. Note that if H is not absolutely irreducible then $N = H$. By Schur's lemma the group $C_{\mathrm{GL}(n,p)}(N)$ has order $p^s - 1$. Both IsAbsolutelyIrreducible and IsSemiLinear will return a centralising matrix $C \in C_{\mathrm{GL}(n,p)}(N)$, of order dividing $p^s - 1$ but not $p^i - 1$ for any $i < s$.

The input to our conjugacy function is two groups $H, K \leq \mathrm{GL}(n,p)$ and a homomorphism ϕ, which is a faithful permutation representation of $\mathrm{GL}(n,p)$. It is assumed that H and K are semilinear.

We start by double-checking that the two groups are of the same dimension, and that they are defined over fields of the same size. To **assert** something in Magma is to declare that it must be true: if it is false, our function will terminate with an error message. This is a useful way of making absolutely sure that our input data is what we think that it is.

```
ConjSemilin := function(H, K, φ)
  n := Degree(H); assert n eq Degree(K);
  p := #BaseRing(H); assert p eq #BaseRing(K);
```

The command $\mathrm{GL}(n,p)@\phi$ returns the image of $\mathrm{GL}(n,p)$ under the map ϕ; which in this case is a permutation representation of $\mathrm{GL}(n,p)$.

```
  glp := GL(n,p)@φ;
```

First we check whether or not the groups are absolutely irreducible. If the groups are not absolutely irreducible then C_1 and C_2 are centralising matrices, and *ext_h* and *ext_k* are the degrees of the field extension, otherwise they will remain undefined. Obviously if one of them is absolutely irreducible and the other is not then they are not conjugate under $\mathrm{GL}(n,p)$.

```
  is_abs_irred_h, C₁, ext_h := IsAbsolutelyIrreducible(H);
  is_abs_irred_k, C₂, ext_k := IsAbsolutelyIrreducible(K);
  if not (is_abs_irred_h cmpeq is_abs_irred_k) then
    return false, _;
  end if;
```

The matrices C_1 and C_2 are returned as elements of the matrix algebra, so we tell Magma to coerce them into $\mathrm{GL}(n,p)$.

```
  if not is_abs_irred_h then
    if not ext_h eq ext_k then
      return false, _;
    end if;
    C₁ := GL(n,p) ! C₁; C₂ := GL(n,p) ! C₂;
  end if;
```

We now compute centralising matrices in the absolutely irreducible case, and check that in this instance the degrees of the field extensions are equal.

```
if is_abs_irred_h then
    C₁ := GL(n, p) ! CentralisingMatrix(H);
    C₂ := GL(n, p) ! CentralisingMatrix(K);
    if not DegreeOfFieldExtension(H) eq
        DegreeOfFieldExtension(K) then
        return false, _;
    end if;
end if;
```

The matrices C_1 and C_2 centralise the largest normal subgroups of H and K respectively which are conjugate to a subgroup of $\mathrm{GL}(n/s, p^s)$. However, it is not guaranteed that they generate the entire centraliser of these normal subgroups, so in particular the two matrices may have different orders even when H and K are conjugate.

```
o₁ := Order(C₁); o₂ := Order(C₂);
gcd := Gcd(o₁, o₂);
C₁ := C₁^(o₁ div gcd); C₂ := C₂^(o₂ div gcd);
```

We now find an element x of $\mathrm{GL}(n, p)$ that conjugates some power of C_2 to C_1. Conjugating by x will map K into $N_{\mathrm{GL}(n,p)}(C_1) = \Gamma\mathrm{L}(n/s, p^s)$. The **break** command will cause us to exit from the **for**-loop as soon as an x is found.

```
c1grp := sub< GL(n, p) | C₁ >;
c2grp := sub< GL(n, p) | C₂ >;
centralisers_conj := false;
for elt in c2grp do
    if Order(elt) eq Order(C₁) then
        centralisers_conj, x := IsSimilar(elt, C₁);
        if centralisers_conj then
            x := GL(n, p) ! x;
            break;
        end if;
    end if;
end for;
```

If the variable *centralisers_conj* is still false then we have not found a power of C_2 that is conjugate to C_1. We therefore cannot use this method to find an element conjugating H to K.

```
if not centralisers_conj then
    return false, _;
end if;
```

We compute the normaliser of *c1grp* in *glp*: this is a group n which is isomorphic to $\Gamma\mathrm{L}(n/s, p^s)$. We then look inside n for an element y that conjugates H to K^x.

```
n := Normaliser(glp, c1grp@φ);
conj_in_gaml, y := IsConjugate(n, H@φ, ((K^x)@φ));
```

If y is defined then $(H@\phi)^y = (K^x)@\phi$. The expression $y@@\phi$ returns the *preimage* of y under ϕ; that is a matrix representing y.

```
    if not conj_in_gaml then
        return false, _ ;
    else
        x₂ := (y@@φ)*x⁻¹;
        return true, x₂ ;
    end if;
end function;
```

We now have a function ConjSemilin which, on being given two conjugate semilinear subgroups of $\mathrm{GL}(n, p)$, will usually find a conjugating element far faster than the general subgroup conjugacy algorithm in Magma V2.9.

Imprimitive groups

As the functions for dealing with imprimitive groups are similar to those dealing with semilinear groups, we give only a brief description of the function ConjImprim(H, K, ϕ). The input groups H and K are matrix groups that are *assumed* to be imprimitive.

1. Run Blocks(H); Blocks(K); several times to determine whether or not H and K have block systems of the same size. This function returns a list of blocks, and is randomised so that it may return different sets of blocks if run more than once.
2. If #Blocks(H) is not equal to #Blocks(K) then return false.
3. Set $b :=$ #Blocks(H).
4. If IsConjugate(Sym(b), BlocksAction(H), BlocksAction(K))) is false then return false, as this tells us that the induced action of H on Blocks(H) is not permutation isomorphic to the induced action of K on Blocks(K).
5. Set Overgroup $:=$ WreathProduct(GL(n div b, p), Sym(b)).
6. For L in $[H, K]$ do
   - Set *blocks_L* := Blocks(L).
   - Let *mat_L* be a matrix whose rows are the basis vectors of *blocks_L*, and set M_L := *mat_L*$^{(-1)}$. Then L^{M_L} is a subgroup of Overgroup.
7. Return IsConjugate(Overgroup@ϕ, $H^{(M\_H)}$@ϕ, $K^{(M\_K)}$@ϕ).

This algorithm is analysed in detail in [23]: given two conjugate imprimitive subgroups $G, H \leq \mathrm{GL}(n, p)$, then it will in general find an $x \in \mathrm{GL}(n, p)$ such that $G^x = H$ in much less time than the general conjugacy algorithm in Magma V2.9.

3.3 The main conjugacy algorithm

In this section we describe our main conjugacy algorithm, IsGLConjugate. Here H and K are two subgroups of $\mathrm{GL}(n, p)$, represented as permutation groups, and ϕ is a homomorphism from the matrix representation of $\mathrm{GL}(n,p)$ to a faithful permutation representation.

```
IsGLConjugate := function(H, K, φ)
  h_mat := H@@φ;
  k_mat := K@@φ;
  n := Degree(h_mat); assert n eq Degree(k_mat);
  p := #BaseRing(h_mat); assert p eq #BaseRing(k_mat);
  glp := GL(n,p)@φ;
```

First we run NotConjTest to try to find a quick proof that H and K are not conjugate. Recall that the only time that NotConjTest will discover that two groups are conjugate is if they are identical, in which case the conjugating element is the identity.

```
a,b := NotConjTest(H, K);
if a then
  if b then
    return true, glp ! 1;
  else
    return false, _;
  end if;
end if;
```

Next we look to see if we can use ConjSemilin. Since IsSemiLinear is not always able to reach a decision, we can only decide at this stage that $H \not\sim_{\mathrm{GL}} K$ if IsSemiLinear returns true for one group and false for the other. Since IsSemiLinear may return a string or a boolean, we use *cmpeq* instead of *eq* to test equality: this command can compare objects of different types.

```
hsemi_lin := IsSemiLinear(h_mat);
ksemi_lin := IsSemiLinear(k_mat);
if not ((hsemi_lin cmpeq ksemi_lin) or (hsemi_lin cmpeq
          "unknown") or (ksemi_lin cmpeq "unknown")) then
  return false, _;
end if;
```

If IsSemiLinear returns true for both groups then we use ConjSemilin to look for a conjugating element. If it finds a conjugating element then we are done. If it fails to find one we do *not* conclude that they are not conjugate, as it could simply be that they are not conjugate subgroups of $\Gamma\mathrm{L}(n/t, p^t)$ for any divisor t of n.

```
if (hsemi_lin cmpeq true) and (ksemi_lin cmpeq true) then
  boolean, conjelt := ConjSemilin(h_mat, k_mat, φ);
  if boolean then return true, conjelt@φ; end if;
end if;
```

Next we check that it is not the case that one group is known to be primitive whilst the other is known to be imprimitive. Once again, IsPrimitive may not always be able to reach a decision.

```
h_imprim := IsPrimitive(h_mat);
k_imprim := IsPrimitive(k_mat);
if not ((h_imprim cmpeq k_imprim) or (h_imprim cmpeq
        "unknown") or (k_imprim cmpeq "unknown")) then
    return false, _;
end if;
```

If both groups are imprimitive then we can apply our additional conjugacy test. Note that we only accept a positive result from this test as a proof that the groups are conjugate, since there is a small possibility that ConjImprim could return false when in fact the two groups are conjugate.

```
if (h_imprim cmpeq false) and (k_imprim cmpeq false) then
    boolean, conjelt := ConjImprim(h_mat, k_mat, φ);
    if boolean then return true, conjelt @ φ; end if;
end if;
```

We are now reduced to standard permutation group methods for finding a conjugating element. These rely on backtrack search, which proved expensive in early tests, so we try to help out a little more. We do this by identifying matching characteristic subgroups of H and K, called HR and KR. If H and K are conjugate then HR and KR will be conjugate, so we test this. Either they are not conjugate, in which case H and K are not conjugate, or we find b with $HR^b = KR$. In this latter case we replace H by H^b and test for the conjugacy of H and K using a backtrack search within the normaliser of KR. The hope is that this normaliser is smaller than the original group, so the search will be faster. Of course, we have to compute the normaliser: this is generally large enough that it may be computed quickly.

Our first attempt at choosing a suitable characteristic subgroup is the soluble radical.

```
HR := Radical(H);
KR := Radical(K);
if #HR ne #KR then return false, _; end if;
if #HR gt 1 and #HR lt #H then
```

If the radicals are not trivial and not the whole group then we use them. To test conjugacy of the radicals we make a recursive call to this function. If the groups are not imprimitive or semilinear, then the recursive call will use the section of the code dealing with soluble groups, described below.

```
a, b := $$(HR, KR, φ);
if not a then return false, _; end if;
H := H^b;
if H eq K then return true, b; end if;
c, d := IsConjugate(Normaliser(glp, KR), H, K);
if c then return true, b*d; else return false, _; end if;
```

If the radicals are trivial then we select the socles as our characteristic subgroups.

```
elif #HR eq 1 then
   HR := Socle(H);
   KR := Socle(K);
   if H ne HR then
      a, b := $$(HR, KR, φ);
      if not a then return false, _; end if;
      H := H^b;
      if H eq K then return true, b; end if;
      c, d := IsConjugate(Normaliser(glp, KR), H, K);
      if c then return true, b∗d; else return false, _; end if;
```

If the groups are equal to their socles then we use the standard conjugacy function in Magma: this is where the recursive call in the previous paragraph may finish, if the socles are neither imprimitive nor semilinear.

```
   else
      c, d := IsConjugate(glp, H, K);
      if c then return true, d; else return false, _; end if;
   end if;
```

If the radicals are equal to the whole group, then the groups are soluble and we use the derived series.

```
   else
      dsH := DerivedSeries(H); dsK := DerivedSeries(K);
      len := #dsH;
      if len ne #dsK then return false, _; end if;
      if exists{i: i in [1..len]|#dsH[i] ne #dsK[i]} then
         return false, _;
      end if;
```

Finally we consider the penultimate term in the derived series. This is abelian and hence has regular orbits. The backtrack search used by Magma contains elements of the method described in [10] and can take advantage of this.

```
      if len le 2 then
         c, d := IsConjugate(glp, H, K);
         if c then return c, d; else return false, _; end if;
      else
         HR := dsH[len −1]; KR := dsK[len −1];
         a, b := IsConjugate(glp, HR, KR);
         if not a then return false, _; end if;
         H := H^b;
         if H eq K then return true, b; end if;
         c, d := IsConjugate(Normaliser(glp, KR), H, K);
         if c then return true, b∗d; else return false, _; end if;
      end if;
   end if;
end function;
```

We have now described an algorithm to determine the conjugacy of subgroups $G, H \leq \mathrm{GL}(n,p)$ that is dramatically faster than the generic subgroup conjugacy algorithm. It operates in three main stages. First it runs **NotConjTest** to try to find a proof that $G \not\sim_{\mathrm{GL}} H$. If this is inconclusive then if G and H are either semilinear or imprimitive, it uses geometric techniques to search quickly for a conjugating element. If this also fails, either because both G and H are primitive and not semilinear or because a conjugating element has not been found, then it finishes with an enhanced version of the standard conjugacy algorithm for permutation groups, which exploits characteristic subgroups of G and H.

4 Maximal irreducible subgroups of $\mathrm{GL}(4,5)$

In this section we describe our construction of the potentially maximal subgroups of $\mathrm{GL}(4,5)$. These will be used as initial data for our main algorithm, which is described in § 5.

Proposition 4.1 [22, Theorem 3.3] *Let G be an irreducible maximal subgroup of $\mathrm{GL}(4,5)$. Then G lies in the following list:*

$$\mathcal{M}_1 = \{\mathrm{GL}(1,5) \wr \mathrm{Sym}(4), \quad \mathrm{GL}(2,5) \wr \mathrm{Sym}(2), \quad \Gamma\mathrm{L}(2,25),$$
$$(4 \circ 2^{1+4}) . \mathrm{Sp}(4,2), \quad (\mathrm{GL}(2,5) \circ \mathrm{GL}(2,5)) : \mathrm{Sym}(2),$$
$$N_{\mathrm{GL}(4,5)}(\mathrm{O}^-(4,5)), \quad N_{\mathrm{GL}(4,5)}(\mathrm{Sp}(4,5)), \quad \mathrm{SL}^{\pm}(4,5)\}.$$

Most of these groups are completely straightforward to construct. We make the group $\mathrm{SL}^{\pm}(4,5)$ by adding a matrix of determinant -1 to the list of generators for $\mathrm{SL}(4,5)$. Since $\mathrm{SL}^{\pm}(4,5)$ is too big for **Magma** V2.9 to compute its subgroups automatically, we must include some of its maximal subgroups, so we append $\mathrm{SL}(4,5)$. However, all other maximal subgroups of $\mathrm{SL}^{\pm}(4,5)$, and all maximal subgroups of $\mathrm{SL}(4,5)$, are contained in other groups in our list which are small enough for **Magma** to compute their maximal subgroups. The first three groups are those with composition factors which are too large for **Magma** V2.9's **MaximalSubgroups()** function. (We note that from **Magma** V2.11, **MaximalSubgroups()** may be successfully applied to all of these groups.)

```
init_seq := [
  GL(4,5),
  sub< GL(4, 5) | SL(4,5), GL(4,5) ! DiagonalMatrix([−1,1,1,1]) >,
  SL(4,5),
  WreathProduct(GL(1,5), Sym(4)),
  WreathProduct(GL(2,5), Sym(2)),
```

The next statement creates the group $(\mathrm{GL}(2,5) \circ \mathrm{GL}(2,5)) : \mathrm{Sym}(2)$. This group resembles a wreath product, but the base group acts on a *tensor* product of subspaces of V.

```
  TensorWreathProduct(GL(2,5), Sym(2)),
```

There is no function in Magma to construct the group $\Gamma L(2, 25) \leq GL(4, 5)$, but WriteOverSmallerField will represent $GL(2, 25)$ as a group over $\mathbb{F}_5$. Its normaliser in $GL(4, 5)$ is precisely $\Gamma L(2, 25)$.

```
Normaliser( GL(4,5), WriteOverSmallerField(GL(2,25), GF(5)) )
];
```

We use the Subgroups function to create the group $(4 \circ 2^{(1+4)}) . \mathrm{Sp}(4, 2)$. We start by making a faithful permutation representation of $GL(4, 5)$, as the Subgroups function does not at present apply to matrix groups. The OrbitAction function also returns a map ϕ from the natural representation of $GL(4, 5)$ to our permutation representation.

```
gl := GL(4,5);
V := RSpace(gl);
φ, glp := OrbitAction(gl, V.1);
```

The normalised 2-group $T \cong 4 \circ 2^{(1+4)}$ has order 64, and when take the quotient of T by the central element $-I$ of order 2, we get an elementary abelian group of order 32. We use this to find T inside the Sylow 2-subgroup of $GL(4, 5)$, and finally obtain the desired group as its normaliser. We set up the quotient of the Sylow 2-group by the central element of order 2. For reasons of time and memory, quotient groups should only be constructed when they are known to be reasonably small.

```
S := Sylow(glp, 2);
Z := Centre(S);
assert exists(x){ x : x in Generators(Z) | Order(x) eq 4 };
Q, onto := quo< S | x² >;
```

The quo constructor returns the quotient group of S by x^2, and the natural map from S to Q. Now we find $T/\langle -I \rangle$: the restrictions that we place on the Subgroups function make it run much more quickly, and there turns out to be only one class of such subgroups in Q. We obtain T by pulling the subgroup of Q back into S, and finish by pulling back the normaliser of T into $GL(4, 5)$.

```
sQ := Subgroups(Q : OrderEqual := 32, IsElementaryAbelian :=true);
T := sQ[1]`subgroup @@ onto;
Append(~init_seq, Normaliser(glp, T)@@φ);
```

Finally, we take the normalisers of the two remaining classical groups. At this stage *init_seq* is $\mathcal{M}_1 \cup \mathcal{M}_2$, as described in §2.1, together with $GL(4, 5)$.

```
init_seq cat:= [ Normaliser(GL(4,5), GOMinus(4,5)),
                 Normaliser(GL(4,5), Sp(4,5)) ];
```

5 The main algorithm

In this section, we describe our main loop for finding the irreducible subgroups of $GL(n, p)$. We give it the list of subgroups that we constructed in §4 as starting data. Whilst creating the full list of subgroups, most of the computing time is taken up with checking for duplicates: that is, conjugacy testing in the general linear group. Thus the algorithms of §3 play a crucial role. When run, the code in this section returns a list $\mathcal{I}$ of conjugacy class representatives of all irreducible subgroups of $GL(n, p)$.

The lists of groups that we use are *wait_perm*, which is the queue of permutation groups whose maximal subgroups are yet to be computed, and *done_mat*, the matrix groups for which we have maximal subgroups. At the end, *done_mat* will contain all the irreducible subgroups of $GL(4, 5)$.

```
done_mat := init_seq[1..3];
wait_perm := [ H@φ : H in init_seq ];
```

After initialising *done_mat* and *wait_perm*, we start going through the queue, computing maximal subgroups and discarding the reducible ones. We include a print statement, as the code is likely to run for some time, and it is helpful to be able to monitor its progress.

```
i := #done_mat;
while i lt #wait_perm do
    i +:= 1;
    Hp := wait_perm[i];
    Append(~done_mat, Hp @@ φ);
    "Expanding nr", i, "Order", #Hp, "Found", #wait_perm;
    max_perm := [ x`subgroup : x in MaximalSubgroups(Hp) |
                    IsIrreducible(x`subgroup @@ φ) ];
```

We now test each new irreducible maximal subgroup for conjugacy with subgroups already found, and check that if a conjugating element is found then it really works. This isn't strictly necessary, but since we are using the first version of IsGLConjugate it is wise to double-check. The **continue** statement tells Magma to go back to the beginning of the **for**-loop and look at the next group if a conjugate of *max_perm*[j] is found in *wait_perm*. If none is found, we append *max_perm*[j] to the queue.

```
    for j in [1..#max_perm] do
        if exists{ H : H in wait_perm | conj and max_perm[j]^x eq H
                where conj, x is IsGLConjugate(max_perm[j], H, φ)} then
            continue j;
        end if;
        Append(~wait_perm, max_perm[j]);
    end for;
end while;
```

After around 16 hours of computing time, we print out the list of subgroups.

"Found", **#done_mat**, "irreds";
done_mat:Magma;

As noted above, we checked that a true result from IsGLConjugate was correct. To confirm that false results were also correct, we checked the final list of primitive groups to be sure that there were no conjugates in the list. We describe the steps taken to check for duplicates in [22].

We find that GL(4,5) has a total of 647 irreducible subgroups, up to conjugacy, and, therefore, there are 647 primitive groups of degree 625 which have soluble socles.

6 Results

The full algorithm, including the construction of maximal subgroups as in § 4, was run for the groups $GL(d,2)$ with $6 \leq d \leq 9$, $GL(d,3)$ with $5 \leq d \leq 6$ and $GL(4,5)$. In all remaining cases with $p^d < 1000$ the Subgroups function in Magma V2.9 was able to compute the full list of subgroups.

In Table 1 we list the numbers of classes of soluble and insoluble irreducible subgroups of $GL(d,p)$ for all natural numbers d and primes p with $p^d < 1000$. In the second row of the table, S stands for soluble and I for insoluble.

Table 1. Number of soluble and insoluble irreducible subgroups of $GL(d,p)$

$d =$	2		3		4		5		6		7		8		9	
	S	I	S	I	S	I	S	I	S	I	S	I	S	I	S	I
$p = 2$	2	0	2	1	10	10	2	1	40	24	2	1	129	109	21	15
3	7	0	9	2	108	37	16	18	324	147						
5	19	3	22	11	509	138										
7	29	4	62	14												
11	42	6														
13	62	6														
17	75	5														
19	77	9														
23	54	4														
29	100	10														
31	114	12														

As the groups are too numerous to list here, they may be found on the web page http://magma.maths.usyd.edu.au/users/colva. The corresponding groups of affine type have been included in Magma's primitive groups database, which now contains all primitive groups of degree less than 1000. These groups are accessed by PrimitiveGroup(*degree, num*), where *degree* is the degree of the group and *num* is its position in the list of primitive groups of that degree. This function returns three things: the primitive permutation group, a string representing the name of the group (possibly empty), and a string representing its O'Nan–Scott class. An irreducible matrix group corresponding to the primitive group G of affine type may be obtained by invoking the function MatrixQuotient(G).

References

1. M. Aschbacher, *On the maximal subgroups of the finite classical groups*, Invent. Math. **76** (1984), 469–514.
2. Wieb Bosma, John Cannon, Catherine Playoust, *The Magma algebra system I: The user language*, J. Symbolic Comput. **24** (1997) 235–265.
   See also the Magma home page at http://magma.maths.usyd.edu.au/magma/.
3. E. R. Bennett, *Primitive groups with a determination of the primitive groups of degree 20*, Amer. J. Math. **34** (1912), 1–20.
4. J. J. Cannon, *An introduction to the group theory language Cayley*, pp. 145–183 in: Michael D. Atkinson (ed.), *Computational Group Theory*, London: Academic Press 1984.
5. J. J. Cannon, D. F. Holt, *Computing maximal subgroups of finite groups*, J. Symbolic Comput. **37** (2004), 589–609.
6. J. D. Dixon, B. Mortimer, *The primitive permutation groups of degree less than 1000*, Math. Proc. Cambridge Philos. Soc. **103** (1988), 213–238.
7. J. D. Dixon, B. Mortimer, *Permutation Groups*, New York: Springer, 1996.
8. B. Eick, B. Höfling, *The solvable primitive permutation groups of degree at most 6560*, LMS J. Comput. Math., bf 6 (2003), 29–39.
9. The GAP Group, *GAP – Groups, Algorithms and Programming, Version 4.3*, 2002. (http://www.gap-system.org).
10. D. F. Holt, *The computation of normalisers in permutation groups*, J. Symbolic Comp. **12** (1991), 499–516.
11. D. F. Holt, C. R. Leedham-Green, E. A. O'Brien, S. Rees, *Computing matrix group decompositions with respect to a normal subgroup*, J. Algebra **184** (1996), 818–838.
12. C. Jordan, *Traitè des Substitutions et des Equations Algébriques*, reprinted by Albert Blanchard, 1957.
13. P. Kleidman, M. Liebeck, *The subgroup structure of the finite classical groups*, Cambridge: Cambridge University Press, 1990.
14. E. N. Martin, *On the imprimitive substitution groups of degree fifteen and the primitive substitution groups of degree eighteen*, Amer. J. Math. **23** (1901), 259–286.
15. G. A. Miller, *Note on the transitive substitution groups of degree 12*, Bull. Amer. Math. Soc. **1-2** (1895), 255–258.
16. G. A. Miller, *List of transitive substitution groups of degree 12*, Quart. J. Pure Appl. Math. **28** (1896), 193–231. Erratum: Quart. J. Pure Appl. Math. **29** (1898), 249.
17. G. A. Miller, *On the primitive substitution groups of degree fifteen*, Proc. London Math. Soc. **28**-1 (1897), 533–544.
18. G. A. Miller, *Sur l'énumeration des groupes primitifs dont le degré est inférieur à 17*, C. R. Acad. Sci. **124** (1897), 1505–1508.

19. G. A. Miller, *On the transitive substitution groups of degrees 13 and 14*, Quart. J. Pure Appl. Math. **29** (1898), 224–249.

20. G. A. Miller, *On the primitive substitution groups of degree 16*, Amer. J. Math. **20** (1898), 229–241.

21. G. A. Miller, *On the transitive substitution groups of degree seventeen*, Quart. J. Pure Appl. Math. **31** (1900), 49–57.

22. C. M. Roney-Dougal, W. R. Unger, *The primitive affine permutation groups of degree less than 1000*, J. Symbolic Comput. **35** (2003), 421–439.

23. C. M. Roney-Dougal, *Conjugacy of subgroups of the general linear group*, Experiment. Math. **13** (2004), 151–163.

24. M. W. Short, *The Primitive Soluble Permutation Groups of Degree less than 256*, Berlin: Springer-Verlag, 1991.

25. C. C. Sims, *Computational methods for permutation groups*, pp. 169–183 in: J. Leech (ed.), *Computational Problems in Abstract Algebra*, Pergamon, 1970.

26. H. Theißen, *Eine Methode zur Normalisatorberechnung in Permutationsgruppen mit Anwendungen in der Konstruktion primitiver Gruppen*. Dissertation, Aachen: Rheinisch Westfälische Technische Hochschule, 1997.

Computer aided discovery of a fast algorithm for testing conjugacy in braid groups

Volker Gebhardt

School of Computing and Mathematics
University of Western Sydney
Sydney, Australia
v.gebhardt@uws.edu.au

1 Introduction

This chapter describes how Magma [3] was used to investigate and understand a phenomenon observed when implementing a conjugacy test for elements of a braid group. These investigations ultimately lead to the discovery of a new invariant of conjugacy classes in braid groups, to an efficient way of computing this invariant, and in particular to a much more powerful conjugacy test than the one which was originally to be implemented [11].

While at the end of this journey stood a purely theoretical result whose proof does not involve any computations, using Magma proved crucial both for recognising the new class invariant and its important properties and for getting ideas as to how to prove these results.

The structure of this article is as follows. In Section 2 we introduce Artin braid groups and outline the known approaches to the conjugacy and conjugacy search problems. Both problems have been known to be solvable for some time, but computations in practice have been seen to be hard or infeasible even for moderate values of the relevant parameters, whence the conjugacy search problem came under consideration as possible basis for public key cryptosystems. In Section 3 we describe how during the process of implementing the established algorithms, evidence emerged for the existence of a much better way of solving the conjugacy and the conjugacy search problems. Sections 4 and 5 describe how Magma was used systematically for gaining insight into the problem and for finding a proof for the conjectured results. In Section 6, finally, we describe one of the cryptographic protocols based on braid groups, describe an attack on this protocol based on conjugacy search and compare the performance of our new approach to that of the established algorithms.

Because of limited space, we can only present very few actual computations. The structures described later in this chapter became apparent only following extensive tests involving many examples. Some of these examples were chosen at random, others emerged in systematic searches for group elements satisfying certain properties.

You are encouraged to do your own experiments. The functions ("intrinsics") written in the Magma programming language for our investigations are provided in the package file braid.m, which must be attached using the command

Attach("braid.m");

to make them available in a Magma session* Note, however, that the established class invariant which forms the starting point for our work, the so-called *super summit set* introduced in Section 2.3, is quite difficult to compute. (This is the very reason why our results are interesting!) So keep the values of braid index and canonical length (see Section 2) small, be patient if computations take a bit longer ... and don't get too upset when you run out of memory occasionally.

For a detailed explanation of the way in which braid groups and their elements can be defined and used in Magma, we refer to the Magma documentation, [12].

2 Background: braid groups and testing conjugacy

This section outlines some facts about Artin braid groups which form the starting point of our investigations. For more details and proofs we refer the interested reader to [1, 4, 7, 8, 10]. We remark that the results described in this section, and indeed the result obtained in [11], can be extended to a more general class of groups, the so-called *Garside groups* [5, 6, 14].

2.1 Artin braid groups

The notion of braid groups was introduced by Artin [1], who considered a sequence B_n $(n = 1, 2, \dots)$ of groups, where B_n, the *braid group on n strings*, is presented on the generators $\sigma_1, \dots, \sigma_{n-1}$ with the defining relations

$$\sigma_i \sigma_j = \sigma_j \sigma_i \qquad (1 \leq i < j < n, \quad j - i > 1)$$
$$\sigma_i \sigma_{i+1} \sigma_i = \sigma_{i+1} \sigma_i \sigma_{i+1} \quad (1 \leq i < n - 1).$$

The elements of B_n can be thought of as operations on a set of n strings in a plane, running "essentially in parallel" from left to right with positions of strings numbered from top to bottom. The generator σ_i corresponds to intertwining the strings at positions i and $i+1$ once, with the string at position i passing over the string at position $i + 1$.

---

*See the Preface to this volume for style conventions regarding Magma code; code appearing in this book is available at http://magma.maths.usyd.edu.au/magma/.

Let B_n^+ denote the set of elements of B_n which can be represented as words in $\sigma_1, \ldots, \sigma_{n-1}$ not containing inverses of generators. It turns out that we obtain a partial ordering on B_n by defining $x \preceq y$ if $x^{-1}y \in B_n^+$ for $x, y \in B_n$. In this case, we say that x *is a divisor of* y. Moreover, for any $x, y \in B_n$ there is a unique element $x \wedge y \in B_n$ such that $x \wedge y \preceq x$, $x \wedge y \preceq y$ and $m \preceq x$ and $m \preceq y$ together imply $m \preceq x \wedge y$ for all $m \in B_n$. Similarly, there is a unique element $x \vee y \in B_n$, such that $x \preceq x \vee y$, $y \preceq x \vee y$ and $x \preceq m$ and $x \preceq m$ together imply $x \vee y \preceq m$ for all $m \in B_n$. That is, B_n is a lattice with respect to the partial ordering $\preceq$ with least common multiple (lcm) and greatest common divisor (gcd) operations given by $\vee$ and $\wedge$, respectively.

The *fundamental element* $\delta = \sigma_1 \vee \cdots \vee \sigma_{n-1}$ and the automorphism $\tau : x \mapsto x^\delta = \delta^{-1}x\delta$ of B_n which conjugation by δ induces play a special role. It can be shown that τ maps the set B_n^+ to itself. Consequently, the partial ordering $\preceq$ is invariant under τ, that is, $x \preceq y$ if and only if $\tau(x) \preceq \tau(y)$ holds for all $x, y \in B_n$.

2.2 Permutation braids and normal form

The elements of B_n^+ which are divisors of δ are called *simple elements* or *permutation braids*. The reason for the latter is that every simple element of B_n is described uniquely by the permutation it induces on the n strings on which B_n acts.[1] We denote the set of simple elements by D.

We can use Magma to check what δ looks like for B_7.

```
>    B := BraidGroup(7);
>    δ := FundamentalElement(B);
>    δ eq LCM({B.i : i in [1..6]}); // this had better be true
        true

>    InducedPermutation(δ);
        (1, 7)(2, 6)(3, 5)
```

It can be shown that, for arbitrary n, the element δ can be represented as the word $(\sigma_1 \cdots \sigma_{n-1})(\sigma_1 \cdots \sigma_{n-2}) \cdots (\sigma_1\sigma_2)\sigma_1$ and that the permutation it induces on the strings is $(1, n)(2, n-1) \cdots (\lfloor n/2 \rfloor, \lceil n/2 \rceil + 1)$.

The key to the study of braid groups is the existence of a unique way of writing any given element as product of simple elements of a certain form. Roughly speaking, uniqueness is achieved by requiring factors to occur as far to the left as possible in the product of simple elements.

Theorem 2.1 *For every $x \in B_n$ there exist unique integers $r \geq 0$ and k and unique simple elements $A_1, \ldots, A_r \in D \setminus \{1, \delta\}$ such that $x = \delta^k A_1 \cdots A_r$ and $A_{i-1}^{-1}\delta \wedge A_i = 1$ for $i = 2, \ldots, r$. Moreover, $A_i = \delta \wedge (\delta^k A_1 \cdots A_{i-1})^{-1}x$ for $i = 1, \ldots, r$.*

---

[1]Because of certain implementation details, Magma internally uses the inverse of this permutation for representing permutation braids, in particular for printing.

We call k the *infimum* of x, denoted $\inf(x)$, r the *canonical length* of x, denoted $\text{len}(x)$, and $r + k$ the *supremum* of x, denoted $\sup(x)$. The product representation $\delta^k A_1 \cdots A_r$ is called the *normal form* of x.

Note that the element $A_{i-1}^{-1} \delta$ is the maximal element which can be multiplied from the right to A_{i-1} such that the product is still simple. Hence the requirement $A_{i-1}^{-1} \delta \wedge A_i = 1$ in Theorem 2.1 means that no divisor of A_i can be moved from the factor A_i to the factor A_{i-1} while keeping all factors simple.

The normal form of any element of B_n^+ can be obtained by repeatedly computing its gcd with δ, as shown below. (Note that this code does not represent the algorithm employed by the Magma command NormalForm.)

```
intrinsic MyNormalForm(x :: GrpBrdElt)  →  RngIntElt, SeqEnum
{
   Input:   a positive element x of a braid group
   Output:
              - inf(x)
              - a sequence of permutations defining A_1,...,A_r
}
   require Id(Parent(x)) le x : "The argument must be positive.";
   δ := FundamentalElement(Parent(x));
   k := 0;
   seq := [ ];
   while not IsId(x) do
      d := GCD(x,δ);
      if d eq δ then
         k +:= 1;
      else
         Append(~seq, InducedPermutation(d)⁻¹);
      end if;
      x := d⁻¹ * x;
   end while;
   return k, seq;
end intrinsic;
```

2.3 Super summit sets and conjugacy testing

The number of conjugates of an element of B_n in general is infinite. We can decide in finite time whether two elements are conjugate if we can compute a finite invariant of the conjugacy class of a given element of B_n. All known deterministic algorithms solving the conjugacy problem work this way and the invariant used in the most efficient established conjugacy tests is a subset of the conjugacy class, namely the set of conjugates having maximal infimum and minimal supremum.

For an element $x \in B_n$ let x^{B_n} denote the set of conjugates of x and define $\inf_s(x) = \max\{\inf(y) : y \in x^{B_n}\}$ and $\sup_s(x) = \min\{\sup(y) : y \in x^{B_n}\}$.

We call $S_x = \{y \in x^{B_n} : \inf(y) = \inf_s(x), \sup(y) = \sup_s(x)\}$ the *super summit set* of x. The set S_x clearly depends only on the conjugacy class of x and S_x is finite, since the number of simple elements of B_n is finite.

Using the normal form of elements introduced in Section 2.2, we define two maps on B_n. Let $\delta^k A_1 \cdots A_r$ be the normal form of $x \in B_n$. If $r = 0$, define $\mathbf{c}(x) = \mathbf{d}(x) = x$, otherwise define $\mathbf{c}(x) = x^{\tau^{-k}(A_1)}$ and $\mathbf{d}(x) = x^{A_r^{-1}}$. We call $\mathbf{c}(x)$ the *cycling* of x and $\mathbf{d}(x)$ the *decycling* of x. It is easy to check that $\mathbf{c}(x^\tau) = \mathbf{c}(x)^\tau$ and $\mathbf{d}(x^\tau) = \mathbf{d}(x)^\tau$ hold for all $x \in B_n$.

We can use cycling and decycling operations to compute a representative $\tilde{x}$ of S_x for given $x \in B_n$.

Theorem 2.2 (i) *For any element of B_n, cycling and decycling neither decrease the infimum nor increase the supremum.*

(ii) *For any $x \in B_n$, a representative $\tilde{x}$ of S_x can be obtained by applying a finite sequence of cycling and decycling operations to x.*

In particular, the super summit set of any element is non-empty. Hence two elements x and y of B_n are conjugate if and only if $S_x = S_y$ or, equivalently, if and only if $S_x \cap S_y \neq \emptyset$.

It is shown in [2] that the infimum and supremum are extremal if they remain unchanged under $n(n-1)/2 - 1$ cycling and decycling operations, respectively. Hence we can compute a representative $\tilde{x}$ of S_x as follows. (We record the conjugating elements for the cycling and decycling operations for later use.)

intrinsic MySuperSummitRepresentative(x :: GrpBrdElt)

$\rightarrow$ GrpBrdElt, GrpBrdElt

```
{
   Input:   an element x of a braid group
   Output:
               - a representative of S_x
               - an element conjugating x to this representative
}
n := NumberOfStrings(Parent(x));
conj := Id(Parent(x));
// maximise infimum
count := n*(n−1)/2 − 1;
inf := Infimum(x);
while count gt 0 and CanonicalLength(x) gt 0 do
   x, c := Cycle(x); count −:= 1;
   conj := conj * c;
   if Infimum(x) gt inf then
      count := n*(n−1)/2 − 1;
      inf := Infimum(x);
   end if;
end while;
```

```
// minimise supremum
count := n*(n−1)/2 − 1;
sup := Supremum(x);
while count gt 0 and CanonicalLength(x) gt 0 do
  x, c := Decycle(x); count −:= 1;
  conj := conj * c;
  if Supremum(x) lt sup then
    count := n*(n−1)/2 − 1;
    sup := Supremum(x);
  end if;
end while;
return x, conj;
end intrinsic;
```

The following "convexity" result allows us to compute the super summit set of an element, starting from a single representative, as the closure with respect to conjugation by simple elements in a finite number of steps.

Theorem 2.3 *Let $x \in B_n$. For arbitrary $y, z \in S_x$ there exist elements $y_0, \ldots, y_t \in S_x$ and elements $c_1, \ldots, c_t \in D$ such that $y_0 = y$, $y_t = z$ and $y_{i-1}^{c_i} = y_i$ for $i = 1, \ldots, t$.*

Given elements $x \in B_n$, $y \in S_x$ and a simple element $s \in D$, the following result from [9] implies the existence of a unique $\preceq$-minimal element $\rho_s = \rho_s(y)$ satisfying $s \preceq \rho_s \preceq \delta$ and $y^{\rho_s} \in S_x$.

Theorem 2.4 *Let $x \in B_n$, $y \in S_x$ and $u, v \in D$. If $y^u \in S_x$ and $y^v \in S_x$ then $y^{u \wedge v} \in S_x$.*

An algorithm for computing $\rho_s(y)$ is described in [9]. The Magma function MinimalElementConjugatingToSuperSummit provides an implementation of this algorithm.

Using the function MySuperSummitRepresentative from above, we can test whether two elements of a braid group are conjugate, and, if they are, compute a conjugating element[2] as follows.

```
intrinsic MyIsConjugate(x :: GrpBrdElt, y :: GrpBrdElt)
                            → BoolElt, RngIntElt, GrpBrdElt
{
  Input:  two elements x and y of a braid group
  Output:
            - true if x and y are conjugate, false otherwise
            - the number of computed super summit elements
            - an element conjugating x to y (if applicable)
}
```

---

[2] The problem of finding a conjugating element for two given elements which are known to be conjugate, is referred to as *conjugacy search problem.*

```
x, c_x := MySuperSummitRepresentative(x);
S := {@ x @};
conj := {@ c_x @};
y, c_y := MySuperSummitRepresentative(y);
if y eq x then
    return true, 1, c_x * c_y⁻¹;
end if;
// close S with respect to conjugation by simple elements
pos := 1;
while pos le #S do
    for s in { MinimalElementConjugatingToSuperSummit(S[pos], a)
                    : a in Generators(Parent(x)) } do
        ns := LeftNormalForm(S[pos]ˢ);
        if ns notin S then
            Include(~S, ns);
            Include(~conj, LeftNormalForm(conj[pos]*s));
            if y eq ns then
                return true, #S, conj[pos] * s * c_y⁻¹;
            end if;
        end if;
    end for;
    pos +:= 1;
end while;
// S is closed with respect to conjugation by simple elements,
// that is, S is the super summit set of x. Since S does not
// contain y, the super summit sets of x and y are distinct.
    return false, #S, _;
end intrinsic;
```

Note that we used Theorem 2.4 to restrict the number of conjugates by simple elements tested in the for-loop. For later use the function also returns the number of elements of S_x which had to be computed.

The functions MySuperSummitRepresentative and MyIsConjugate above are a simplified version of the conjugacy test which was to be implemented for Magma. Section 3 describes what happened during tests of the implementation.

2.4 Limitations for conjugacy testing and conjugacy search

Unfortunately, the conjugacy test and the search for conjugating elements as outlined in Section 2.3 have severe limitations in practice. If the arguments to MyIsConjugate are not conjugate, the entire super summit set S_x of the first argument has to be computed in any case. For conjugate arguments, the same is true in the worst case, while on average one would expect that about half of the elements of S_x have to be computed.[3] Hence both in the worst case and in the average case, time and memory requirements are proportional to $|S_x|$.

The best proven bound for $|S_x|$ is exponential both in n and in len(x). While it is conjectured that for fixed n a polynomial bound in len(x) exists, $|S_x|$ appears to grow exponentially in n, that is, the complexity of conjugacy testing using the function MyIsConjugate is exponential in n. In practice, computations are hard or infeasible even for moderate values of n and len(x).

Recently, interest has grown in braid groups as a possible basis for public key cryptosystems, and it is this hardness of (variations of) the conjugacy problem, on which the security of most of the proposed cryptosystems is based. In Section 6 we will present an example of a protocol for key exchange over an insecure channel using braid groups.

3 Coming across another class invariant

This section describes how the conjugacy test from Section 2 behaved in practice and how this behaviour pointed towards another invariant of conjugacy classes and, more importantly, to the fact that this invariant might be an interesting thing to look at.

3.1 A small surprise

Taking a look at the function MyIsConjugate, one would expect that if the first argument x is fixed, the number of computed super summit elements varies according to the result of the conjugacy test. If the second argument is not conjugate to x, the entire super summit set S_x of x must be computed before the negative result can be established. If, however, y is conjugate to x then the super summit representative $\tilde{y}$ of y computed in the fourth line of the function MyIsConjugate is an element of S_x and the test can be aborted with a positive result as soon as $\tilde{y}$ is encountered when computing S_x.

For a uniform distribution of $\tilde{y}$ in S_x, one would expect that for conjugate arguments, on average, $|S_x|/2$ super summit elements would have to be computed before $\tilde{y}$ is encountered. Moreover, the case $\tilde{y} = \tilde{x}$, that is, $\tilde{y}$ is found as the first element in the computation of S_x, should be expected to occur in roughly $1/|S_x|$ of the tests.

Both completely failed to be the case. Instead, the typical behaviour looked similar to the following. (We will return to this example later in this section.)

```
>    Attach("braid.m");
>    B := BraidGroup(7);
>    x := RandomCFP(B, 0, 1, 5, 10);
```

---

[3] We will see in Section 3.1 that this is not quite true. The qualitative dependency on $|S_x|$, however, is as claimed.

```
>    #SuperSummitSet(x);
        578

>    count₁ := 0;
>    sum := 0;
>    for i := 1 to 1000 do
>      _, c := MyIsConjugate(x, x^Random(B));
>      if c eq 1 then
>        count₁ +:= 1;
>      end if;
>      sum +:= c;
>    end for;
>    count₁; // expectation: approx. 1.7
        402

>    Round(sum/1000); // expectation: approx. 290
        126

>    Round( (sum−count₁)/(1000−count₁) );
        210
```

In particular, the frequency with which the elements $\tilde{x}$ and $\tilde{y}$ coincided was remarkable. Note that in the example above this happens in about 40% of the cases, whereas for a uniform distribution on S_x one would expect it to happen in a mere 0.2% of the cases! Clearly, the super summit elements produced by the function MySuperSummitRepresentative are far from uniformly distributed. Instead, there is obviously a strong bias towards a relatively small subset of the super summit set.

Moreover, the average number of computed super summit elements is much smaller than expected. However, if one discards the cases for which $\tilde{x} = \tilde{y}$, the average number of computed super summit elements for the remaining cases is not too far off the expected value of $|S_x|/2$ in the example above. Hence, unless $\tilde{x} = \tilde{y}$, the conjugacy test does not seem to benefit from the additional structure described above.

In the light of the remarks from Section 2.4, using the set of elements whose existence was suggested by the experimental results for conjugacy testing instead of the super summit set seemed extremely attractive. To be able to do this, we had to establish that this set is indeed a class invariant and find a way of computing it efficiently.

3.2 Identifying the smaller invariant

Taking a closer look at the function MySuperSummitRepresentative and recalling Theorem 2.2 (i) revealed the identity of the smaller class of elements mentioned above.

Theorem 2.2 (i) implies that the cycling operation induces a map from the super summit set S_x of an element $x \in B_n$ to itself. More formally, the cycling operation endows S_x with the structure of a directed graph Γ_x with set of vertices S_x and set of edges $\{(y, \mathbf{c}(y)) : y \in S_x\}$. Since $\mathbf{c}(x^\tau) = \mathbf{c}(x)^\tau$ holds for all $x \in B_n$, τ induces an automorphism of Γ_x.

Let U_x be the union of the vertex sets of all circuits of Γ_x. We call U_x the *ultra summit set* of x. Clearly U_x is invariant under τ.

As S_x is finite, applying repeated cycling operations to any element of S_x must eventually produce an element of U_x. If S_x is known, the set U_x can be computed as follows.

```
intrinsic Circuits(S :: SetIndx)  → SetIndx
{
  Input:
       the super summit set of an elt x as an indexed set
  Output:
       a set containing the vertex sets of the circuits
       of the graph Gamma_x
}
  C := {@ @};
  seen := [ false : s in S ];
  while exists(pos){ i : i in [1..#seen] | not seen[i] } do
     // follow trajectory till we arrive at previously seen elt
     T := [ S| ];
     while not seen[pos] do
        seen[pos] := true;
        Append(~T, S[pos]);
        pos := Index(S, Cycle(S[pos]));
     end while;
     // if periodic part of trajectory is new, add it to C
     if S[pos] in T then
        Include(~C, {@ T[i] : i in [Index(T, S[pos])..#T] @});
     end if;
  end while;
  return C;
end intrinsic;
```

As S_x depends only on the conjugacy class of x, the same holds for U_x. Since U_x is a non-empty set of conjugates of x, two elements x and y are conjugate if and only if $U_x = U_y$ or, equivalently, if and only if $U_x \cap U_y \neq \emptyset$.

The function MySuperSummitRepresentative proves maximality and minimality of the infimum and supremum, respectively, of a potential super summit element by checking that these values are invariant under a certain number of iterated cycling and decycling operations. In particular, the returned super summit element is the image of another super summit element under a series of cycling and decycling operations. Hence it seemed conceivable that this process produces elements with a bias towards U_x.

Indeed tests confirmed that super summit elements computed by the function MySuperSummitRepresentative extremely often, though not always, actually are ultra summit elements. Moreover, a comparison of $|S_x|$ and $|U_x|$ for random elements $x \in B_n$ for various values of n suggested that, in general, U_x is much smaller than S_x. For the example of Section 3.1 the results look like this:

```
>    Sx := SuperSummitSet(x);
>    Ux := &join(Circuits(Sx));
>    #Sx;

        578

>    #Ux;

         6

>    countU := 0;
>    for i := 1 to 1000 do
>      y := MySuperSummitRepresentative(x^Random(B));
>      if y in Ux then
>        countU +:= 1;
>      end if;
>    end for;
>    countU;

        1000
```

3.3 Looking for a way of computing ultra summit sets

Establishing that the ultra summit set U_x is an invariant of the conjugacy class of x and that two elements x and y are conjugate if and only if $U_x \cap U_y \neq \emptyset$ was, however, in itself not sufficient for more efficient conjugacy testing. With the results described in Section 3.2, we could compute U_x only from its definition, using the function Circuits. Since this required S_x to be known, it was at least as impractical as the conjugacy test from Section 2.3. In order to make use of our new invariant, we needed a better way of computing it.

Searching for an idea, it was natural to look at the way the super summit set is computed. In Section 2.3 we established that the super summit set S_x of an element x can be obtained, starting with a single element $\tilde{x} \in S_x$, as the closure with respect to minimal simple elements. The reason why this works lies in the properties of the super summit set described in Theorems 2.3 and 2.4, which can be summarised as follows.

Theorem 3.1 *Let $x \in B_n$, $y \in S_x$ and $u, v \in B_n^+$. If $y^u \in S_x$ and $y^v \in S_x$ then $y^{u \wedge v} \in S_x$.*

The equivalence of this property to Theorems 2.3 and 2.4 follows easily from the observation that in the above situation $z \in S_x$ is equivalent to $z^\delta \in S_x$. We remark that an analogous property is satisfied by the set of all conjugates of x lying in B_n^+ [9].

Being optimistic, we were wondering whether a property analogous to Theorem 3.1 could be shown for ultra summit sets too. Our hope was to be able to compute a representative of U_x by applying repeated cycling operations to an element of S_x and then to compute the closure with respect to conjugation by minimal simple elements, in a way analogous to the approach used in the function MyIsConjugate for computing super summit sets.

This was asking for a lot. Before setting out on the quest for a proof of this claim (or at this stage "hope"), we decided to spend some time looking for a counter-example.[4] We performed extensive tests using the functions SatisfiesConvexity and SatisfiesGCD shown below.

The following function SatisfiesConvexity tests whether a given set S satisfies a "convexity property" analogous to Theorem 2.3. Note that we only check whether every element of S can be reached from the first element of S by a chain of elements in S which are linked by conjugation with simple elements. This is sufficient to test the "convexity property" of S, if S is closed with respect to conjugation with δ. To see this, observe that for any $x, y \in S$ and $s \in D$ satisfying $x^s = y$, we have $x = y^{s^{-1}} = (y^{(s^{-1}\delta)})^{\delta^{-1}}$, where $s^{-1}\delta \in D$ and $y^{(s^{-1}\delta)} = x^\delta \in S$ according to the assumptions.

```
intrinsic SatisfiesConvexity(S :: SetIndx)  → BoolElt
{
    Input:   a set S which is closed under tau
    Output:  whether S satisfies the "convexity" property
}
    B := Universe(S);
    D := [ B | s : s in Sym(NumberOfStrings(B)) ]; // simple elts.
    seen := [ false : i in [1..#S] ];
    // mark elements reachable from S[1] by a chain of elements
    // of S linked by conjugation with simple elements
    new := { 1 };
    seen[1] := true;
    while #new gt 0 do
      ExtractRep(∼new, ∼i);
      u := S[i];
      for d in D do
        idx := Index(S, u^d);
        // u^d is in S if and only if idx is positive
        if idx gt 0 and not seen[idx] then
```

---

[4]Unlike attempting to prove a wrong theorem, attempting to find counter-examples to a correct theorem wastes only *CPU* time. Moreover, working on proving a claim is the more enjoyable, the more you believe in its correctness.

```
        Include(~new, idx);
          seen[idx] := true;
        end if;
      end for;
    end while;
    return forall{ i : i in [1..#S] | seen[i] };
  end intrinsic;
```

The following function SatisfiesGCD tests whether a given set S satisfies the "gcd" property analogous to Theorem 2.4.

```
intrinsic SatisfiesGCD(S :: SetIndx) → BoolElt
{
  Input:   a set S of elements
  Output:  whether S satisfies the "gcd" property
}
  B := Universe(S);
  D := [ B | s : s in Sym(NumberOfStrings(B)) ]; // simple elts.
  for s in S do
    for i := 1 to #D−1 do
      if s^D[i] in S then
        for j := i+1 to #D do
          if s^D[j] in S then
            if s^LeftGCD(D[i],D[j]) notin S then
              return false;
            end if;
          end if;
        end for;
      end if;
    end for;
  end for;
  return true;
end intrinsic;
```

The results for the example from Section 3.1 are as follows:

SatisfiesConvexity(Ux);

> true

SatisfiesGCD(Ux);

> true

Extensive tests with many randomly chosen elements of braid groups on a different number of strings failed to produce a counter-example, giving us enough confidence to put forward the following conjecture.

Conjecture 3.2 *Let $x \in B_n$, $y \in U_x$ and $u, v \in B_n^+$. If $y^u \in U_x$ and $y^v \in U_x$ then $y^{u \wedge v} \in U_x$.*

4 On the way to a proof

The next step after starting to believe in Conjecture 3.2 was to get ideas as to how it could be proved. Experiments turned out to be very helpful for finding the way to a proof. Nevertheless, as might be expected, a few turns were taken which in the end led only to dead ends. We will skip these wrong turns and instead focus on the path which was successful ... and on the experimental signposts on the way.

4.1 Linking cycling and conjugation

The first obvious thing to look at was the lengths of the circuits. Initial examples suggested that all circuits in the graph Γ_x for a given element x have the same length. Moreover, the circuits were either all invariant under the graph automorphism induced by τ or all came in pairs of circuits interchanged by this automorphism. This suggested some sort of transitive operation on the set of circuits, which might have made life very easy.

```
>    B₁ := BraidGroup(7);
>    x := RandomCFP(B₁, 0, 1, 5, 10);
>    Sx := SuperSummitSet(x);
>    #Sx;
          578

>    Cx := Circuits(Sx);
>    [ #p : p in Cx ];
          [ 3, 3 ]

>    {@ z^FundamentalElement(B₁) : z in Cx[1] @} eq Cx[2];
          true
```

However, a systematic search for counter-examples to this ideal situation soon exhibited slightly more complicated behaviour as the one shown below, thereby ruling out a transitive action on the circuits induced by a graph automorphism.

```
>    B₂ := BraidGroup(8);
>    y := B₂ ! <"Artin", 1, [Sym(8)|(1,8)(3,5,7,6,4),(2,4)(3,6,7),
>        (2,7,6,5,4,3)],0>; // element found by systematic search
>    Sy := SuperSummitSet(y);
>    #Sy;
          3928

>    Cy := Circuits(Sy);
>    [ #p : p in Cy ];
          [ 5, 5, 10, 10 ]
```

> $\{@ \ z^{\mathsf{FundamentalElement}(B_2)} : z \ \textbf{\textit{in}} \ Cy[1] \ @\} \ \textbf{\textit{eq}} \ Cy[2];$

  true

> $\{@ \ z^{\mathsf{FundamentalElement}(B_2)} : z \ \textbf{\textit{in}} \ Cy[3] \ @\} \ \textbf{\textit{eq}} \ Cy[4];$

  true

As Conjecture 3.2 is all about conjugating elements lying in circuits, another natural question to ask was how circuits behave under conjugation by simple elements. Not surprisingly, conjugation by a given element does not respect circuits.

> $\{@ \ z^{(B_1.1)} : z \ \textbf{\textit{in}} \ Cx[1] \ @\} \ \textbf{\textit{in}} \ [\ Cx[1], Cx[2] \ ];$

  false

Counting the number of simple elements conjugating elements of one circuit into another circuit nevertheless revealed some structure. For a fixed enumeration $C_1, \ldots, C_k$ of the circuits, we define $A_{i,j}$ to be the number of triples (y, z, s) such that $y \in C_i$, $z \in C_j$ and $s \in D$. The resulting "adjacency matrix" A is computed by the function SimpleElementAdjacency.

```
intrinsic SimpleElementAdjacency(C :: SetIndx)  → Mtrx
{
  Input:
      an indexed set of sets of elements of a braid group B
  Output:
      the adjacency matrix A
}
  if #C eq 0 then
    return Matrix(Integers(), 0, 0, [ ]);
  end if;
  B := Universe(C[1]);
  D := [ B | s : s in Sym(NumberOfStrings(B)) ]; // simple elts.
  A := Matrix(Integers(), #C, #C, [ 0 : I In [1..(#C)²] ]);
  Ux := &join(C);
  for i := 1 to #C do
    for b in D do
      for p in C[i] do
        pb := pᵇ;
        if pb in Ux and exists(j){ k : k in [1..#C] | pb in C[k] } then
          A[i,j] +:= 1;
        end if;
      end for;
    end for;
  end for;
  return A;
end intrinsic;
```

The results for the examples from the beginning of this section are as follows.

```
>    [ #p : p in Cx ];
        [ 3, 3 ]
```

```
>    SimpleElementAdjacency(Cx);
            [6 6]
            [6 6]
```

```
>    [ #p : p in Cy ];
        [ 5, 5, 10, 10 ]
```

```
>    SimpleElementAdjacency(Cy);
            [ 30   30   60   60]
            [ 30   30   60   60]
            [ 60   60  120  120]
            [ 60   60  120  120]
```

In all examples, $A_{i,j}$ was a multiple of $|C_i|$. In other words, although conjugation by a fixed simple element does not respect circuits, conjugation "by all simple elements" in some sense seemed to do so.

This led to the question as to whether for given $z \in C_i$ and $s \in D$ satisfying $z^s \in C_j$, there always exists an element $t \in D$ such that $\mathbf{c}(z^s) = \mathbf{c}(z)^t$.

Closer investigation suggested that this was indeed the case and that it even seemed to hold in a more general context, namely when both z and z^s were super summit elements. However, if either z or z^s failed to be super summit elements, there were counter-examples. We also observed that the elements t above were in general not unique and that non-uniqueness was strongly correlated with the existence of circuits of different sizes.

```
    intrinsic CheckTransport(x :: GrpBrdElt, s :: GrpBrdElt)  → RngIntElt
    {
      Input:
          elements x and s of a braid group
      Output:
          number of simple elts t with Cycle(x^s) eq Cycle(x)^t
    }
      B := Parent(x);
      D := [ B | s : s in Sym(NumberOfStrings(B)) ]; // simple elts.
      return #{ t : t in D | Cycle(x)^t eq Cycle(x^s) };
    end intrinsic;
```

For the examples from the beginning of this section, we obtain the following.

```
>    s := B₁ ! Sym(7) ! (1,4,7,3,5)(2,6);
>    IsSuperSummitRepresentative(Sx[1]ˢ);
            false
```

```
>   CheckTransport(Sx[1], s);

        0

>   IsSimple(s⁻¹*FundamentalElement(B₁));

        true

>   CheckTransport(Sx[1]ˢ, s⁻¹*FundamentalElement(B₁));

        0

>   IsSuperSummitRepresentative(Sy[1]^{B₂.1});

        true

>   CheckTransport(Sy[1], B₂.1);  // no uniqueness

        2

>   time forall{ <z,s> : z in Sx, s in Sym(7) |
>       not IsSuperSummitRepresentative(z^{(B₁ ! s)})
>       or CheckTransport(z, B₁ ! s) eq 1 };  // note uniqueness!

        true
        Time: 24181.140
```

4.2 Defining the transport map

For a super summit element x consider $D_x = \{s \in D : x^s \in S_x\}$, that is, the set of simple elements s for which x^s is also a super summit element. The experimental results described in the preceding section hint at the existence of a map $\varphi_x : D_x \to D_{\mathbf{c}(x)}$. In fact, this map turned out to be the crucial tool in the proof of Conjecture 3.2.

For any $x \in B_n$ define $\mathrm{LF}(x) = \delta \wedge \delta^{-\inf(x)}x$, the *leading factor* of x. The element $\mathrm{LF}(x)$ is simple and it is either trivial or the first (non-fundamental) factor occurring in the normal form of x as introduced in Theorem 2.1.

For a super summit element x and an element $s \in D_x$ we define the *transport of s along $x \mapsto \mathbf{c}(x)$* as $\varphi_x(s) = \tau^{-\inf(x)}(\mathrm{LF}(x))^{-1} s \tau^{-\inf(x)}(\mathrm{LF}(x^s))$. From this definition it directly follows that $\varphi_x(s)$ satisfies $\mathbf{c}(x^s) = \mathbf{c}(x)^{\varphi_x(s)}$.

If we could show that $\varphi_x(s)$ is simple, we could immediately conclude that $\varphi_x(s) \in D_{\mathbf{c}(x)}$, as x^s and hence $\mathbf{c}(x^s)$ are super summit elements. Tests suggested that this is indeed the case.

```
intrinsic MyTransport(x :: GrpBrdElt, s :: GrpBrdElt) → GrpBrdElt
{
    Input:   elements x and s of a braid group, such that s
        is simple and both x and x^s are super summit elts
    Output:
        the transport of s
}
```

```
    k := −Infimum(x);
    δ := FundamentalElement(Parent(x));
    lfx := GCD(δ, δ^k * x);
    lfxs := GCD(δ, δ^k * (x^s));
    return (lfx^d)^−1 * s * lfxs^d where d := δ^k;
end intrinsic;
```

The result for one of the examples from the beginning of this section is shown here.

```
>    time forall{ <z, s> : z in Sx, s in Sym(7) |
>            not IsSuperSummitRepresentative(z^(B_1 ! s))
>            or IsSimple(MyTransport(z, B_1 ! s)) };

    true
    Time: 1648.570
```

Conjecture 4.1 *For a super summit element x and an element $s \in D_x$, the transport satisfies $\varphi_x(s) \in D$, that is, $\varphi_x(s) \in D_{\mathbf{c}(x)}$.*

Assume that Conjecture 4.1 holds. Given an ultra summit element x and $s \in D_x$, we can iterate the transport map, that is, we can define $s^{(0)} = s$ and $s^{(i+1)} = \varphi_{\mathbf{c}^i(x)}(s^{(i)})$ for $i \geq 0$. Since x is an ultra summit element, there exists an integer $N > 0$ such that $\mathbf{c}^N(x) = x$, whence $s^{(N)} \in D_x$. However, $s^{(N)}$ may be different to $s^{(0)} = s$ and this is necessarily the case if $\mathbf{c}^{(N)}(x^s) \neq x^s$, that is, if the lengths of the circuits containing x and x^s are distinct. Hence Conjecture 4.1, if correct, would also explain the observed connection between different circuit lengths and the existence of more than one simple element t satisfying $\mathbf{c}(x^s) = \mathbf{c}(x)^t$ mentioned at the end of Section 4.1.

4.3 Proving properties of the transport map

In order to prove Conjecture 4.1, it was necessary to understand how, for a given super summit element x and $s \in D_x$, the simple factors in the normal forms of x and x^s are related. Since both x and x^s are super summit elements, there are integers k and r as well as simple elements $A_1, \ldots, A_r$ and $\bar{A}_1, \ldots, \bar{A}_r$ such that the normal forms of x and x^s are $\delta^k A_1 \cdots A_r$ and $\delta^k \bar{A}_1 \cdots \bar{A}_r$, respectively.

From Theorem 2.1 follows that there are unique elements $s_1 \ldots, s_{r+1} \in B_n^+$ such that $s_1 = \tau^k(s)$, $s_{r+1} = s$ and $\bar{A}_i = s_i^{-1} A_i s_{i+1}$ for $i = 1, \ldots, r$. Note that $\varphi_x(s) = \tau^{-k}(s_2)$. We refer to $s_1 \ldots, s_{r+1}$ as *moving factors* for the conjugation of x by s.

Given x and s as above, the sequence $[s_1, \ldots, s_{r+1}]$ can be computed using the following function.

```
intrinsic MovingFactors(x :: GrpBrdElt, s :: GrpBrdElt) → SeqEnum
{
    Input:
```

```
        an element x and a simple element s such
        that x and x^s are super summit elements
    Output:
        the sequence [s_1,...,s_(r+1)] as defined above
}
```

require IsSuperSummitRepresentative(x)
    and IsSuperSummitRepresentative(x^s) :
        "x and x^s should be super summit elements";
B := Parent(x);
δ := FundamentalElement(B);
mov := [s^d **where** d := $\delta^{\text{Infimum}(x)}$];
xs := $\delta^{(-\text{Infimum}(x))} * x^s$;
x := $\delta^{(-\text{Infimum}(x))} * x$;
r := CanonicalLength(x);
for i := 1 **to** r **do**
  Ai := GCD(δ, x);
  Asi := GCD(δ, xs);
  Append($\sim mov$, NormalForm($Ai^{-1} * mov[\#mov] * Asi$));
  x := $Ai^{-1} * x$;
  xs := $Asi^{-1} * xs$;
end for;
return mov;
end intrinsic;

Experiments suggested that the moving factors respect both the partial ordering $\preceq$ and the gcd operation. More precisely, for a super summit element x with $\text{len}(x) = r$ and elements $s, t \in D_x$ with moving factors $s_1 \ldots, s_{r+1}$ and $t_1 \ldots, t_{r+1}$, respectively, the implications

$$s \preceq t \quad \Rightarrow \quad s_i \preceq t_i \quad \text{for} \quad i = 1, \ldots, r+1 \tag{1}$$

and

$$s \wedge t = 1 \quad \Rightarrow \quad s_i \wedge t_i = 1 \quad \text{for} \quad i = 1, \ldots, r+1 \tag{2}$$

seemed to hold. Since D_x is closed with respect to the gcd operation, (1) and (2) together are equivalent to

$$u = s \wedge t \quad \Rightarrow \quad u_i = s_i \wedge t_i \quad \text{for} \quad i = 1, \ldots, r+1$$

where $u \in D_x$ and $u_1 \ldots, u_{r+1}$ denote the moving factors for u.

intrinsic TestMovingFactors(x :: GrpBrdElt) $\rightarrow$ BoolElt
  { Return whether moving factors for conjugates of x
  satisfy (1) and (2) }
  require IsSuperSummitRepresentative(x) :
       "x should be a super summit element";
B := Parent(x);

```
r := CanonicalLength(x);
Dx := [ B | s : s in Sym(NumberOfStrings(B))
                      | IsSuperSummitRepresentative(x^(B ! s)) ];
for s in Dx do
  for t in Dx do
    u := GCD(s, t);
    movs := MovingFactors(x, s);
    movt := MovingFactors(x, t);
    movu := MovingFactors(x, u);
    if exists{ i : i in [1..r+1] | movu[i] ne GCD(movs[i], movt[i]) }
      then return false;
    end if;
  end for;
end for;
return true;
end intrinsic;
```

Extensive tests similar to the one below did not produce a counter-example.

```
>   Sx := SuperSummitSet(RandomCFP(BraidGroup(6), 0, 1, 5, 10));
>   TestMovingFactors(Sx[1]);

        true

>   #Sx;

        228

>   time forall{ y : y in Sx | TestMovingFactors(y) };

        true
        Time: 4079.730
```

At this point, the "right" conjectures had been found and we were ready to complete the proof of Conjecture 3.2.

The properties (1) and (2) of the moving constituents could be proved directly. Having done that, Conjecture 4.1 followed from the easy observation that the moving constituents for $s = \delta$ satisfy $s_i = \delta$ for $i = 1, \ldots, r+1$.

With the transport map and its properties following from (1) and (2) established, Conjecture 3.2 could be proven. It was sufficient to show that for a super summit element x satisfying $\mathbf{c}(x) \in U_x$, the existence of simple elements u and v satisfying $x^u, x^v \in U_x$ and $u \wedge v = 1$ implies that x is an ultra summit element. This was done in two steps. First it was shown that repeated transport of u and v is periodic, that is, that there is an integer M such that $u^{(M)} = u$, $v^{(M)} = v$, $\mathbf{c}^M(x^u) = x^u$, $\mathbf{c}^M(x^v) = x^v$, and $\mathbf{c}^{M+1}(x) = \mathbf{c}(x)$. Then it was possible to use (2) and the information $u \wedge v = 1$ to deduce from this $\mathbf{c}^M(x) = x$, that is, $x \in U_x$. For details we refer to [11].

5 Computing minimal simple elements

Having established the correctness of Conjecture 3.2, we knew that for every ultra summit element x and every simple element s there exists a uniquely defined $\preceq$-minimal element $c_s = c_s(x)$ satisfying $s \preceq c_s \preceq \delta$ and $x^{c_s} \in U_x$.

In particular, the ultra summit set of an element can be computed, starting from a single representative, as the closure with respect to conjugation by simple elements. In order to devise an efficient version of this process similar to the computation of the super summit set in the function **MyIsConjugate** from Section 2.3, we still had to find a way of computing $c_s(x)$ for given elements x and s. The transport map introduced in Section 4 again turned out to be the crucial tool.

Let x be an ultra summit element and $s \in D$. Since, as we saw in (2), the transport map respects the gcd operation, there is a unique $\preceq$-minimal element $\pi_x(s) \in D_x$ satisfying $s \preceq \varphi_x(\pi_x(s))$. We call $\pi_x(s)$ the *pullback of s along* $x \mapsto \mathbf{c}(x)$. It is not hard to see that $\pi_x(s)$ can be computed as in the following function **PullbackBraid**; for a proof we refer to [11].

```
intrinsic PullbackBraid(x :: GrpBrdElt, s :: GrpBrdElt) → GrpBrdElt
{ Return the pullback of s along x -> c(x). }
  B := Parent(x);
  require s in B :
    "Elements do not belong to a common braid group";
  LeftNormalForm(~x);
  require IsSuperSummitRepresentative(x) :
    "First argument must be a super summit element";
  require IsSimple(s) : "Second argument must be simple";
  δ := FundamentalElement(B);
  τ := δ^Infimum(x);
  τ₁ := τ*δ;
  cfp := [ B | cf : cf in CFP(x)[3] ]; // the simple factors
  A1s := cfp[1]⁻¹*δ;
  m := sᵀ; // this should be moving into first simple factor
  // ...ensure that the product of the first simple factor of
  // the conjugate and m is simple
  u₁ := (A1s⁻¹*LCM(A1s,m))ᵗ where t := τ₁⁻¹;
  // ...and that m is a divisor of the second simple factor
  u₂ := m;
  for i := 2 to #cfp do
    u₂ := cfp[i]⁻¹*LCM(cfp[i],u₂);
  end for;
  u := MinimalElementConjugatingToSuperSummit(x, LeftLCM(u₁,u₂));
  return LeftNormalForm(u);
end intrinsic;
```

Given an ultra summit element x and $s \in D$, the element $c_s(x)$ can be computed using repeated pullback and transport operations along the circuit containing x. We refer to [11] for details.

The following intrinsics are provided in the package file braid.m. The function MyMinimalElementConjugatingToSuperSummit computes the element $c_s(x)$ for a given ultra summit element x and $s \in D$, the function MyUltraSummit-Representative returns for given $x \in B_n$ a representative $\tilde{x}$ of U_x and the function MyIsConjugate_Ultra provides a conjugacy test analogous to the one in the function MyIsConjugate from Section 2.3, using ultra summit sets instead of super summit sets.

6 An application: key recovery

In this section we consider a protocol using braid groups for facilitating a key exchange over an insecure channel, which was proposed in [13] and is in some respects similar to the Diffie–Hellman key exchange. We also describe an attack on this protocol based on computing conjugating elements and compare the performance of the conjugacy test based on our new ultra summit sets to the one based on super summit sets.

6.1 Key exchange using braid groups

We outline the key exchange protocol proposed in [13]. Let B be a braid group acting on $l + r$ strings with generators $\sigma_1, \ldots, \sigma_{l+r-1}$ and consider the subgroups B_L generated by $\sigma_1, \ldots, \sigma_{l-1}$ and B_R generated by $\sigma_{l+1}, \ldots, \sigma_{l+r-1}$, respectively. Clearly B_L and B_R commute.

Let $x \in B$ be a publicly known, sufficiently complicated braid. Two users, obviously named Alice and Bob, can exchange a common key over an insecure channel as follows. Alice chooses a random element $a \in B_L$, which she keeps secret and sends the normal form of x^a to Bob. Sending the normal form is crucial, as it disguises the structure of x^a as conjugate of x by a. Similarly, Bob chooses a random element $b \in B_R$ which he keeps secret and sends the normal form of x^b to Alice. Alice then can compute the normal form of $K_A = (x^b)^a$, whereas Bob can compute the normal form of $K_B = (x^a)^b$. Since a and b commute, $K_A = x^{ba} = x^{ab} = K_B$ and its normal form can, in suitable form, be used as a shared secret.

In order to recover the key, an eavesdropper needs to deduce x^{ab} from the known braids x, x^a and x^b. The security of the key depends crucially on the difficulty of determining the conjugating elements a or b knowing x and x^a or x and x^b. Note the similarity to the discrete logarithm problem in the context of the Diffie–Hellman key exchange.

Using the implementation of braid groups in Magma, a small example might look like this.

```
>    // public data: we use a product of 8 simple elements in
>    // a braid group on 9 strings
>    l := 5;
>    r := 4;
>    len := 8;
>    B_L := BraidGroup(l);
>    B_R := BraidGroup(r);
>    B := BraidGroup(l+r);
>    x := Random(B, 0, 0, len, len);
>    // identify B_L and B_R with subgroups of B
>    f := hom< B_L → B | [ B_L.i  → B.i : i in [1..l−1] ] >;
>    g := hom< B_R → B | [ B_R.i  → B.(l+i) : i in [1..r−1] ] >;
>    a := Random(B_L); // Alice: keeps a secret, sends y₁ to Bob
>    y₁ := NormalForm(x^{f(a)});
>    b := Random(B_R); // Bob: keeps b secret, sends y₂ to Alice
>    y₂ := NormalForm(x^{g(b)});
>    K_A := Eltseq(NormalForm(y₂^{f(a)})); // Alice extracts shared key
>    K_B := Eltseq(NormalForm(y₁^{g(b)})); // Bob extracts shared key
>    K_A eq K_B;
         true
```

6.2 An attack using conjugacy search

In order to recover the braid a, and hence the key, from the openly available information x and x^a, finding *any* element $c \in B$ conjugating x to x^a is not necessarily sufficient. The conjugating element c can be used to recover the key x^{ab}, if and only if the commutator $[b^{-1}, c^{-1}] = bcb^{-1}c^{-1}$ commutes with x. This is in particular the case if $c \in B_L$.

To keep things easy, we ignore this complication here and look for conjugating elements in B anyway; we will see that, provided a conjugating element can be found, the chances of a successful attack are still quite good.[5]

Although extremely small values were chosen for the parameters in the above example, computing a conjugating element using the super summit set approach is rather painful.

```
>    // attack using super summit sets
>    time _, nr_s, c_s := MyIsConjugate(x, y₁);
         Time: 1255.620

>    nr_s;
         134188
```

---

[5] This may be different if the braid x is chosen with more care but, again, we want to keep things easy.

```
>    y₂^{c_s} eq y₂^{f(a)};
```
$$y_2^{c\_s} \text{ eq } y_2^{f(a)};$$
>   true

```
>    K_recover_s := Eltseq(NormalForm(y₂^{c_s}));
>    K_recover_s eq K_A;
```
$$\text{K\_recover\_s := Eltseq(NormalForm}(y_2^{c\_s}));$$
$$\text{K\_recover\_s eq K\_A};$$
>   true

Nevertheless, we were successful in recovering the key after computing 134 188 super summit elements.

For real cryptographic applications much larger parameters would be used, vastly increasing the size of the super summit set and thus spoiling the attack.[6]

Now we try, for the same example, to recover the key using the conjugacy test based on ultra summit sets.

```
>    // attack using ultra summit sets
>    time _, _, c_u := MyIsConjugate_Ultra(x, y₁);
```
>   Time: 0.750

```
>    y₂^{c_u} eq y₂^{f(a)};
```
$$y_2^{c\_u} \text{ eq } y_2^{f(a)};$$
>   true

```
>    K_recover_u := Eltseq(NormalForm(y₂^{c_u}));
>    K_recover_u eq K_A;
```
$$\text{K\_recover\_u := Eltseq(NormalForm}(y_2^{c\_u}));$$
$$\text{K\_recover\_u eq K\_A};$$
>   true

The reason for this enormous difference in performance is that the ultra summit set is *much* smaller than the super summit set.

```
>    #UltraSummitSet(x);
```
>   24

The Magma function UltraSummitSet used here is based on a more sophisticated implementation of the ideas presented in this chapter.

In fact, we can relatively easily compute the ultra summit sets of randomly chosen braids for parameter values which have been suggested as appropriate for cryptographic purposes in some publications. Here we compute the ultra summit sets of randomly chosen products of between 1 000 and 2 000 simple elements in the braid group on 100 strings. The super summit sets of such braids are, in general, far too large to be computable.

```
>    B := BraidGroup(100);
>    time #UltraSummitSet(Random(B, 0, 1, 1000, 2000));
```
>   2040
>   Time: 816.610

---

[6]At least that is the idea of the proposed key exchange. However, as we saw in Section 3.1, even the conjugacy test using the function MyIsConjugate may work much more often than one might expect from the size of the super summit set!

```
>    time #UltraSummitSet(Random(B, 0, 1, 1000, 2000));
     2670
     Time: 111.380
```

Note that if we are able to compute the ultra summit set of a braid x, we can, in particular, decide conjugacy to x and compute conjugating elements using the method of the function **MyIsConjugate_Ultra**. Consequently, the key exchange protocol described in Section 6.1 based on conjugates of x can be attacked as described above.

It may be possible to prevent attacks based on ultra summit sets by choosing the braid x carefully, that is, in such a way that its ultra summit set is large and has a complicated structure. Whether suitable braids with this property exist, and how they might be constructed if they do, remains to be seen ...

References

1. E. Artin, *Theory of braids*, Ann. of Math. (2) **48** (1947), 101–126.
2. Joan S. Birman, Ki Hyoung Ko, Sang Jin Lee, *The infimum, supremum, and geodesic length of a braid conjugacy class*, Adv. Math. **164**-1 (2001), 41–56.
3. Wieb Bosma, John Cannon, Catherine Playoust, *The Magma algebra system I: The user language*, J. Symbolic Comput. **24** (1997), 235–265.
   See also the Magma home page at http://magma.maths.usyd.edu.au/magma/.
4. Patrick Dehornoy, *A fast method for comparing braids*, Adv. Math. **125**-2 (1997), 200–235.
5. Patrick Dehornoy, *Groupes de Garside*, Ann. Sci. École Norm. Sup. (4), **35**-2 (2002), 267–306.
6. Patrick Dehornoy, Luis Paris, *Gaussian groups and Garside groups, two generalisations of Artin groups*, Proc. London Math. Soc. (3) **79**-3 (1999), 569–604.
7. Elsayed A. El-Rifai, H. R. Morton, *Algorithms for positive braids*, Quart. J. Math. Oxford Ser. (?) **45**-180 (1994), 479–497.
8. David B. A. Epstein, James W. Cannon, Derek F. Holt, Silvio V. F. Levy, Michael S. Paterson, William P. Thurston, *Word processing in groups*, chapter 9, Boston: Jones and Bartlett Publishers, 1992.
9. Nuno Franco, Juan González-Meneses, *Conjugacy problem for braid groups and Garside groups*, J. Algebra, **266**-1 (2003), 112–132.
10. F. A. Garside, *The braid group and other groups*, Quart. J. Math. Oxford Ser. (2), **20** (1969), 235–254.
11. Volker Gebhardt, *A new approach to the conjugacy problem in Garside groups*, Journal of Algebra **292**-1 (2005), 282–302.
12. Volker Gebhardt, *Braid groups*, Chapter 31, pp. 963–1014 in: John Cannon, Wieb Bosma (eds.), *Handbook of Magma Functions*, Version 2.11, Volume **3**, Sydney, 2004.
13. Ki Hyoung Ko, Sang Jin Lee, Jung Hee Cheon, Jae Woo Han, Ju-sung Kang, Choonsik Park, *New public-key cryptosystem using braid groups*, pp. 166–183 in: *Advances in cryptology—CRYPTO 2000 (Santa Barbara, CA)*, Lecture Notes in Comput. Sci. **1880**, Berlin: Springer, 2000.
14. Matthieu Picantin, *The conjugacy problem in small Gaussian groups*, Comm. Algebra, **29**-3 (2001), 1021–1039.

Searching for linear codes with large minimum distance

Markus Grassl

Institut für Algorithmen und Kognitive Systeme
Arbeitsgruppe Prof. Beth, Fakultät für Informatik, Universität Karlsruhe
Am Fasanengarten 5
76 128 Karlsruhe, Germany
grassl@ira.uka.de

Summary. There are many tables which summarise bounds on the parameters of error-correcting codes. We are undertaking a project to find constructions for codes with large minimum distance. In the course of the project, many algorithms to construct and search for good codes have been devised, and some are presented here. In particular, a very efficient algorithm for computing the minimum distance of a code has been developed.

1 Introduction

1.1 Coding theory

Coding theory, in general terms, deals with the protection of information against errors. The foundations of the field as a part of information theory were laid by Shannon in 1948 [11]. While information theory is often concerned with asymptotic properties, there is of course also the necessity to devise methods for error detection and error correction for a given system. The origins of this branch of coding theory can be seen in the work of Hamming [8]. The main principle for error protection is to systematically add redundancy to the information. This is equivalent to using only a subset of all possible messages to encode information. If the possible messages are words of equal (finite) length over a finite alphabet, the resulting encoding is called a block code. Quite often, the solution to the combinatorial problem of constructing a good block code is found with the aid of algebraic structures, such as finite fields, groups and group representations, algebraic geometry, or designs. For explorations in the field of algebraic codes, Magma [3] is a very useful tool as it integrates all these areas of discrete mathematics.

1.2 Mathematical background

For given length n and cardinality M, a central problem in coding theory is to construct a block code $C = (n, M, d)_q \subset \mathcal{A}^n$ over a (finite) alphabet $\mathcal{A}$ such that the minimum distance d is as large as possible. The *minimum distance* of a block code is defined as

$$\mathrm{d_{min}}(C) := \min\{\mathrm{dist}(\boldsymbol{c}, \boldsymbol{d})\colon \boldsymbol{c}, \boldsymbol{d} \in C \mid \boldsymbol{c} \neq \boldsymbol{d}\}, \tag{1}$$

where the *Hamming distance* $\mathrm{dist}(\boldsymbol{c}, \boldsymbol{d})$ of two words $\boldsymbol{c}, \boldsymbol{d} \in \mathcal{A}^n$ is the number of positions where $\boldsymbol{c}$ and $\boldsymbol{d}$ differ. We consider the special case where $C = [n, k, d]_q$ is a linear block code over a field $\mathbb{F}_q$, i.e., C is a k-dimensional subspace of $\mathbb{F}_q^n$. Then the minimum distance of C equals the *minimum weight* of the non-zero codewords in C and is given by

$$\begin{aligned}
\mathrm{d_{min}}(C) &= \min\{\mathrm{dist}(\boldsymbol{c}, \boldsymbol{d})\colon \boldsymbol{c}, \boldsymbol{d} \in C \mid \boldsymbol{c} \neq \boldsymbol{d}\} \\
&= \min\{\mathrm{dist}(\boldsymbol{c} - \boldsymbol{d}, \boldsymbol{0})\colon \boldsymbol{c}, \boldsymbol{d} \in C \mid \boldsymbol{c} \neq \boldsymbol{d}\} \\
&= \min\{\mathrm{wgt}(\boldsymbol{c})\colon \boldsymbol{c} \in C \mid \boldsymbol{c} \neq \boldsymbol{0}\},
\end{aligned} \tag{2}$$

where the *Hamming weight* $\mathrm{wgt}(\boldsymbol{c})$ of a vector $\boldsymbol{c}$ is the number of non-zero coefficients. One advantage of linear codes over general block codes is that a linear code $C = [n, k]$ can be compactly represented by either a *generator matrix* $G \in \mathbb{F}_q^{k \times n}$ or a *parity check matrix* $H \in \mathbb{F}_q^{(n-k) \times n}$. Given a generator matrix G, the code C is defined as the row span of G, and given a parity check matrix H, the code is defined as the kernel of H^t.

Brouwer has published lower and upper bounds on the minimum distance of linear codes over small fields [4]. The on-line version of those tables is updated frequently. In addition to the bounds, some information on the construction of the codes and some references are provided, but in general generator matrices are not given. So in 1999, a Magma project was initiated to find explicit constructions for *good linear codes*, i.e., codes whose minimum distance achieves the lower bounds in the table. In many cases it was not only possible to find a code which achieves the published lower bound, but we have even found codes with better lower bounds. A complete table for binary codes was released in Magma V2.8 in July 2001, followed by tables of codes over $\mathbb{F}_4$ and $\mathbb{F}_3$ in V2.10 (April 2003) and V2.11 (May 2004), respectively [5, Section 115.13, 'Best Known Linear Codes']. Some results have already been published [7], while an updated version of the tables for codes over $\mathbb{F}_q$, $q = 2, 3, 4, 5, 7, 8, 9$, is in preparation.

In the following, some of the techniques used to search for the codes are described. A core problem in this process is to compute the minimum distance. Vardy [14] has shown that for linear binary codes, computing the minimum weight is in general an NP-hard problem. As the length of the codes increases, one cannot expect to be able to compute the minimum weight explicitly. Instead, one has to use particular constructions for linear codes which allow us to deduce at least a lower bound on the minimum weight. Important examples of such codes are cyclic codes, in particular BCH codes and Reed–Solomon codes, Goppa codes (a good reference for these codes, and coding theory in general, is e.g. [10]), and codes from algebraic geometry [13]. However, some bounds in Brouwer's table are derived from non-constructive counting arguments (such as the Gilbert–Varshamov bound). In order to find codes whose minimum distance achieves these lower bounds, we need to compute the minimum weight of randomly chosen codes or use techniques to construct new codes from explicitly known codes. Those constructions yield bounds on the minimum weight of the resulting code when the minimum weight of the constituent codes is known. Therefore, we need good algorithms to compute the minimum weight, a topic which is discussed in the next section.

2 Computing the minimum weight

In this section, we present algorithms for computing the minimum weight of linear codes. First, we consider general linear codes, then an algorithm for cyclic codes is discussed, followed by some alternative methods for special cases.

2.1 General linear codes

In the sequel we describe the algorithm that is used in Magma [5, Section 115.8.1, 'The Minimum Weight'] to compute the minimum weight of a general linear code, together with some improvements. The algorithm, which was refined during the search for good linear codes, is based on an algorithm by Zimmermann (see [15] and [2, Algorithmus 1.3.6]) who improved an algorithm by Brouwer.

The main idea of the algorithm is to enumerate the codewords in such a way that one not only obtains an upper bound on the minimum weight of the code via the minimum of the weight of the words that have been encountered, but also lower bounds on the minimum weight. For this, we need the concept of information sets.

Definition 2.1 (Information Set). *Let $C = [n, k, d]_q$ be a linear code of length n and dimension k over $\mathbb{F}_q$. A subset $\mathcal{I} \subseteq \{1, 2, \ldots, n\}$ of size $|\mathcal{I}| = k$ is called an* information set *if the corresponding columns in a generator matrix G of C are linearly independent. Then there also exists a* systematic generator matrix G_{syst} of C such that the columns of G_{syst} specified by $\mathcal{I}$ form an identity matrix.*

Without loss of generality, let $\mathcal{I} = \{1, 2, \ldots, k\}$ be an information set. Then G_{syst} is of the form $G_{\text{syst}} = (I|A)$. Using this systematic generator matrix to encode a vector $i \in \mathbb{F}_q^k$, we get:

Lemma 2.2. *Let* $G_{\mathrm{syst}} = (I|A)$ *and* $i \in \mathbb{F}_q^k$. *Then* $c := iG = (i, iA)$ *and hence* $\mathrm{wgt}(c) \geq \mathrm{wgt}(i)$.

This observation directly leads to our first algorithm:

Algorithm 2.3 (One Information Set).
$d_{\mathrm{lb}} := 1;$
$d_{\mathrm{ub}} := n - k + 1;$
$w := 1;$
while $w \leq k$ and $d_{\mathrm{lb}} < d_{\mathrm{ub}}$ do
$\quad d_{\mathrm{ub}} := \min(d_{\mathrm{ub}}, \min\{\mathrm{wgt}(iG) \colon i \in \mathbb{F}_q^k \mid \mathrm{wgt}(i) = w\});$
$\quad d_{\mathrm{lb}} := w + 1;$
$\quad w := w + 1;$
end while;
return $d_{\mathrm{ub}};$

In this algorithm, the words $i \in \mathbb{F}_q^k$ are enumerated in order of increasing Hamming weight w. Having enumerated all words with $\mathrm{wgt}(i) \leq w$, from Lemma 2.2 it follows that the weight d_{lb} of the codewords that have not been encountered is at least $\mathrm{wgt}(c) \geq \mathrm{wgt}(i) > w$. The enumeration process terminates when all words have been enumerated or a codeword is found whose weight does not exceed the lower bound. As presented, Algorithm 2.3 compares the upper and lower bound only after having computed the minimum weight of all codewords iG with fixed weight $\mathrm{wgt}(i) = w$. Of course, we can exit the while-loop as soon as a codeword is encountered whose weight does not exceed the lower bound.

Nevertheless, if the minimum weight of the code is at least $k + 1$, all q^k codewords have to be enumerated. The only way to avoid this is to find a way to compute an improved lower bound d_{lb}. For this, we use the fact that any set of k linearly independent codewords can be used as rows of a generator matrix of C, and hence we may use different generator matrices in the enumeration process.

Below, let $\mathcal{I}_1, \mathcal{I}_2, \ldots, \mathcal{I}_m$ be a collection of pairwise disjoint information sets of the code C with corresponding systematic generator matrices G_j.

Algorithm 2.4 (Disjoint Information Sets).
$d_{\mathrm{lb}} := 1;$
$d_{\mathrm{ub}} := n - k + 1;$
$w := 1;$
while $w \leq k$ and $d_{\mathrm{lb}} < d_{\mathrm{ub}}$ do
$\quad$ for $j := 1$ to m do
$\quad\quad d_{\mathrm{ub}} := \min(d_{\mathrm{ub}}, \min\{\mathrm{wgt}(iG_j) \colon i \in \mathbb{F}_q^k \mid \mathrm{wgt}(i) = w\});$
$\quad$ end for;
$\quad d_{\mathrm{lb}} := m(w + 1);$
$\quad w := w + 1;$
end while;
return $d_{\mathrm{ub}};$

Having encoded all words of weight $\mathrm{wgt}(i) \leq w$ using the systematic generator matrix G_j corresponding to the information set $\mathcal{I}_j$, we know that for the remaining codewords there are at least $w+1$ non-zero elements at the positions in $\mathcal{I}_j$. As the information sets are mutually disjoint, we get the lower bound $d_{\mathrm{lb}} = m(w+1)$.

Note that Algorithm 2.4 can also be used to verify lower and upper bounds on the minimum weight, which is the function of the Magma intrinsics VerifyMinimumWeightLowerBound and VerifyMinimumWeightUpperBound. In order to prove that $d_{\min} \geq d_0$, all words $i \in \mathbb{F}_q^k$ of weight $\mathrm{wgt}\, i \leq \lceil d_0/m - 1 \rceil :=$ w_0 have to be encoded using the m different generator matrices. So the total number of codewords considered is

$$\sum_{w=1}^{\lceil d_0/m-1 \rceil} m \binom{k}{w} (q-1)^{w-1}, \tag{3}$$

where we have used the fact that it is sufficient to consider only *normalised* vectors, i.e., the first non-zero position of i is 1. Compared to Algorithm 2.3, for each fixed weight w the number of codewords considered in Algorithm 2.4 is m times as large. On the other hand, the lower bound on the minimum weight grows m times faster so that the maximal weight w_0 of the vectors i in the enumeration process is considerably smaller. As the sum (3) is mainly dominated by the last term, increasing m results in an overall saving.

In the ideal situation, a code $C = [n,k]_q$ has $\lfloor n/k \rfloor$ disjoint information sets. Unfortunately, for a general linear code with systematic generator matrix $G_1 = (I|A)$, the rank of the matrix A can be less than k which implies that there is no information set $\mathcal{I}_2$ with $\mathcal{I}_1 \cap \mathcal{I}_2 = \emptyset$. In this situation, we can only obtain a *partial information set* $\mathcal{I}_2'$ of size $r_2 := |\mathcal{I}_2'| = \mathrm{rank}\, A$. In order to obtain an information set $\mathcal{I}_2 \supseteq \mathcal{I}_2'$, $k - r_2$ elements of $\mathcal{I}_1$ have to be included. Although $\mathcal{I}_1$ and $\mathcal{I}_2$ are not disjoint, using both G_1 and the systematic generator matrix G_2 corresponding to $\mathcal{I}_2$ in the enumeration process yields an improved lower bound: encoding all vectors $i \in \mathbb{F}_q^k$ with $\mathrm{wgt}(i) \leq w$, the remaining codewords have at least $w + 1$ non-zero symbols at the coordinate positions in each of $\mathcal{I}_1$ and $\mathcal{I}_2$. As the information sets overlap, some positions are counted twice, so the resulting lower bound is $2(w+1) - |\mathcal{I}_1 \cap \mathcal{I}_2|$. This can be generalised to an ordered list $(\mathcal{I}_1, \mathcal{I}_2, \ldots, \mathcal{I}_m)$ of information sets. To compute the lower bound, we define the *relative rank* r_j of $\mathcal{I}_j$ as

$$r_j := k - \left| \mathcal{I}_j \cap \bigcup_{l=1}^{j-1} \mathcal{I}_l \right|,$$

i.e., r_j equals the number of positions in the information set $\mathcal{I}_j$ that are not contained in any information set $\mathcal{I}_l$ with $l < j$. Similarly as in the case $m = 2$ discussed earlier, the matrix G_j contributes at most $(w+1) - (k - r_j)$ to the lower bound when encoding words of weight w with the corresponding systematic generator matrices G_j. This improvement, due to K.-H. Zimmermann [15], allows the use of additional systematic generator matrices when computing the minimum weight, resulting in faster growth of the lower bound (see Algorithm 2.5).

Algorithm 2.5 (Overlapping Information Sets).

$d_{\mathrm{lb}} := 1;$
$d_{\mathrm{ub}} := n - k + 1;$
$w := 1;$
while $w \leq k$ and $d_{\mathrm{lb}} < d_{\mathrm{ub}}$ do
  for $j := 1$ to m do
    $d_{\mathrm{ub}} := \min\big(d_{\mathrm{ub}}, \min\{\mathrm{wgt}(iG_j): i \in \mathbb{F}_q^k | \mathrm{wgt}(i) = w\}\big);$
  end for;
  $d_{\mathrm{lb}} := \sum_{j=1}^{m} \max(0, (w+1) - (k - r_j));$
  $w := w + 1;$
end while;
return $d_{\mathrm{ub}};$

The method of computing a sequence of information sets employed by Magma proceeds as follows. Without loss of generality, the first generator matrix may be assumed to have the form $G_1 = (I|A_1)$. Using Gaussian elimination on G_1, the submatrix A_1 is transformed into row echelon form

$$A_1' = \begin{pmatrix} I_{r_2} & B_{12} \\ 0 & 0 \end{pmatrix}.$$

For the whole generator matrix we obtain the form

$$G_2 = \left(\begin{array}{cc|cc} B_{11} & 0 & I_{r_2} & B_{12} \\ B_{21} & I_{k-r_2} & 0 & 0 \end{array} \right),$$

where the vertical bar indicates the first k positions. Note that r_2 is the rank of A_1. The next generator matrix G_3 is computed by transforming the submatrix B_{12} of G_2 into row echelon form and so forth. The process ends when the rank of the submatrix is zero or the information sets cover all positions, i.e., $\bigcup_j \mathcal{I}_j = \{1, 2, \ldots, n\}$. For the sequence of information sets obtained by this method, the sequence r_j is non-increasing, and sometimes we even get $r_j = 1$. A generator matrix with small r_j makes little contribution to the lower bound and can quite often be omitted. Given the sequence $(r_1, r_2, \ldots, r_m)$ of relative ranks, one can compute the sequence of lower bounds d_{lb} in Algorithm 2.5 and the weight w_0 for which the lower bound crosses the actual upper bound (see Algorithm 2.6). All matrices with $r_j < k - w_0$ can be omitted as they never contribute to the lower bound. This is one of our improvements over the algorithm of Zimmermann mentioned before. In addition to the sequence of lower bounds, we can also compute the total number of vectors that are considered in Algorithm 2.5. So one can estimate the total running time of the algorithm in advance or based on the timing information provided by the verbose output of Magma (see Example 2.7 below). To obtain that additional information, we have to use the command SetVerbose("Code", 1); which enables the conditional print statements **vprintf** Code: "..."; (see Algorithm 2.6)*

Algorithm 2.6 (Work Factor).

```
>    intrinsic WorkFactor(F :: FldFin, n :: RngIntElt, k :: RngIntElt,
>                         d :: RngIntElt, ranks :: [RngIntElt]) → RngIntElt
>    {
>      Computes max. number of codewords considered in Magma's
>      algorithm to compute the minimum weight of linear codes.
>    }
>      w:=0;
>      repeat
>        w +:= 1;
>        lb := &+[ w+1−(k−x): x in ranks | w+x−k ge 0 ];
>      until lb ge Minimum(d, n−k+1) or w ge k;
>      w_0 := w;
>      vprintf Code: "maximal weight: %o\n", w_0;
>      ranks_used := [ x : x in ranks | x ge k−w_0 ];
>      vprintf Code: "relative ranks used: %o\n", ranks_used;
>      work := 0;
>      for w := 1 to w_0 do
>        work +:= #ranks_used*Binomial(k, w)*(#F−1)^(w−1);
>        lb := &+[ w+1−(k−x) : x in ranks_used | x ge k−w ];
>        vprintf Code: "w=%o: lower bound: %o,
>                count: %o=2^%o\n", w, lb, work, Ilog2(work);
>      end for;
>      return work;
>    end intrinsic;
```

The running time of Algorithm 2.5 is minimal when the overlap between the information sets is minimal. So in the most favourable situation, there are $m = \lceil n/k \rceil$ information sets with relative ranks $r_1 = \ldots = r_{m-1} = k$ and $r_m = n - (m-1)k$. When the algorithm used in Magma fails to produce such a sequence, this does not imply that such a sequence does not exist. Applying a (random) permutation to the code—which does not change the minimum weight of the code—might help, as shown by the following example.[1]

Example 2.7 (Comparison of different strategies).

```
>    SetVerbose("Code", 1);
>    c:=BKLC(GF(2), 77, 15);
```

---

*See the Preface to this volume for style conventions regarding Magma code; code appearing in this book is available at http://magma.maths.usyd.edu.au/magma/.

[1]Note that the heuristic to apply a random permutation in order to get a better sequence of generator matrices is now implemented in Magma.

```
>    ResetMinimumWeightBounds(c);
>    MinimumWeight(c);
```

```
Code MinimumWeightZimmermann: length 77,
dimension 15, not cyclic
Lower Bound: 1 , Upper Bound: 63
Finished constructing sets, time taken: 0.009
Ranks: 15 15 15 15 11 6
Maximum Projected r: 12
Starting Search...
r: 1, i: 0, lower: 1, upper: 63, time: 0.009
New codeword identified of weight 33, time 0.009
New codeword identified of weight 29, time 0.009
r: 1, i: 1, lower: 1, upper: 29, time: 0.009
r: 1, i: 2, lower: 1, upper: 29, time: 0.009
r: 1, i: 3, lower: 1, upper: 29, time: 0.009
r: 1, i: 4, lower: 1, upper: 29, time: 0.009
r: 1, i: 5, lower: 1, upper: 29, time: 0.009
New Maximum Projected r: 6
Deleting non-contributing rank 6 matrix
New Ranks: 15 15 15 15 11
Continuing Search...
r: 2, i: 0, lower: 8, upper: 29, time: 0.009
```

verbose output partially deleted

```
r: 6, i: 4, lower: 26, upper: 29, time: 0.039
Final Results: lower = 31, upper = 29,
  Total time: 0.039
29
```

In this example, we are using a code from the database of codes in Magma which is accessed via the intrinsic **BKLC**. As the minimum weight of the codes in the database is already known, we have to clear this information using the command **ResetMinimumWeightBounds** in order to enforce the computation of the minimum weight. The statement **SetVerbose("Code", 1)**; sets the verbose level for the coding theory functions to one. Then Magma provides additional information about intermediate results.

At the beginning, the sequence of relative ranks is printed:

```
Ranks: 15 15 15 15 11 6
```

Based on the current upper bound, given by the Singleton bound $d \leq n-k+1$, the projection for the maximal weight is given by

```
Maximum Projected r: 12
```

Then the search starts encoding all vectors of weight $r = 1$, using the different generator matrices indexed by i. Every time when a codeword is found whose weight is less than the actual upper bound, the new upper bound is reported:

```
Starting Search...
r: 1, i: 0, lower: 1, upper: 63, time: 0.009
New codeword identified of weight 33, time 0.009
New codeword identified of weight 29, time 0.009
```

When all generator matrices have been used to encode the words of weight one, the lower bound is updated.[2] From the actual upper bound it is deduced that the weight to be considered is at most 6. As the relative rank of the last generator matrix is only 6, it will never make a contribution to the lower bound and can be deleted.

```
New Maximum Projected r: 6
Deleting non-contributing rank 6 matrix
New Ranks: 15 15 15 15 11
Continuing Search...
r: 2, i: 0, lower: 8, upper: 29, time: 0.009
```

The search proceeds with no further decrease of the upper bound until the lower bound has crossed the upper bound.

Applying a randomly chosen permutation to the code, we obtain 5 generator matrices with full rank, compared to only 4 for the original code. The progress of the algorithm is as follows:

Example 2.7 (Continued).

```
>    perm := Sym(77) ! (1, 64, 4, 66, 36, 59, 65, 73, 12, 18, 31, 45, 57, 17)
>      (2, 55, 68, 42, 56, 24, 39, 47, 54, 33, 30, 8, 60, 40, 71, 21, 58, 16, 46, 61, 50,
>       26, 9, 27, 28, 43, 67, 69, 75, 38, 19, 76, 35, 25, 6, 74, 13, 53, 3, 29, 14, 20,
>       10, 37, 77, 32, 63, 15, 5)(7, 23, 22, 52, 11, 44, 49, 62, 70, 48, 72, 34, 51);
>    c_2 := c^{perm};
>    ReootMinimumWeightBounds(c_2);
>    MinimumWeight(c_2);
```

```
Code MinimumWeightZimmermann: length 77,
dimension 15, not cyclic
Lower Bound: 1 , Upper Bound: 63
Finished constructing sets, time taken: 0.000
Ranks: 15 15 15 15 15 2
Maximum Projected r: 12
Deleting non-contributing rank 2 information set
New Ranks: 15 15 15 15 15
Starting Search...
r: 1, i: 0, lower: 1, upper: 63, time: 0.009
New codeword identified of weight 30, time 0.009
New codeword identified of weight 29, time 0.009
r: 1, i: 1, lower: 1, upper: 29, time: 0.009
```

---

[2]Note that the lower bound can be updated every time when all words of weight r have been encoded with a single generator matrices (inside the loop j in Algorithm 2.5). This improvement is now included in Magma.

```
r: 1, i: 2, lower: 1, upper: 29, time: 0.009
r: 1, i: 3, lower: 1, upper: 29, time: 0.009
r: 1, i: 4, lower: 1, upper: 29, time: 0.009
New Maximum Projected r: 5
r: 2, i: 0, lower: 10, upper: 29, time: 0.009
```

verbose output partially deleted

```
r: 5, i: 4, lower: 25, upper: 29, time: 0.019
Final Results: lower = 30, upper = 29,
   Total time: 0.019
29
```

The running time for the permuted code is only about half the running time for the original code, as predicted by Algorithm 2.6:

> WorkFactor(GF(2), 77, 15, 29, [15, 15, 15, 15, 11, 6]);

```
maximal weight: 6
relative ranks used: [ 15, 15, 15, 15, 11 ]
w=1: lower bound: 8, count: 75=2^6
w=2: lower bound: 12, count: 600=2^9
w=3: lower bound: 16, count: 2875=2^11
w=4: lower bound: 21, count: 9700=2^13
w=5: lower bound: 26, count: 24715=2^14
w=6: lower bound: 31, count: 49740=2^15
49740
```

> WorkFactor(GF(2), 77, 15, 29, [15, 15, 15, 15, 15, 2]);

```
maximal weight: 5
relative ranks used: [ 15, 15, 15, 15, 15 ]
w=1: lower bound: 10, count: 75=2^6
w=2: lower bound: 15, count: 600=2^9
w=3: lower bound: 20, count: 2875=2^11
w=4: lower bound: 25, count: 9700=2^13
w=5: lower bound: 30, count: 24715=2^14
24715
```

For this code, even using Algorithm 2.3 with only one information set would be faster than using the non-optimal information sets, as the code has $2^{15} = 32768$ codewords.

2.2 Cyclic codes

A cyclic code $C = [n, k, d]$ is a block code that is invariant under cyclic shifts of the codewords, i.e., if $c = (c_1, \ldots, c_n)$ is a codeword, then $(c_n, c_1, \ldots, c_{n-1})$ is a codeword as well. For a cyclic code $C = [n, k, d]$, if there is an information set formed by k consecutive positions, there are always $m = \lfloor n/k \rfloor$ disjoint information sets, e.g., $\mathcal{I}_1 = \{1, \ldots, k\}$, $\mathcal{I}_2 = \{k + 1, \ldots, 2k\}$, etc. What is more, the codewords which are obtained by encoding a vector $i \in \mathbb{F}_q^k$ using the generator matrices corresponding to these information sets can all be obtained by a cyclic shift of only one of those codewords. As a permutation of the coordinate positions does not change the Hamming weight, it is sufficient to use only one generator matrix in the encoding process, but we can use all m information sets to compute the lower bound. The cyclic automorphism group allows further improvements to the lower bound (see [6] and [9, Theorem 1]):

Lemma 2.8. *For any codeword of length n having weight u there is an information set containing not more than ku/n non-zero positions.*

Proof. Starting with an arbitrary information set $\mathcal{I}_0$, we obtain n different information sets $\mathcal{I}_j := \{i + j \bmod n : i \in \mathcal{I}_0\}$ by cyclic shifts. Every symbol of the codeword is in exactly k of those information sets. Hence the total number of non-zero elements in those n information sets is ku. Therefore the average weight in one information set is ku/n, and in at least one of them the weight must not exceed the average weight. $\square$

After all words $i \in \mathbb{F}_q^k$ of weight wgt $i \leq w$ have been considered, the remaining codewords will have at least $w + 1$ non-zero coefficients in the information set. Applying Lemma 2.8, this implies a lower bound $d_{\text{lb}} = \lceil (w+1)n/k \rceil$. Similar to Algorithm 2.6, we can pre-compute the resulting sequence of lower bounds and project the total number of vectors that have to be considered (see Algorithm 2.9).

Algorithm 2.9 (Cyclic Work Factor).

```
intrinsic CyclicWorkFactor(F :: FldFin, n :: RngIntElt, k :: RngIntElt,
                                        d :: RngIntElt) → RngIntElt
{
    Computes max. number of codewords considered in Magma's
    algorithm to compute the minimum weight of a cyclic code
}
    w0 := Floor(d*k/n);
    vprintf Code: "maximal weight: %o\n", w0;
    work := 0;
    w := 0;
    lb := 1;
    repeat
      w +:= 1;
      work +:= Binomial(k, w)*(#F−1)^(w−1);
      lb := Ceiling(n/k*(w+1));
      vprintf Code: "w=%o: lower bound: %o,
                count: %o=2^%o\n", w, lb, work, Ilog2(work);
    until lb ge Minimum(d, n−k+1) or w ge k;
    return work;
end intrinsic;
```

The following example demonstrates the faster growth of the lower bound for cyclic codes. We consider a code $C = [85, 16, 44]_4$ over $\mathbb{F}_4$. The sequence of lower bounds and the number of vectors to be considered reads

```
>    WorkFactor(GF(4), 85, 16, 44, [16,16,16,16,16,5]);
         maximal weight: 8
         relative ranks used: [ 16, 16, 16, 16, 16 ]
         w=1: lower bound: 10, count: 80=2^6
         w=2: lower bound: 15, count: 1880=2^10
         w=3: lower bound: 20, count: 27080=2^14
         w=4: lower bound: 25, count: 272780=2^18
         w=5: lower bound: 30, count: 2041820=2^20
         w=6: lower bound: 35, count: 11771540=2^23
         w=7: lower bound: 40, count: 53470340=2^25
         w=8: lower bound: 45, count: 194203790=2^27
         194203790
```

for a code that is not known to be cyclic, and

```
>    CyclicWorkFactor(GF(4), 85, 16, 44);
         maximal weight: 8
         w=1: lower bound: 11, count: 16=2^4
         w=2: lower bound: 16, count: 376=2^8
         w=3: lower bound: 22, count: 5416=2^12
         w=4: lower bound: 27, count: 54556=2^15
         w=5: lower bound: 32, count: 408364=2^18
         w=6: lower bound: 38, count: 2354308=2^21
         w=7: lower bound: 43, count: 10694068=2^23
         w=8: lower bound: 48, count: 38840758=2^25
         38840758
```

for a cyclic code. Despite the fact that the lower bound for a cyclic code grows faster, the maximal weight that needs to be considered is $w_{\max} = 8$ for both cases. Nevertheless, the algorithm tailored for cyclic codes is five times faster because the general algorithm uses five different generator matrices while one generator matrix suffices for cyclic codes. The relative rank of the sixth generator matrix is only five. As it makes no contribution to establish the lower bound, it can be discarded.

The verbose output of the actual algorithm (see Example 2.10) reveals an additional saving. As the code is even, i.e., it has only codewords of even weight, all odd lower bounds can be increased by one. Therefore the weight w that needs to be considered in the cyclic case is at most $w_{\max} = 7$, while there is no such advantage for the general algorithm.

Example 2.10 (Cyclic Code).

```
>    c := BKLC(GF(4), 85, 16);
>    ResetMinimumWeightBounds(c);
>    MinimumWeight(c);
```

```
Code MinimumWeightZimmermann: length 85, dimension 16, cyclic
Lower Bound: 1 , Upper Bound: 70
BCHBound gives the lower bound = 22
Using congruence d mod 2 = 0
Maximum Projected r: 16
Starting Search...
lower: 22, upper: 70, r: 1, time: 0.010
New codeword identified of weight 50, time 0.010
New codeword identified of weight 48, time 0.010
lower: 22, upper: 48, r: 2, time: 0.010
New codeword identified of weight 46, time 0.010
lower: 22, upper: 46, r: 3, time: 0.010
New codeword identified of weight 44, time 0.010
lower: 22, upper: 44, r: 4, time: 0.020
lower: 28, upper: 44, r: 5, time: 0.060
lower: 32, upper: 44, r: 6, time: 0.400
lower: 38, upper: 44, r: 7, time: 2.120
Final Results: lower = 44, upper = 44,
   Total time: 9.660
44
```

In general, if one knows that all weights in the code are divisible by a constant t, the lower bound can always be increased to the next multiple of t. In Magma this is implemented for binary and quaternary codes for which one can easily check whether they are even ($t = 2$) or doubly-even ($t = 4$, implemented for binary codes only), as well as for ternary codes ($t = 3$).

2.3 The MacWilliams transform

In some cases there may be alternatives to using Algorithm 2.5 for computing the minimum weight. We have already seen in Example 2.7 that sometimes using only a single information set might be better than using many information sets with large overlaps. If we use only one information set for the code $C = [77, 15, 29]_2$, the algorithm terminates after having enumerated all codewords. In the minimum weight algorithm, the codewords of the form iG are enumerated with increasing weight $\mathrm{wgt}(i)$. Instead, we can choose a different order for enumerating all codewords which yields a faster algorithm. As a by-product, the complete weight distribution of the code can be computed using the intrinsic **WeightDistribution**. For codes with high rate $R = k/n > 1/2$, the dual code

$$C^{\perp} := \{x : x \in \mathbb{F}_q^n \mid \forall c \in C : x \cdot c = 0\}$$

has relatively few codewords. Then the computation of the weight distribution can be done using the MacWilliams transform (see, e.g., [10]) which relates the weight distribution of the code to that of its dual. Using Algorithms 2.6 and 2.9 we can decide a priori which algorithm should be used. Algorithm 2.11 integrates the different strategies into one function:

Algorithm 2.11 (Minimum Weight Computation).

```
intrinsic new_MinimumWeight(C :: Code) → RngIntElt
{
   Computes minimum distance of the code either by Magma's
   internal algorithm or via the weight distribution
}
   F := Alphabet(C);
   n := Length(C);
   k := Dimension(C);
   if not assigned C`MinimumWeight then
     work_wd := #F^Minimum(k, n−k);
     if IsCyclic(C) then
       work := CyclicWorkFactor(F, n, k, n−k+1);
     else
       ranks := [ k : i in [1..n−k−1 by k] ] cat [n mod k];
       work := WorkFactor(F, n, k, n−k+1, ranks);
     end if;
     if work_wd lt work then
       vprint Code: "using the weight distribution";
       wd := WeightDistribution(C);
     end if;
   end if;
   return MinimumWeight(C);
end intrinsic;
```

First the algorithm checks whether the minimum weight of the code is already known, i.e., if the attribute C`MinimumWeight has been assigned. If not, the total number of vectors **work** that have to be considered is estimated. If this number is larger than the number of codewords in the code or in the dual code, the weight distribution of the code is computed. Note that Magma automatically uses the MacWilliams transform to determine the weight distribution of a code if the dual code has less codewords than the code. Additionally, the weight distribution is internally stored. This enables the intrinsic MinimumWeight to derive the minimum weight directly from the weight distribution of the code if it is already known.

Of course the Singleton bound $d \leq n - k + 1$ used to compute the work factor in Algorithm 2.11 should be replaced by a better upper bound if available, e.g., by sampling some codewords. Otherwise, the running time of Algorithm 2.5 might be overestimated. But even with this weak upper bound the new algorithm is better, as demonstrated by the following example.

Example 2.12 (Hybrid Algorithm).

```
>   c := BKLC(GF(4), 100, 90);
>   ResetMinimumWeightBounds(c);
```

```
>    new_MinimumWeight(c);

          maximal weight: 10
          relative ranks used: [ 90 ]
          w=1: lower bound: 2, count: 90=2^6
          w=2: lower bound: 3, count: 12105=2^13
          w=3: lower bound: 4, count: 1069425=2^20
          w=4: lower bound: 5, count: 70059555=2^26
          w=5: lower bound: 6, count: 3629950263=2^31
          w=6: lower bound: 7, count: 154925305353=2^37
          w=7: lower bound: 8, count: 5601558088593=2^42
          w=8: lower bound: 9, count: 175128003466938=2^47
          w=9: lower bound: 10, count: 4808850843808368=2^52
          w=10: lower bound: 11, count: 117408315864105117=2^56
          using the weight distribution
          Checking if the code is MDS:
```

verbose output partially deleted

```
          Code is not MDS
          5
```

In order to establish the lower bound $d_{lb} = 5$, one would have to enumerate approximately 4^{13} codewords, while the dual code has only 4^{10} codewords. Turning off the verbose mode, the timings are:

```
>    SetVerbose("Code",0);
>    c := BKLC(GF(4), 100, 90);
>    ResetMinimumWeightBounds(c);
>    time MinimumWeight(c);

          5
          Time: 23.690

>    ResetMinimumWeightBounds(c);
>    time new_MinimumWeight(c);

          5
          Time: 1.320
```

3 Constructing new codes from old ones

In this section, we will briefly discuss some constructions which enable us to build new codes from explicitly known codes. In most of the cases, the construction also yields a lower bound on the minimum weight of the resulting code based on the minimum weight (resp. bounds) of the constituent codes.

3.1 Construction X and variations

The first family of constructions uses two or more codes of the same length where one code contains the other codes which have a higher minimum weight. The simplest case starts with two nested codes and a third auxiliary code:

Lemma 3.1 (Construction X (cf. [12])). *Let $C_1 = [n, k_1, d_1]_q$ and $C_2 = [n, k_2, d_2]_q$ be two linear codes of length n over $\mathbb{F}_q$ with $C_2 \subset C_1$ and $k_2 < k_1$. Using a third code $C_3 = [n', k_1 - k_2, d_3]_q$, one obtains a code $C = [n+n', k_1, d]_q$ where $d \geq \min\{d_2, d_1 + d_3\}$.*

Proof (Sketch). The generator matrix G of C is of the form

$$G = \begin{pmatrix} G_{12} & G_3 \\ G_2 & 0 \end{pmatrix},$$

where G_2 and G_3 are generator matrices of C_2 and C_3, respectively, and G_{12} and G_2 together form a generator matrix of G_1. Partitioning the code C_1 into cosets with respect to the subcode C_2 shows that the codewords of C are either a juxtaposition $(c_1|c_3)$ of non-zero codewords of C_1 and C_3 if c_1 is in a proper coset of C_2, or they are of the form $(c_2|0)$, where $c_2 \in C_2$. $\square$

For a chain of three codes the construction can be generalised as follows:

Lemma 3.2 (Construction X3 (cf. [10])). *Let $C_1 = [n, k_1, d_1]_q$, $C_2 = [n, k_2, d_2]_q$, and $C_3 = [n, k_3, d_3]_q$ be three linear codes with $C_3 \subset C_2 \subset C_1$ and $k_3 < k_2 < k_1$. Using auxiliary codes $C_4 = [n', k_2 - k_3, d_4]_q$ and $C_5 = [n'', k_1 - k_2, d_5]_q$, one obtains a code $C = [n + n' + n'', k_1, d]_q$ where $d \geq \min\{d_3, d_2 + d_4, d_1 + d_5\}$.*

Proof (Sketch). The generator matrix G of C is of the form

$$G = \begin{pmatrix} G_{12} & 0 & G_5 \\ G_{23} & G_4 & 0 \\ G_3 & 0 & 0 \end{pmatrix},$$

where the generator matrices G_1 and G_2 of C_1 and C_2 are decomposed in the obvious way. $\square$

Another variation of Construction X uses two arbitrary subcodes of a code and two auxiliary codes. The parameters of the new code additionally depend on the intersection of the two subcodes.

Lemma 3.3 (Construction XX (cf. [1])). *Let $C_1 = [n, k_1, d_1]_q$, $C_2 = [n, k_2, d_2]_q$, and $C_3 = [n, k_3, d_3]_q$ be three linear codes with $C_2 \subset C_1$, $k_2 < k_1$, and $C_3 \subset C_1$, $k_3 < k_1$. Furthermore, let $C_4 = [n, k_4, d_4] := C_2 \cap C_3$. Using auxiliary codes $C_5 = [n', k_1 - k_2, d_5]_q$ and $C_6 = [n'', k_1 - k_3, d_6]_q$, one obtains a code $C = [n + n' + n'', k_1, d]_q$ where $d \geq \min\{d_4, d_2 + d_6, d_3 + d_5, d_1 + d_5 + d_6\}$.*

Proof (Sketch). The four codes of length n form the following lattice:

$$
\begin{array}{ccc}
 & C_1 & \\
C_5 \nearrow & & \nwarrow C_6 \\
C_2 & & C_3 \, . \\
C_6 \nwarrow & & \nearrow C_5 \\
 & C_4 &
\end{array}
$$

In a first step, we apply Construction X to the codes $C_2 \subset C_1$ and C_5. The resulting code can be partitioned such that the first n coordinates form cosets of C_3 in C_1. With respect to this decomposition, we apply again Construction X using the auxiliary code C_6. So we append to any codeword of C_1 that is in a proper coset with respect to both C_2 and C_3 a non-zero codeword of C_5 and a non-zero codeword of C_6. Codewords which lie in a proper coset of C_4 in C_2 are extended by a non-zero codeword of C_5, and similarly for cosets of C_4 in C_3. Only the codewords of C_4 will be padded by zeroes. $\square$

There are even more involved variations of Construction X, which are labelled X3a, X3u, X4, X6, etc. (see [4]). All these constructions start from a lattice of codes which have the same length, but the subcodes have a higher minimum weight. So when searching for good codes, it is helpful to find codes which have good subcodes or supercodes as well.

3.2 Construction Y1

While Construction X and its variations increase the length of a code in order to enlarge the minimum weight, we now consider a construction which decreases both the length and the dimension of a code.

Lemma 3.4 (Construction Y1 (cf. [10])). *Let $C = [n, k, d]_q$ denote a linear code over $\mathbb{F}_q$. If the dual code $C^\perp$ contains a word $v \in C^\perp$ of weight w, then there is a code $C' = [n - w, k' \geq k - w + 1, d' > d]_q$.*

Proof. The code C' is obtained from C via shortening, i.e., taking the subcode of C which consists of all codewords that have zeroes at the positions where v is non-zero, and then deleting these w positions. Obviously, the minimum weight of the resulting code is at least d. As $v \in C^\perp$ is a codeword of the dual code there exists a parity check matrix H of C which has v as its first row. After deleting all coordinates where v is non-zero, the first row of the resulting matrix H' is zero. Hence the rank of H' is at most $n - k - 1$. Thus the dimension of the code C' is at least $k - w + 1$.

4 Searching for good codes

During the search for good linear codes that establish or even improve the lower bounds in Brouwer's tables, many constructions and techniques have been used. In the following, we illustrate some of the most successful approaches, as well as a very special case.

4.1 Cyclic codes

Cyclic codes play a central rôle in our search, for many reasons. Firstly, the BCH bound provides a lower bound on the minimum weight that is easy to obtain. Secondly, the algorithms for computing the minimum weight perform best for cyclic codes. Additionally, the search space for cyclic codes is relatively small. And finally, cyclic codes naturally give rise to lattices of codes that are contained in each other.

A cyclic code is defined by a generator polynomial $g(x) \in \mathbb{F}_q[x]$ which is a divisor of $x^n - 1$, where $\gcd(n, q) = 1$. Hence the set of all cyclic codes of length n corresponds to all factors of $x^n - 1$, or equivalently to subsets of the set of all irreducible factors. So provided that $x^n - 1$ does not have too many factors, we may compute all cyclic codes of length n from the factorisation of $x^n - 1$. The following Magma program returns all cyclic codes of given length:

```
intrinsic AllCyclicCodes(F :: FldFin, n :: RngIntElt)  →  SetEnum
{
    Returns the set of all cyclic codes of  length n over F
}
    P := PolynomialRing(F);
    L_f := {f[1]:f in Factorisation(P.1^n−1)};
    if #L_f gt 16 then
        printf "WARNING: there are 2^%o codes\n",#L_f;
    end if;
    L_c := { CyclicCode(n, &*y): y in Subsets(L_f) };
    return L_c;
end intrinsic;
```

Note that Magma can generate all cyclic codes from the set *L_f* of all irreducible factors of $x^n - 1$ in a single statement:

```
L_c:= { CyclicCode(n, &*y) : y in Subsets(L_f) };
```

The intrinsic **Subsets** returns all subsets of the argument. The product of all elements in each subset is computed using the reduction operator &*y.

Given a set of codes, one can simply test whether there exists a code of given dimension that establishes or even improves a given lower bound. Using Algorithm 2.9 one can exclude codes from the search for which the running time would be too high to establish the lower bound. Using the following Magma program, we find a cyclic code $[73, 25, 24]_3$:

```
>    cand := AllCyclicCodes(GF(3), 73);
>    time exists(c){ c : c in cand | Dimension(c) eq 25
>                        and VerifyMinimumWeightLowerBound(c, 24) };
```

```
Code MinimumWeightZimmermann: length 73,
dimension 25, cyclic
Lower Bound: 1 , Upper Bound: 49
```

```
BCHBound gives the lower bound = 13
Maximum Projected r: 25
Starting Search...
lower: 13, upper: 49, r: 1, time: 0.469
New codeword identified of weight 35, time 0.469
New codeword identified of weight 28, time 0.469
New codeword identified of weight 26, time 0.469
lower: 13, upper: 26, r: 2, time: 0.469
New codeword identified of weight 20, time 0.469
Final Results: lower = 13, upper = 20,
   Total time: 0.469
Code MinimumWeightZimmermann: length 73,
dimension 25, cyclic
Lower Bound: 1 , Upper Bound: 49
BCHBound gives the lower bound = 19
Maximum Projected r: 25
Starting Search...
lower: 19, upper: 49, r: 1, time: 0.470
New codeword identified of weight 36, time 0.470
New codeword identified of weight 31, time 0.470
New codeword identified of weight 25, time 0.470
lower: 19, upper: 25, r: 2, time: 0.470
New codeword identified of weight 24, time 0.470
lower: 19, upper: 24, r: 3, time: 0.470
lower: 19, upper: 24, r: 4, time: 0.470
lower: 19, upper: 24, r: 5, time: 0.510
lower: 19, upper: 24, r: 6, time: 0.810
lower: 21, upper: 24, r: 7, time: 2.890
Final Results: lower = 24, upper = 24,
   Total time: 14.350
true
Time: 14.830
```

Note that for the first code tested the upper bound drops to $d_{ub} = 20$ after only half a second when the lower bound is $d_{lb} = 13$. As we are looking for a code with minimum distance 24, the computation can be aborted without computing the true minimum distance of that code. So the intrinsic VerifyMinimumWeightLowerBound proves to be very useful when searching for good codes.

Using a similar technique, we found, for example, a code $[85, 43, 21]_4$ which improves the lower bound $d_{\text{Brouwer}} = 18$ by three. The verification of the lower bound took almost 78 days on a 2.5 GHz Linux PC:[3]

```
>   F<a> := GF(4);
>   g := PolynomialRing(F) ! [1, 1, 1, a, a, 0, a², a, a, 1, 0, 0, 1, 1, 0, 1, 1, a, a²,
>             a², a², a², a, a, 1, a², 1, a, 0, 0, 0, a², a, 1, 0, a, 1, a, 0, a², 1, 1, 1]
```

---

[3]In Magma V2.11, an improved codeword enumeration algorithm has been implemented, reducing the projected running time on the same PC to approx. 19 days.

```
>    c := CyclicCode(85, g);
>    MinimumWeight(c);
```

```
        Code MinimumWeightZimmermann: length 85,
        dimension 43, cyclic
        Lower Bound: 1 , Upper Bound: 43
        BCHBound gives the lower bound = 8
        Maximum Projected r: 42
        Starting Search...
        lower: 8, upper: 43, r: 1, time: 0.009
        New codeword identified of weight 34, time 0.009
        New codeword identified of weight 29, time 0.009
        New codeword identified of weight 28, time 0.009
        lower: 8, upper: 28, r: 2, time: 0.009
        New codeword identified of weight 25, time 0.009
        lower: 8, upper: 25, r: 3, time: 0.009
        New codeword identified of weight 24, time 0.009
        New codeword identified of weight 23, time 0.009
        New codeword identified of weight 22, time 0.029
        lower: 8, upper: 22, r: 4, time: 0.029
        New codeword identified of weight 21, time 0.149
        lower: 10, upper: 21, r: 5, time: 0.539
        lower: 12, upper: 21, r: 6, time: 12.639
        lower: 14, upper: 21, r: 7, time: 242.569
        lower: 16, upper: 21, r: 8, time: 3916.620
        lower: 18, upper: 21, r: 9, time: 53859.480
        lower: 20, upper: 21, r: 10, time: 642562.619
        Final Results: lower = 21, upper = 21,
           Total time: 6729809.280
```

Note that more than 2^{22} codewords have been enumerated per second. The original lower bound in Brouwer's table was proven in only about 15 hours. The comparison of the different methods to compute the minimum weight clearly demonstrates the advantage of cyclic codes:

method	approx. running time
cyclic code	78 days
overlapping information sets	155 days
one information set	$7.8 \cdot 10^7$ years
weight distribution of the dual code	$9.8 \cdot 10^{11}$ years

Having spent so much time on a single code, we would also like to find subcodes and supercodes which are suitable for Construction X and its variations. In order to get subcodes or supercodes of a given cyclic code, we multiply the generator polynomial by another irreducible factor of $x^n - 1$, respectively, divide by one of the factors. The following functions implement this simple approach:

intrinsic CyclicSubcodes(C :: Code) $\rightarrow$ SetEnum
{ Returns set of maximal proper cyclic subcodes of C }
  if Dimension(C) **eq** 0 **then**
    return {};
  else
    n := Length(C);
    g := GeneratorPolynomial(C);
    g_0 := Parent(g).1^n-1;
    g_1 := g_0 **div** g;
    L_c := { CyclicCode(n, $g*f[1]$) : f **in** Factorisation(g_1) };
    return L_c;
  end if;
end intrinsic;

intrinsic CyclicSupercodes(C :: Code) $\rightarrow$ SetEnum
{ Returns the set of all cyclic codes
for which C is a maximal proper subcode }
  n := Length(C);
  if Dimension(C) **eq** n **then**
    return {};
  else
    g := GeneratorPolynomial(C);
    L_c := { CyclicCode(n, g **div** $f[1]$) : f **in** Factorisation(g) };
    return L_c;
  end if;
end intrinsic;

Table 1. Codes derived from the cyclic code $[85, 43, 21]_4$.

construction	ingredients	resulting code	d_{Brouwer}
X	$[85, 40, 22]_4 \subset [85, 43, 21]_4$, $[3, 3, 1]_4$	$[88, 43, 22]_4$	20
X	$[85, 43, 21]_4 \subset [85, 47, 18]_4$, $[5, 4, 2]_4$	$[90, 47, 20]_4$	18
X	$[85, 43, 21]_4 \subset [85, 47, 18]_4$, $[7, 4, 3]_4$	$[92, 47, 21]_4$	19
X3	$[85, 40, 22]_4 \subset [85, 43, 21]_4 \subset [85, 46, 18]_4$ $[3, 3, 1]_4$, $[6, 3, 4]_4$	$[94, 46, 22]_4$	21
X	$[85, 43, 21]_4 \subset [85, 49, 16]_4$, $[11, 6, 5]_4$	$[96, 49, 21]_4$	20

Using this approach, we found the following chain of cyclic codes:

$$[85, 40, 22]_4 \subset [85, 43, 21]_4 \subset [85, 47, 18]_4 \subset [85, 49, 16]_4.$$

In addition to these cyclic codes, we can take a (non-cyclic) subcode of $[85, 46, 18]_4 \subset [85, 47, 18]_4$ which contains the cyclic code $[85, 43, 21]_4$ to refine the chain. Then applying Construction X or Construction X3 to some of these codes, we obtain new codes which improve the original lower bounds d_{Brouwer} in Brouwer's table. The new codes are summarised in Table 1.

4.2 Other techniques

From a particular code $C = [n, k, d]_q$ one can derive bounds on the minimum weight of many other codes obtained by simple constructions such as shortening, puncturing, lengthening by padding with zeroes, and taking subcodes. If the code is explicitly given by a generator matrix one can apply even more constructions, as demonstrated in the sequel.

The algorithm implemented in Magma for computing the minimum weight stores a codeword whose weight equals the actual upper bound d_{ub}. In order to test whether we can apply Construction Y1 to a code $C = [n, k, d]_q$ to obtain a code $C' = [n - w, k - w + 1, d' \geq d]_q$, we have to test whether there is a codeword of weight at most w in the dual code $C^\perp$. This is achieved by the following program:

```
intrinsic TestConstructionY1(C :: Code, w :: RngIntElt) → Code
{ Searches for a non-zero codeword of weight
  at most w in the dual code of C. If such a word
  exists, Construction Y1 is applied to C, else
  the code C is returned. }
  D := Dual(C);
  if VerifyMinimumWeightUpperBound(D, w) then
    v := D`MinimumWeightUpperBoundWord;
    C₁ := ShortenCode(C, Support(v));
  else
    C₁ := C;
  end if;
  return C₁;
end intrinsic;
```

The intrinsic VerifyMinimumWeightUpperBound returns true if and only if the algorithm for computing the minimum weight has found a codeword whose weight does not exceed the prescribed bound w. The support of such a word, which is stored in the attribute D`MinimumWeightUpperBoundWord, can be used for Construction Y1.

As in the case of Example 2.7, a (random) permutation of the code might help as this results in the codewords being enumerated in a different order. Hence a codeword of low weight might be encountered sooner. Using this technique, we found ternary codes $[162, 122, 14]_3$ and $[162, 127, 12]_3$, starting from BCH codes $[242, 201, 14]_3$ and $[242, 206, 12]_3$, respectively.

As a final example, we describe the search for a ternary code $[168, 153, 6]_3$ which is listed in Brouwer's table. The recipe for constructing such a code given in [4, Sect. 4.1.9.(x)], however, contains an error, as the required chain of ternary codes does not exist. As many codes in Brouwer's table are derived from this code, we search for an alternative construction. So far, we have not been completely successful, but we were able to construct most of the missing codes. The main idea, which will be described in the sequel, is 'to invert shortening', i.e., to start with a code $C = [n, k, d]$ that has the same redundancy $n-k$ and minimum distance as the missing code, and then increase both its length and dimension.

The longest ternary code with $n - k \leq 15$ and $d \geq 6$ in our list is a code $C_{146} = [146, 132, 6]_3$ which is constructed as follows:

```
>    P<x> := PolynomialRing(GF(9));
>    w := Roots(x²−x−1)[1,1];
>    g := x⁷+w*x⁶+w⁶*x⁵+w³*x⁴+w⁷*x³+w²*x²+w⁵*x+2;
>    C_146_132_6 := ConcatenatedCode(
>              CyclicCode(73, g), UniverseCode(GF(3), 2) );
```

As in Example 2.7, the minimum weight computation is much faster when using the dual code:

```
>    time MinimumWeight(C_146_132_6);

     6
     Time: 305.970


>    ResetMinimumWeightBounds(C_146_132_6);
>    time new_MinimumWeight(C_146_132_6);

     6
     Time: 1.520
```

Using a cyclic code $[73, 67, 4]_9$ which contains the cyclic code $[73, 66, 6]_9$ over $\mathbb{F}_9$, we construct a supercode $[146, 134, 4]_3 \supset [146, 132, 6]_3$:

```
>    C_146_134_4 := ConcatenatedCode(
>              CyclicCode(73, g div (x−1)), UniverseCode(GF(3),2) );
```

Next we construct all codes between those two codes:

```
>    subcodes := { sub< C_146_134_4 | C_146_132_6, x >:
>         x in CodeComplement(C_146_134_4, C_146_132_6) | x ne 0 };
>    L_d := { MinimumWeight(c) : c in subcodes };
>    subcodes:Minimal;

         {
             [146, 133, 5] Linear Code over GF(3),
             [146, 133, 5] Linear Code over GF(3),
             [146, 133, 4] Linear Code over GF(3),
             [146, 133, 4] Linear Code over GF(3)
         }
```

As there are two codes with minimum weight 5, we can apply Construction XX and obtain a code $C_{148} = [148, 134, 6]_3$.

```
>     subcodes := [ c : c in subcodes | MinimumWeight(c) eq 5 ];
>     C_2 := subcodes[1];
>     C_3 := subcodes[2];
>     C_4 := UniverseCode(GF(3), 1);
>     C_148_134_6 := ConstructionXX(C_146_134_4, C_2, C_3, C_4, C_4);
>     new_MinimumWeight(C_148_134_6);

      6
```

Note that this code improves the lower bound $d_{\text{Brouwer}} = 5$ in Brouwer's table. As the redundancy $n - k = 14$ of this code is not too large, we can compute the distance distribution of the cosets:

```
>     time CosetDistanceDistribution(C_148_134_6);

      [ <0, 1>, <1, 296>, <2, 43512>, <3, 2788308>,
        <4, 1950852> ]
      Time: 25.550
```

As there is a coset with distance 4, the code can be enlarged to a code $[148, 135, 4]_3$. Then, applying Construction X to those codes, we obtain a code $C_{150} = [150, 135, 6]_3$. As there are many cosets with distance 4 we can use a probabilistic search to find a suitable vector: we pick a random vector x from the vector space complement of C_{148} in $\mathbb{F}_3^{148}$ and test if the code generated by C_{148} and x has at least minimum weight 4:

```
>     cc := CodeComplement(Generic(C_148_134_6), C_148_134_6);
>     repeat
>        c_1 := LinearCode< GF(3), 148 | C_148_134_6, Random(cc) >;
>     until c_1 ne C_148_134_6 and VerifyMinimumWeightLowerBound(c_1, 4);
>     C_150_135_6 := ConstructionX(
>                   c_1,C_148_134_6, RepetitionCode(GF(3), 2) );
```

As the code $C_{150} = [150, 135, 6]_3$ already has the same redundancy $n - k = 15$ as the code we are looking for, we now have to search for a coset with distance at least 5. If such a coset exists, we obtain a new code $C_1 = [n, n - 14, 5]_3$. Applying Construction X to the codes C_1, $C = [n, n - 15, 6]_3$, and the trivial code $[1, 1, 1]_3$, we obtain a code $C' = [n+1, (n+1) - 15, 6]_3$ and we can repeat the process. Simplified, the search strategy can be implemented as follows:

```
>     c:=C_150_135_6;
>     while CoveringRadius(c) ge 5 do
>        cc := CodeComplement(Generic(c), c);
>        time repeat
>           c_1 := LinearCode< GF(3), Length(c) | c, Random(cc) >;
>        until c_1 ne c and VerifyMinimumWeightLowerBound(c_1, 5);
>        c := ConstructionX(c_1, c, UniverseCode(GF(3), 1) );
>     end while;
```

Using such an approach, we found a code $C_{167} = [167, 152, 6]_3$. As the covering radius of this code is only 4, we had to stop the process and did not obtain the missing code. A different choice of the coset at distance 5 in each step might have been better. But we face the problem that at the beginning of our search, there are too many cosets to consider all of them. As the length of the codes obtained by this strategy increases, the number of cosets with distance 5 decreases (see Table 2). Hence the probability of finding a coset with distance 5 becomes very small. So instead of the random search we have to use a different strategy, such as tabu search, i.e., excluding the vectors that have been considered, or even a systematic search of the whole space of cosets of size 3^{15}. So we must have both—fast computers with a lot of memory and luck.

Table 2. Coset distance distribution of the codes obtained when searching for a putative $[168, 153, 6]_3$ code.

code	number of cosets with distance d					
	$d = 0$	$d = 1$	$d = 2$	$d = 3$	$d = 4$	$d = 5$
$[150, 135, 6]_3$	1	300	44700	2962424	9747998	1593484
$[151, 136, 6]_3$	1	302	45300	3049698	10626544	627062
$[152, 137, 6]_3$	1	304	45904	3136130	10900844	265724
$[153, 138, 6]_3$	1	306	46512	3223548	10980832	97708
$[154, 139, 6]_3$	1	308	47124	3310682	10950806	39986
$[155, 140, 6]_3$	1	310	47740	3397578	10886486	16792
$[156, 141, 6]_3$	1	312	48360	3484094	10809124	7016
$[157, 142, 6]_3$	1	314	48984	3570960	10725586	3062
$[158, 143, 6]_3$	1	316	49612	3657340	10640240	1398
$[159, 144, 6]_3$	1	318	50244	3743836	10553882	626
$[160, 145, 6]_3$	1	320	50880	3830068	10467324	314
$[161, 146, 6]_3$	1	322	51520	3915716	10381192	156
$[162, 147, 6]_3$	1	324	52164	4003250	10293118	50
$[163, 148, 6]_3$	1	326	52812	4089298	10206440	30
$[164, 149, 6]_3$	1	328	53464	4175102	10119996	16
$[165, 150, 6]_3$	1	330	54120	4259258	10035190	8
$[166, 151, 6]_3$	1	332	54780	4346504	9947288	2
$[167, 152, 6]_3$	1	334	55444	4431990	9861138	0

5 Conclusion

Our search for codes with large minimum distance revealed many new codes which improve the lower bound compared to Brouwer's tables [4]. Even for binary codes, which are the most intensely studied and for which earlier results of our project have already been included in Brouwer's table, we could improve the minimum distance for about 3 percent of all codes up to length 256 (see Fig. 1 a)).

The most dramatic progress has been made for codes over $\mathbb{F}_4$. Almost two thirds of the codes of length 89, about one third of all codes up to length 100, and more then 9 percent of all codes up to length 256 have been improved (see Fig. 1 b)). Most of the codes are derived from cyclic codes and applying Construction X and its variations to chains of cyclic codes, in particular to cyclic codes of length $n = 85$.

We have not yet found constructions for all codes in Brouwer's tables. Many of those which are missing are those for which the lower bound is based on non-constructive counting arguments. But we are confident that the various techniques developed will help to find more good codes.

alphabet	max. length	new codes number	fraction
$\mathbb{F}_2$	256	980	2.98 %
$\mathbb{F}_3$	243	2098	7.08 %
$\mathbb{F}_4$	256	3110	9.45 %
$\mathbb{F}_5$	130	1250	14.68 %
$\mathbb{F}_7$	50	86	6.75 %
$\mathbb{F}_8$	130	738	8.67 %
$\mathbb{F}_9$	130	389	4.57 %

a) Total number of new codes.

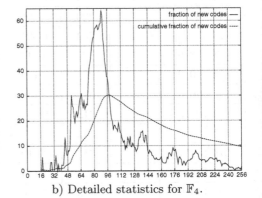

b) Detailed statistics for $\mathbb{F}_4$.

Fig. 1. Summary of the number of new codes which improve the lower bound on the minimum distance compared to Brouwer's on-line table [4] as of February 3, 2004.

6 Acknowledgements

First of all, I would like to thank all members of the Magma group who have been involved in this project, in particular John Cannon, Damien Fisher, Allan Steel, and Greg White. Part of this work was performed while visiting the Computational Algebra Group within the School of Mathematics and

Statistics of the University of Sydney whose financial support is particularly acknowledged.

Special thanks go to my colleagues at IAKS for providing computing power to search for good codes.

References

1. W. O. Alltop, *A Method for Extending Binary Linear Codes*, IEEE Transactions on Information Theory **30** (1984), 871–872.
2. A. Betten, H. Fripertinger, A. Kerber, A. Wassermann, K.-H. Zimmermann, *Codierungstheorie: Konstruktionen und Anwendungen linearer Codes*, Berlin: Springer, 1998.
3. Wieb Bosma, John Cannon, Catherine Playoust, *The Magma algebra system I: The user language*, J. Symbolic Comput. **24** (1997), 235–265.
   See also the Magma home page at http://magma.maths.usyd.edu.au/magma/.
4. A. E. Brouwer, *Bounds on the Size of Linear Codes*, in: V. S. Pless, W. C. Huffman (eds), *Handbook of Coding Theory*, Amsterdam: Elsevier, 1998, pp. 295–461. On-line version at http://www.win.tue.nl/~aeb/voorlincod.html.
5. J. Cannon, A. Steel, G. White, *Linear Codes over Finite Fields*, Chapter 115 in: John Cannon, Wieb Bosma (eds.), *Handbook of Magma Functions*, Version 2.11, Volume **8**, Sydney, 2004, pp. 3521–3596.
6. C.-L. Chen, *Computer Results on the Minimum Distance of Some Binary Cyclic Codes*, IEEE Transactions on Information Theory **16** (1970), 359–360.
7. M. Grassl, *New Binary Codes from a Chain of Cyclic Codes*, IEEE Transactions on Information Theory **47** (2001), 1178–1181.
8. R. W. Hamming, *Error detecting and error correcting codes*, The Bell System Technical Journal **29** (1950), 147–160.
9. F. P. Kschischang, S. Pasupathy, *Some Ternary and Quaternary Codes and Associated Sphere Packings*, IEEE Transactions on Information Theory **38** (1992), 227–246.
10. F. J. MacWilliams, N. J. A. Sloane, *The Theory of Error-Correcting Codes*, Amsterdam: North-Holland, 1977.
11. C. E. Shannon, *A mathematical theory of communication*, Bell System Technical Journal **27** (1948), 379–423 and 623–656.
12. N. J. A. Sloane, S. M. Reddy, C.-L. Chen, *New Binary Codes*, IEEE Transactions on Information Theory **18** (1972), 503–510.
13. M. A. Tsfasman, S. G. Vladut, T. Zink, *Modular curves, Shimura curves, and Goppa codes, better than Varshamov–Gilbert bound*, Mathematische Nachrichten **109** (1982), 21–28.
14. A. Vardy, *The Intractability of Computing the Minimum Distance of a Code*, IEEE Transactions on Information Theory **43** (1997), 1757–1773.
15. K.-H. Zimmermann, *Integral Hecke Modules, Integral Generalized Reed–Muller Codes, and Linear Codes*, Tech. Rep. **3-96**, Technische Universität Hamburg–Harburg, 1996.

Colouring planar graphs

Paulette Lieby [*]

Autonomous Systems and Sensing Technologies
NICTA Canberra Laboratory
Mail : Locked Bag 8001, Canberra ACT 2601, Australia
Paulette.Lieby@anu.edu.au

Summary. There is no known polynomial time 4-colouring algorithm for planar graphs, except the algorithms that can be derived from the proofs of Appel *et al.* and Robertson *et al.* An alternative way to colour planar graphs is to investigate the problem of finding nowhere-zero k-flows in graphs. We demonstrate how Magma can be used to find such flows in graphs.

1 Introduction

The four-colour problem asks whether every planar graph is 4-colourable. The proof of the four-colour theorem by Appel, Haken, and Koch [1, 3] relies on the existence in a planar graph of a set of unavoidable configurations which must be shown to be reducible. A configuration in a graph G is reducible if it can be replaced by a smaller configuration such that a 4-colouring of G can be deduced from a 4-colouring of the smaller configuration. Latest amendments to this proof [2] show that 1256 configurations must be shown to be reducible. Finding unavoidable configurations and proving their reducibility requires computer assistance in the form of thousands of hours of computation time. A later proof by Robertson, Sanders, Seymour, and Thomas [9], using the same approach as Appel *et al.*, involves less then 700 configurations but it still requires computer aid.

There is no known polynomial time 4-colouring algorithm for planar graphs, except the algorithms that can be derived from the proofs of Appel *et al.* and Robertson *et al.* Short of doing this, one way to colour planar graphs is to investigate the problem of finding nowhere-zero k-flows in graphs: Theorem 4.1 below asserts that there is an equivalence between k-colouring a planar graph and finding a nowhere-zero k-flow in a graph.

---

[*] The research for this paper was carried out while the author was a member of the Magma group, School of Mathematics and Statistics, University of Sydney, Sydney, Australia.

In this chapter we demonstrate how Magma [4] can be used to find nowhere-zero k-flows in graphs. It is an open problem to find good heuristics to construct such flows. Any advance in this respect might, in addition to producing a reasonably efficient 4-colouring algorithm for planar graphs, give new leads to a shorter proof of the four-colour theorem. The chapter is organized as follows. In Section 2 we define k-flows; in Section 3 we define planar graphs and the concept of duality for planar graphs, and show how, using Magma one constructs the dual of a planar graph. Section 4 establishes the equivalence between the problem of k-colouring a planar graph and finding a nowhere-zero k-flow in a graph. Section 5 discusses a possible way to find nowhere-zero k-flows; finally, Section 6 provides a simple way to test, in Magma, for a nowhere-zero k-flow.

2 k-Flows in graphs

In the whole of this chapter, unless otherwise stated, G is a general graph, that is, it might have loops and multiple edges. The vertex set of G is denoted by V and its edge set by E. Edges in G may be given an orientation D.

Assume that $G = (V, E, D)$ is a graph whose edges are oriented. Let u be an edge from u to v. If e is an edge without orientation, we write $e = \{u, v\}$. If e is an edge oriented from u to v, we write $e = [u, v]$, in which case we also write $-e = [v, u]$. We often write $e \in (E, D)$ to mean that e is an edge of E with an orientation given by D.

Given $u \in V$, define E_u^+ to be the set of oriented edges in (E, D) with tail u and E_u^- to be the set of oriented edges with head u. That is, E_u^+ and E_u^- are the sets of edges out-going from u and in-going to u respectively.

For $X \subseteq V$, let $\overrightarrow{E}(X, \overline{X}) = \{e : e = [u, v] \mid u \in X, v \notin X\} \cup \{-e : e = [v, u] \mid u \in X, v \notin X\}$. That is, $\overrightarrow{E}(X, \overline{X})$ is the set of all edges incident with u and taken as out-going from u, irrespective of the current edge orientation.

Let H be an additive abelian group and assume that the edges in E have an orientation D. Then a *H-circulation* or a *H-flow* is a mapping $f : (E, D) \to H$ such that the following two conditions hold:

1. Skew symmetry: $f(e) = -f(-e) \ \forall \ e \in (E, D)$,
2. Flow conservation: $\sum_{e \in E_u^+} f(e) = \sum_{e \in E_u^-} f(e) \ \forall \ u \in V$.

It follows that $\sum_{e \in \overrightarrow{E}(X, \overline{X})} f(e) = 0$ for all $X \subseteq V$.

Consider the additive group of integers $\mathbb{Z}$. A *k-flow* is a $\mathbb{Z}$-flow such that $0 \le |f(e)| < k$ for all $e \in E$. The *support* $S(f)$ of a flow f is the set of edges e such that $f(e) \ne 0$. A *nowhere-zero k-flow* is a k-flow with support E.

Two important and non-trivial results are worth noting.

Theorem 2.1. *If H and H' are two finite abelian groups of equal order, then a graph G has a nowhere-zero H-flow if and only if it has a nowhere-zero H'-flow.*

That is, the existence of a nowhere-zero flow only depends on the order of the group, and not on the group itself.

Theorem 2.2. *A graph G has a nowhere-zero k-flow if and only if it has a nowhere-zero Z_k-flow.*

A comprehensive and well-written introduction to the theory of flows in graphs and to the duality between k-flows and k-colouring for planar graphs can be found in [6]. It is now time to turn to the latter problem.

3 Planar graphs

A *plane* graph is a bijective mapping of vertices and edges into the plane such that vertices map to points and edges map to open arcs between the images of their end-vertices so that none of the resulting open arcs contain a point which is an image of a vertex or which belongs to another open arc. Let us call the union of these images the *drawing* D of the plane graph.

Let $G = (V, E)$ be a plane graph and let m be the mapping into the plane just described. In the plane, define points x and y to be equivalent if there is a path from x to y. The equivalence classes of $\mathbb{R}^2 \setminus D$ are called the *regions* of the plane; their pre-images in G under m are called the *faces* of G. Thus a plane graph has always an *outer* face which corresponds to the unbounded region in the plane under m. It may also have *inner* faces corresponding to bounded regions in the plane.

An *planar embedding* of a graph G is an isomorphism between G and a plane graph. A graph is *planar* if it has a planar embedding.

If $G = (V, E)$ is a plane graph then its *dual* graph $G^* = (V^*, E^*)$ is defined to be the graph whose vertex set V^* consists of the faces of G and where the face f_i is adjacent to the face f_j, $(f_i, f_j \in V^*)$, if and only if f_i and f_j share a common edge of G. There is thus a natural bijection between the edges of G and the edges of G^*.

Given an orientation D of the edges in E, we want to find an orientation D^* of the edges in E^* establishing a bijection ϕ between (E, D) and (E^*, D^*) as follows: Let $e = [u, v]$ be an oriented edge in G bordering the faces f_1 and f_2. Then, for $f_x, f_y \in \{f_1, f_2\}$, let $\phi(e) = [f_x, f_y]$ be the oriented edge in G^* such that $\phi(e)$ can be obtained from e by an anti-clockwise rotation.

Note that we could have chosen a clockwise rotation instead, the main point being that the orientation of $\phi(e)$ is obtained in a consistent fashion from the orientation of e. See Figure 1 for an illustration.

It is not difficult to see that $\phi(-e) = -\phi(e)$ for each $e \in (E, D)$. If P is a path $v_0, v_1, \ldots, v_l$ in G define $\overrightarrow{P} = \{[v_i, v_{i+1}] \mid 0 \le i < l\}$. We say that $\overrightarrow{P}$ is a path *with orientation*. (Note that if $e \in \overrightarrow{P}$, then e might not be in (E, D), in which case $-e$ is in (E, D)). A *cycle* C is a closed path $v_0, v_1, \ldots, v_l, v_0$ and a cycle with orientation $\overrightarrow{C}$ is defined in the same way as $\overrightarrow{P}$.

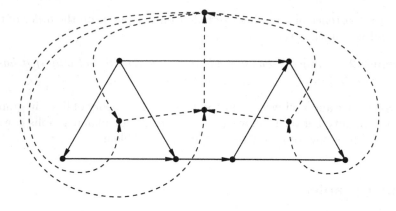

Fig. 1. G and its dual G^*

Assume that G^* has a H-flow f. Define the mapping $g : (E, D) \to H$ as $g(e) = f(\phi(e))$ for $e \in (E, D)$. For $\overrightarrow{P}$ a path with orientation in G, define $g(\overrightarrow{P}) = \sum_{e \in \overrightarrow{P}} g(e)$.

Lemma 3.1. *Let G be a plane graph and G^* its dual. Assume that G^* has a H-flow and define $g : (E, D) \to H$ as above. Let $\overrightarrow{C}$ be any cycle with orientation in G. Then $g(\overrightarrow{C}) = 0$.*

Proof. Let $\overrightarrow{C}$ be a cycle with orientation in the plane graph G. If X is the set of faces lying inside $\overrightarrow{C}$, then $X \subset V^*$. It is not difficult to see that the set of edges $\{\phi(e) : e \in \overrightarrow{C}\}$ is the set $\overrightarrow{E}(X, \overline{X})$ or $\{-e : e \in \overrightarrow{E}(X, \overline{X})\}$ depending on the orientation of $\overrightarrow{C}$. But then we have seen that $\sum_{e \in \overrightarrow{E}(X, \overline{X})} f(e) = 0$. □

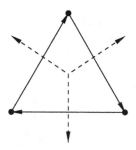

Fig. 2. Lemma 3.1

Figure 2 illustrates the situation in Lemma 3.1 for $\overrightarrow{C}$ bordering a face.

For the remainder of this section we turn to Magma and show* how to construct the dual of the plane graph G as depicted in Figure 3. We start by building G as an undirected graph made of two triangles joined by two edges. We verify that the edge connectivity of G is 2. The Magma function **join** returns a graph whose vertex set and edge set are the union of the vertex set and edge set respectively of the operand graphs.

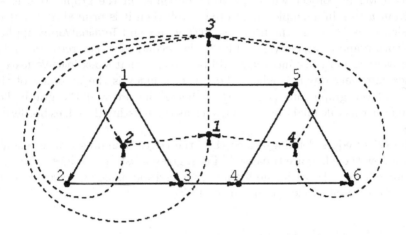

Fig. 3. G and G^* as built in the Magma example

```
>   G := CompleteGraph(3);
>   G := G join G;
>   AddEdge(~G, VertexSet(G) ! 1, VertexSet(G) ! 5);
>   AddEdge(~G, VertexSet(G) ! 3, VertexSet(G) ! 4);
>   assert IsKEdgeConnected(G, 2);
>   assert IsPlanar(G);
>   E := EdgeSet(G);
>   F := Faces(G);
>   F;
        [
            [ {1, 5}, {5, 4}, {4, 3}, {3, 1} ],
            [ {1, 3}, {3, 2}, {2, 1} ],
            [ {1, 2}, {2, 3}, {3, 4}, {4, 6}, {6, 5}, {5, 1} ],
            [ {5, 6}, {6, 4}, {4, 5} ]
        ]
```

A few comments are in order regarding the code segment above. The planarity algorithm is an implementation of a linear algorithm due to Boyer and

---

*See the Preface to this volume for style conventions regarding Magma code; code appearing in this book is available at http://magma.maths.usyd.edu.au/magma/.

Myrvold [5]. This algorithm is particularly interesting as it is significantly easier to implement than previously known linear planarity testers. Moreover, if the graph is not planar, then it isolates a Kuratowski subgraph, also in linear time.

When the graph G is planar, the faces of the embedding found by the algorithm are listed as sequences of undirected edges of G. In Magma a graph edge is a Magma object whose printout depends on the graph. If e is an edge from u to v in a simple undirected graph then it is printed as $\{u, v\}$ or equivalently as $\{v, u\}$. In the first case InitialVertex and TerminalVertex applied to e return u and v respectively, while in the second case they return v and u respectively. In most instances it would be the case that Index(InitialVertex(e)) < Index(TerminalVertex(e)), where the function Index is the mapping of the vertices of the graph into $\{1, \ldots, n\}$, n being the order of the graph. For convenience this rule might be broken, such as when listing the edges bordering a face.

The list of edges defining a face (i.e., the edges bounding a face) should indicate how the edges are traversed. Thus, given a listing of edges defining a face, any edge $e = \{u, v\}$ in the listing is understood as being traversed from u to v. For example, take the first face of G:

```
>    F[1];
```

```
     [ {1, 5}, {5, 4}, {4, 3}, {3, 1} ]
```

The edges defining $F[1]$ are traversed as follows: The first edge is traversed from 1 to 5, the second from 5 to 4, etc.

With this in mind, we can record for each face of G the orientation of the edges defining it. We say that an edge $\{u, v\}$ is forward if $u < v$, and backward otherwise. This is shown below: *Ds* shows, for each face, the orientation of the edges that border it (forward edges are indicated by 1, backward edges by -1).

```
>    Ds := [ [ 1 : x in [ 1..#F[i] ] ] : i in [ 1..#F ] ];
>    for i in [ 1..#F ] do
>      for j in [ 1..#F[i] ] do
>        if Index(InitialVertex(F[i][j])) gt
>           Index(TerminalVertex(F[i][j])) then
>          Ds[i][j] := -1;
>        end if;
>      end for;
>    end for;
>    Ds;
       [
         [ 1, -1, -1, -1 ],
         [ 1, -1, -1 ],
         [ 1, 1, 1, 1, -1, -1 ],
         [ 1, -1, 1 ]
       ]
```

We construct the dual G^* of G as a multidigraph. For each edge e of G that lies in, say, faces f_1 and f_2 we add the appropriate edge between vertex 1 and vertex 2 with a direction consistent with the orientation of e in f_1 as given by $Ds[1]$. Note that taking the orientation of e to be consistent with the orientation of e in f_2 as given by $Ds[2]$ achieves the same result. This is illustrated by the code segment below; we do not exhibit G^* but rather the bijection Bij between the (undirected) edges of G with the (directed) edges in G^*.

```
>    nstar := #F;
>    Gstar := MultiDigraph< nstar | >;
>    Bij := [ EdgeSet(Gstar) | ];
>    Fs := [ SequenceToSet(f) : f in F ];
>    for u in [1..nstar−1] do
>      for v in [u+1..nstar] do
>        M := Fs[u] meet Fs[v];
>        for e in M do
>          p := Position(F[u], e);
>          q := Position(Edges(G), e);
>          if Ds[u][p] eq 1 then
>            Gstar, edge :=
>              AddEdge(Gstar, VertexSet(Gstar) ! u, VertexSet(Gstar) ! v);
>            ChangeUniverse(∼Bij, EdgeSet(Gstar));
>            Bij[q] := edge;
>          else
>            Gstar, edge :=
>              AddEdge(Gstar, VertexSet(Gstar) ! v, VertexSet(Gstar) ! u);
>            ChangeUniverse(∼Bij, EdgeSet(Gstar));
>            Bij[q] := edge;
>          end if;
>        end for;
>      end for;
>    end for;
>    for i := 1 to Size(G) do
>      "  ", EdgeSet(G).i, "  ", Bij[i];
>    end for;
             {1, 2}    < [3, 2], 6 >
             {1, 3}    < [2, 1], 1 >
             {1, 5}    < [1, 3], 2 >
             {2, 3}    < [3, 2], 5 >
             {3, 4}    < [3, 1], 3 >
             {4, 5}    < [4, 1], 4 >
             {4, 6}    < [3, 4], 7 >
             {5, 6}    < [4, 3], 8 >
```

Since G^* is a multidigraph it may have multiple edges between the same end-vertices. In our example, G^* has edge $< [3, 2], 6 >$ and edge $< [3, 2], 5$, both being directed edges from 3 to 2. There is thus a need to distinguish between both edges: This is achieved by identifying the edges with their respective index in the graph's adjacency list, here 6 for the first edge, and 5 for the second. This index uniquely identifies an edge in a Magma multi(di)graph and is guaranteed to do so during the graph's (and the edge's) lifetime; it is accessed via the Index function.

We now turn to the duality between flows and colouring in planar graphs.

4 k-Flows and k-colouring in planar graphs

The following theorem is due to Tutte (1954). A *k-colouring* of a graph $G = (V, E)$ is a mapping $c : V \to \{0, \dots, k-1\}$ so that no two adjacent vertices of G are assigned the same integer/colour.

Theorem 4.1. *Let G be a plane graph and G^* its dual. Then G has a k-colouring if and only if G^* has a nowhere-zero k-flow.*

Proof. Without loss of generality we can assume that G is connected. Further, we can assume that G has no loops so that G^* has no bridge. (A *bridge* is an edge whose removal leaves the resulting graph disconnected.) The special case in which G has bridges is left as an exercise to the reader. We will assume that G has no bridges so that G^* has no loops. Assume that G^* has a nowhere-zero k-flow, then G^* has a nowhere-zero Z_k-flow f (Theorem 2.2). As before, define $g : (E, D) \to Z_k$ by $g(e) = f(\phi(e))$, where as usual ϕ gives the bijection between the edges of G and the edges of G^*. We construct a k-colouring for G as follows.

Let T be a depth-first search tree of G; it is a spanning tree of G. Relabel the vertices of G according to the tree order, that is the order in which the vertices have been visited when constructing T. An edge of G is a *tree edge* if it is in T, it is a *back edge* otherwise. Let r be the root vertex of T and colour r with 0: $c(r) = 0$. For any other vertex u, we set $c(u) = g(\overrightarrow{P})$ where $\overrightarrow{P}$ is the path (with orientation) from r to u in T. We claim that $c(u)$ is a k-colouring of G.

Let $e = \{u, v\}$ be any edge of G and assume that $u < v$ in the tree order. Assume that e is a tree edge. Then

$$c(v) - c(u) = g([u, v]) = f(\phi([u, v])) \neq 0$$

since f is nowhere-zero. Assume that e is a back edge and let $\overrightarrow{P}$ be the $u \to v$ path in T. Then, by Lemma 3.1,

$$c(v) - c(u) = g(\overrightarrow{P}) = -g([v, u]) = -f(\phi([v, u])) \neq 0.$$

This shows that a nowhere-zero k-flow in G^* induces a k-colouring in G. We leave it to the reader to prove the converse.   □

5 Finding nowhere-zero k-flows

Unfortunately it is not a trivial matter to find nowhere-zero k-flows in graphs. The decision problem is in general NP-hard.

Following Kochol [8], one interesting avenue seems to be to attempt to find orientations of the edges of the graph so that for each edge e in G its flow $f(e)$ is strictly positive. Using this idea we developed an algorithm for finding k-flows in graphs, but it would require better heuristics than the ones currently in place to achieve a reasonable running time.

A *positive orientation* of a nowhere-zero Z-flow f in a graph $G = (V, D)$ is an orientation D of the edges of G such that $f(e) > 0$ for all $e \in (E, D)$. It is not difficult to see that D is a positive orientation of a flow f if and only if there is a path with orientation from u to v for every ordered pair $(u, v) \in V \times V$, $u \neq v$: Simply take any $X \subset V$ such that $u \in X$ and $v \notin X$ and recall that $\sum_{e \in \vec{E}(X, \overline{X})} f(e) = 0$. Since we require that $f(e) > e$ for each edge in G, it must be the case that there are edges out-going from X *as well as* edges in-going into X.

A *totally cyclic* orientation D of G is an orientation where every edge $e \in (E, D)$ lies in a cycle with orientation as defined in Section 3. From the previous paragraph we now can conclude that D is a positive orientation of a nowhere-zero Z-flow in G if and only if it is totally cyclic.

We will show how to construct a totally cyclic orientation in Magma. As before, assume that G is connected and has no bridges and no loops: It is easy to see that a graph with a bridge cannot have a nowhere-zero flow, and we again leave it to the reader to consider the case where G has loops. Let T be a (spanning) depth-first search tree of G, and again assume that the vertices of G are labelled according to the tree order, i.e. according to the order in which they have been discovered while performing the depth-first search. One obtains a totally cyclic orientation of G as follows: Let $e = \{u, v\}$ any edge of G with $u < v$ in the tree order. If e is a tree edge then orient e as $[u, v]$, if e is a back edge, then orient e as $[v, u]$.

Returning to our previous example, the problem is to find a nowhere-zero k-flow in G^*, the dual of the plane graph G. Following the above discussion, the first problem is to find a totally cyclic orientation of G^*. We proceed by finding a depth-first search spanning tree of the underlying simple graph S of G^*. (We could also have taken the underlying multigraph of G^*.)

```
>    S := UnderlyingGraph(Gstar);
>    T, _, _, DFS := DepthFirstSearchTree(VertexSet(S) ! 1);
>    DFS;

     [ 1, 4, 3, 2 ]
```

The fourth output argument *DFS* returned by DepthFirstSearchTree is the tree order: *DFS*[i] is the order of the vertex i of S. Let G^{**} be the graph with the same vertex set V^* and edge set E^* as G^* but whose orientation is totally cyclic. To obtain G^{**} it is enough to decide, for any $e \in E^*$, whether e is a tree edge or a back edge in T, and then orient e according to the tree order in *DFS*.

```
EG := EdgeSet(Gstar);
VT := VertexSet(T);
visited := [ false : i in [1..#VT] ];
GDstar := MultiDigraph< Order(Gstar) | >;
for e in EG do
   u := InitialVertex(e);
   v := TerminalVertex(e);
   i := DFS[Index(u)];
   j := DFS[Index(v)];
   BE := false;
   if VT! u adj VT! v then
   // is possibly a tree edge
      if i lt j and not visited[j] then
         visited[j] := true;
         AddEdge(~GDstar, VertexSet(GDstar) ! u, VertexSet(GDstar) ! v);
      elif j lt i and not visited[i] then
         visited[i] := true;
         AddEdge(~GDstar, VertexSet(GDstar) ! v, VertexSet(GDstar) ! u);
      else
         // must be a back edge (parallel edge)
         BE := true;
      end if;
   else
      // must be a back edge
      BE := true;
   end if;
   if BE then
      if i lt j then
         AddEdge(~GDstar, VertexSet(GDstar) ! v, VertexSet(GDstar) ! u);
      else
         AddEdge(~GDstar, VertexSet(GDstar) ! u, VertexSet(GDstar) ! v);
      end if;
   end if;
end for;
assert Size(GDstar) eq Size(Gstar);
```

One can easily verify that the orientation of G^{**} is totally cyclic. From here on, one must test if this orientation allows a feasible nowhere-zero k-flow in G^*. If this is not the case, then one finds another totally cyclic orientation, and tests it again. This is where good heuristics are required; as mentioned earlier, some serious work still needs to be done in this respect. For now we concentrate on demonstrating how one can (easily) test if an orientation allows for a feasible nowhere-zero k-flow. This is the object of the next section.

6 Testing for nowhere-zero k-Flows

We demonstrate how one can test for nowhere-zero k-flows using Magma. Assume $G = (V, E)$ is a graph with edge orientation D. To test if G has a nowhere-zero k-flow we will test if G has a feasible flow f with $0 < f(e) < k$ for all $e \in (E, D)$.

The following well known result states the conditions under which $G = (V, E, D)$ has a feasible nowhere-zero k-flow. For any $X \subseteq V$ define $E_X^+ = \{e \in (E, D) : e \in E_u^+ \mid u \in X\}$ and $E_X^- = \{e \in (E, D) : e \in E_u^- \mid u \in X\}$.

Theorem 6.1. *Let D be an orientation of a graph $G = (V, E)$ and b_e, c_e be integers so that $b_e \le c_e$ for every $e \in E$. Then there exists a Z-flow f in G such that $b_e \le f(e) \le c_e$ for every $x \in (E, D)$ if and only if*

$$\sum_{e \in E_X^+} b_e \le \sum_{e \in E_X^-} c_e$$

for every $X \subseteq V$.

That is, $G = (V, E, D)$ has a feasible nowhere-zero k-flow if and only if $\sum_{e \in E_X^+} \le \sum_{e \in E_X^-}(k-1)$ for every $X \subseteq V$. But instead of verifying that condition directly, we turn G into a network and compute its maximum flow.

Let G^{**} be a totally cyclic multidigraph with orientation as given in Figure 4. Note that G^{**} has *not* the same orientation as the one computed in Section 3. We verify that G^{**} has a nowhere-zero 3-flow.

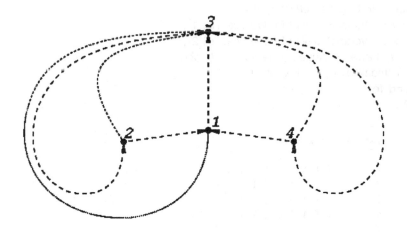

Fig. 4. An orientation of G^{**} with a nowhere-zero 3-flow

We start by turning G^{**} into a network N: A *network* is a multidigraph whose edges are assigned a capacity. Choose two arbitrary vertices of N to be the source s and the sink t. A flow in a network is similar to a H-circulation or H-flow in a graph, except for the fact that flow conservation does not hold for the source s and the sink t.

Define a *maximum flow* f in N to be the maximum amount of flow that can be pushed from the source s to the sink t, where for every edge e in N with capacity $c(e)$ it holds that $0 \leq f(e) \leq c(e)$. In Magma we have implemented two network-flow algorithms: The well-known Dinic algorithm and an algorithm based on the push-relabel method. For most practical purposes the latter is the most efficient and is the one called by default.

In the code segment below, G^{**} is converted into a network N: Any (directed) edge of G^* is a (directed) edge in N. To each such edge in N we assign capacity $k - 2 = 1$. Moreover, we establish the bijection ϕ between the edges in G^{**} and those in N.

```
>    k := 3;
>    GDstar;

          Multidigraph
          Vertex  Neighbours

            1        3 3 ;
            2        1 3 ;
            3        4 2 ;
            4        3 1 ;

>    N := Network< Order(GDstar) | >;
>    φ := [ ];
>    for e in EdgeSet(GDstar) do
>       u := VertexSet(N) ! EndVertices(e)[1];
>       v := VertexSet(N) ! EndVertices(e)[2];
>       N, newe := AddEdge(N, u, v, k−2);
>       φ[Index(e)] := Index(newe);
>    end for;
>    N;

          Network
          Vertex  Neighbours

            1        3 [ 1 ] 3 [ 1 ] ;
            2        3 [ 1 ] 1 [ 1 ] ;
            3        2 [ 1 ] 4 [ 1 ] ;
            4        1 [ 1 ] 3 [ 1 ] ;
```

When adding edges to a multi(di)graph (here a network) it may be convenient to have access to the edge just added since there may exist several parallel edges between the same end-vertices: This is achieved via the second return argument of the function AddEdge which returns the newly added edge. The function Index applied to this edge returns its index in the adjacency list of the network N and thus uniquely identifies the edge.

We see that every edge of N has capacity 1 as indicated by the square brackets. We add two additional vertices to N, the source s and the sink t. For each vertex u in N such that $u \neq s$ and $u \neq t$, we add the following edges to N:

- a (directed) edge from s to u with capacity $c = $ in-degree(u),
- a (directed) edge from u to t with capacity $c = $ out-degree(u).

```
>   N +:= 2;
>   N;
```

```
          Network
          Vertex   Neighbours

          1          3 [ 1 ] 3 [ 1 ] ;
          2          3 [ 1 ] 1 [ 1 ] ;
          3          2 [ 1 ] 4 [ 1 ] ;
          4          1 [ 1 ] 3 [ 1 ] ;
          5          ;
          6          ;
```

```
>   s := VertexSet(N) ! (Order(N) − 1);
>   t := VertexSet(N) ! Order(N);
>   s, t;
```

```
          5 6
```

```
>   for u in VertexSet(N) do
>     if not u eq s and not u eq t then
>       c := InDegree(u);
>       AddEdge(∼N, s, u, c);
>       c := OutDegree(u);
>       AddEdge(∼N, u, t, c);
>     end if;
>   end for;
>   N;
```

```
          Network
          Vertex   Neighbours

          1          6 [ 2 ] 3 [ 1 ] 3 [ 1 ] ;
          2          6 [ 2 ] 3 [ 1 ] 1 [ 1 ] ;
          3          6 [ 2 ] 2 [ 1 ] 4 [ 1 ] ;
          4          6 [ 2 ] 1 [ 1 ] 3 [ 1 ] ;
          5          4 [ 1 ] 3 [ 4 ] 2 [ 1 ] 1 [ 2 ] ;
          6          ;
```

At this point one applies a network-flow algorithm that computes the maximum flow in N. Assume that f is the maximum flow in N. Then it should be easy to see that G^{**} has a nowhere-zero k-flow if and only if every edge in N

out-going from the source s or in-going into the sink t is saturated; that is, if for each such edge e, $f(e) = c(e)$. This is indeed the case in our example.

```
>    s := VertexSet(N) ! s;
>    t := VertexSet(N) ! t;
>    F := MaximumFlow(s, t);
>    for e in Edges(N) do
>      if EndVertices(e)[1] eq s
>          or EndVertices(e)[2] eq t then
>        e, Capacity(e), Flow(e);
>      end if;
>    end for;
```

```
< [1, 6], 10 > 2 2
< [2, 6], 12 > 2 2
< [3, 6], 14 > 2 2
< [4, 6], 16 > 2 2
< [5, 1], 9 > 2 2
< [5, 2], 11 > 1 1
< [5, 3], 13 > 4 4
< [5, 4], 15 > 1 1
```

To get the nowhere-zero k-flow F in G^{**} it is then enough to set $F(e) = f(e)+1$ for every edge e in G^{**} (note how the bijection ϕ is used to that effect). Then $0 < F(e) < k$ as expected.

```
>    E := EdgeSet(N);
>    for e in EdgeSet(GDstar) do
>      e, Flow(E.ϕ[Index(e)]) + 1;
>    end for;
```

```
< [1, 3], 2 > 1
< [1, 3], 1 > 1
< [2, 1], 4 > 1
< [2, 3], 3 > 1
< [3, 4], 6 > 2
< [3, 2], 5 > 2
< [4, 3], 8 > 1
< [4, 1], 7 > 1
```

The graph G^{**} given in this example is depicted in Figure 5 below, together with the plane graph G we constructed in Section 3. The dotted arrows for G^{**} are the edges whose direction is reversed with respect to the original dual G^* of G as depicted in Figure 3. One computes the nowhere-zero k-flow in G^* from the nowhere-zero k-flow in G^{**} by keeping track of these edge inversions. At this point, colouring the vertices of G is an easy task, which we do not demonstrate here. Note that in our example G is 3-colourable.

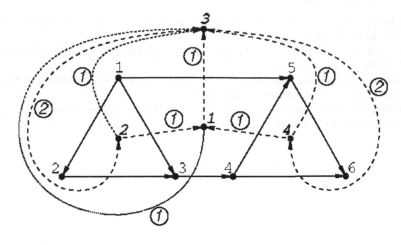

Fig. 5. A nowhere-zero 3-flow for G^{**}

7 Conclusion

We have seen that, in order to be effective, the approach taken in this chapter to colour planar graphs requires good heuristics in finding a totally cyclic orientation for which a nowhere-zero 4-flow is feasible. While it is easy to find all totally cyclic orientations of a graph it is certainly not practical since this results in an $O(2^m)$ implementation where m is the dimension of the cycle space of the graph.

One way to find suitable totally cyclic orientations could be to ensure that the difference between the minimum and maximum of the in-degree and out-degree of the vertices of the graph is kept to a minimum. Another approach would be to examine the cuts of the graph. A cut is a set of edges whose removal disconnects the graph. We tried the first approach with mixed results. The second approach seems promising. We leave it as an open problem.

References

1. K. Appel, W. Haken, *Every planar map is four colorable. I. Discharging*, Illinois J. Math. **21** (1977), 429–490.
2. K. Appel, W. Haken, (with the collaboration of J. Koch), *Every planar map is four colorable*, Providence, RI: American Mathematical Society, 1989.
3. K. Appel, W. Haken, J. Koch, *Every planar map is four colorable. II. Reducibility*, Illinois J. Math. **21** (1977), 491–567.
4. Wieb Bosma, John Cannon, Catherine Playoust, *The Magma algebra system I: The user language*, J. Symbolic Comput. **24** (1997), 235–265.
   See also the Magma home page at http://magma.maths.usyd.edu.au/magma/.
5. J. Boyer, W. Myrvold, *Simplified O(n) Planarity Algorithms*, J. Algorithms, submitted, 2001.

6. R. Diestel, *Graph Theory*, Second Edition. Graduate Texts in Mathematics **173**, Berlin: Springer, 2000.

7. C. Godsil, G. Royle, *Algebraic Graph Theory*. Graduate Texts in Mathematics **207**, Berlin: Springer, 2001.

8. M. Kochol, *Polynomials Associated with Nowhere-Zero Flows*, Journal of Combinatorial Theory (B) **84** (2002), 260–269.

9. N. Roberston, D.P. Sanders, P.D. Seymour, R. Thomas, *The Four-Colour Theorem*, Journal of Combinatorial Theory (B) **70** (1997), 2–44.

Appendix: The Magma language

Geoff Bailey

School of Mathematics and Statistics
The University of Sydney
Sydney, Australia

Introduction

The following is a brief introduction to the Magma language. It is not a comprehensive description of the language, nor is it a programming tutorial (although, inevitably, it has much in common with one). Instead, it is intended to describe the basic language features that account for the vast majority of code written in the Magma language. For a more in-depth treatment, see the language section of the Magma Handbook [1].

Some of the examples below will use language features that are not explained until a later point; it is hoped in each case that the meaning will be clear enough for this not to cause problems. A certain familiarity with basic programming concepts such as expressions and statements will be assumed.

1 Basics

1.1 Statements and printing

Throughout the book, code fragments will use a format similar to the example below. Lines beginning with > (the Magma prompt) indicate input to Magma (the prompt is *not* part of the input), while lines without the prompt indicate output produced by the preceding instructions.

```
>    1 + 1
>    + 1;
     3
```

This trivial example already illustrates two important features of the language. Firstly, statements in Magma must end with a semicolon. Thus, no output occurred after the first line of input, as the statement was still incomplete at that point. Secondly, when a Magma statement is an expression, then the result of evaluating that expression will be printed. If multiple expressions occur (separated by commas) in a single statement, then they will each be printed, separated by spaces.

> "You are number", 6;

      You are number 6

Note the use of a *string* (text surrounded by double quotes) in this example. A string prints as the text it contains.

There are also two explicit print statements in Magma.

    print *value,..., value;*
    printf *format-string, value,..., value;*

The **print** statement prints its values, exactly as would occur if the keyword **print** were omitted. It is thus rarely seen, although its use can improve code clarity. (It is also useful when debugging, as it can be easily searched for and removed once the problem has been solved.)

The **printf** statement behaves similarly to the corresponding C function. The string argument is printed, with two types of substitutions made. Each instance of '%o' in the string is replaced by the output from printing the corresponding value argument to **printf** (the first instance uses the first value argument, and so on). Also, two-character strings beginning with backslash are changed in the usual manner. For example, '\t' is replaced with a tab character and '\n' is replaced with a newline character.

> **printf** "%o green %o...\n", 99, "bottles";

      99 green bottles...

> **print** "...hanging on the wall";

      ...hanging on the wall

1.2 Comments

A *comment* is a way of specifying that some specific text should be ignored by Magma. There are two types of comments in Magma, matching those of the C and C++ languages.

    // comment text
    / comment text */*

The short form // applies to one line only; all text following the // on the same line is ignored. The long form /* ... */ can extend over multiple lines; all text between the /* and the */ is ignored.

> 1 + 1 // ; this semicolon is ignored
> /* As will be the one on the next line
> +1;
> */
> + 1; // Now the statement is finally finished

      3

1.3 Intrinsic functions

A *function* is an object that uses some values (the *arguments*) to perform certain operations and produce some other values (the *return values*). A *procedure* is similar except that it has no return values. In Magma, a function or procedure is called by typing its name, followed by the values of its arguments (if any) enclosed within parentheses.

> name(*arguments*)

If there are multiple arguments then commas are used to separate them. Magma contains a great many inbuilt, or intrinsic, functions and procedures, which are called *intrinsics* for short.

The names of intrinsics are usually descriptive of their purpose. For instance, the intrinsic **GreatestCommonDivisor** computes the greatest common divisor of its two arguments.

```
>    GreatestCommonDivisor(91, 224);
        7
```

Many intrinsics have synonyms that reflect different spellings, usages, or convenient shorthand forms. For example, the names **GCD** and **Gcd** are synonyms of **GreatestCommonDivisor**.

Functions may have more than one return value. One example is **Quotrem**, which returns the quotient and remainder (respectively) arising from Euclidean division of its arguments.

```
>    Quotrem(16, 3);
        5 1
```

If a function returns multiple values but not all are used then the unused return values are silently discarded. For example, if a function call is part of an expression then only the first return value is used.

```
>    3*Quotrem(16, 3);
        15
```

Functions and procedures are described further in a later section which also describes how it is possible to create new functions and procedures, and how procedures may modify their arguments.

1.4 Arithmetic

Magma includes the standard arithmetic operators $+$, $-$, and $*$, as well as parentheses () for grouping.

```
>    (2 + 3)*(4 − 1);
        15
```

There are two versions of division, however, depending on what sort of object one is working with. The operators **div** and **mod** apply when working over

Euclidean rings (such as $\mathbb{Z}$), and return the quotient and remainder (respectively) of the division.

```
>    16 div 3, 16 mod 3;
        5 1
```

The operator / applies when working over fields, where the (exact) quotient is returned. It can also be used when working over rings; in this case the result will be returned as an element of the appropriate field of fractions.

```
>    16.0 / 3.0;
        5.3333333333333333333333333333
>    16 / 3;
        16/3
```

Note that this latter example took two elements from the Euclidean ring $\mathbb{Z}$ but returned a result in its field of fractions $\mathbb{Q}$.

```
>    Parent(16);
        Integer Ring
>    Parent(16 / 3);
        Rational Field
```

(The intrinsic **Parent** returns the structure that contains the specified object.)

The other important arithmetic operator is ^, which is used for exponentiation. However, in this book exponentiation has been indicated in the conventional way as a superscript and the ^ is not shown. So, whereas one would type 4^2 into Magma, in the examples it would appear as follows.

```
>    4²;
        16
```

1.5 Variables, assignment, mutation, and **where**

A *variable* is a name that may have a value associated with it. Variables can be given a value by using an assignment statement.

```
    name := value;
    name,..., name := values;
```

(As is common with interpreted languages, variables in Magma do not need to be declared; they are defined by use and the type of a variable is that of the value it currently contains.)

```
>    seconds := 24*60*60; // seconds per day
>    seconds;
        86400
```

Magma also has *mutation* operations, which are a convenient shorthand for the most common kind of variable modification. These have the form

> variable *op*:= value;

and are equivalent to the longer form *variable* := *variable op* value. Here most binary operators may be used in place of *op*.

```
>    seconds *:= 365 + 1/4; // Average seconds per year
>    seconds;
         31557600
>    seconds /:= 12; // Average seconds per month
>    seconds;
         2629800
```

The special symbol _ may appear on the left hand side of an assignment statement instead of a variable, and indicates that the value that would have been assigned should be discarded instead. It is most commonly used to get rid of unwanted return values from functions.

```
>    p := 17;
>    x := 12;
>    _,_,xinv := XGCD(p, x); // Extended GCD
>    x*xinv mod p;
         1
```

There is a special kind of 'temporary' assignment available in Magma, using the **where** clause. There are two minor variations of this clause.

> expression **where** name := value
> expression **where** name **is** value

In each case the expression is evaluated as though *name* were a variable with the given value. The previous value of the variable (if any) is unchanged by this process; the changed value applies only while evaluating the expression.

```
>    x*xinv mod p where p is 13;
         7
>    p;
         17
```

The use of **where** is particularly desirable when a lengthy or computationally expensive sub-expression occurs multiple times within the one expression.

```
>    Modexp(3, p-1, p) where p is 2^n + 1 where n is 2^5;
         3029026160
```

(The intrinsic Modexp(x, n, m) efficiently computes x^n **mod** m.) Note the use of more than one **where** in this example.

1.6 Generators and generator assignment

Many structures in Magma are generated by a few special elements. These *generators* can be retrieved with the dot operator.

   structure . integer

The integer must be a valid generator number for the given structure. These numbers range from 1 to the number of generators.[1]

```
>   K := QuadraticField(5); // sets K to Q[√5], with generator √5
>   α := K.1;
>   α;

        K.1

>   α²−5;

        0
```

Since it is extremely common to want to use the generators after defining the structure, Magma provides a convenient way to assign these generators to variables when the structure is created.

   variable<names> := structure;

This construct sets the variables specified by *names* to the corresponding generators of the structure. It also has another important effect: The names provided will be used when printing elements of the structure.

```
>   P<x> := PolynomialRing(Integers());
>   (x+1)*(x−1);

        x^2 - 1

>   u := x;
>   (u³−1) div (u−1);

        x^2 + x + 1
```

1.7 Coercion

Given two mathematical structures, it is often the case that there is a natural choice of map from one to the other. In Magma, the act of applying this map is called *coercion*, and is achieved by using the coercion operator ' ! '.

   structure ! element

The appropriate map is applied to the element, producing a value lying in the specified structure. In Magma terminology, the element has been *coerced* into the structure.

---

[1] In certain cases other values are allowed; for instance, if the structure is a group then the negative numbers yield the inverses of the generators and zero gives the identity element.

```
>   K := GF(7); // GF(q) produces the finite field with q elements
>   x := K ! 23;
>   x;
        2
>   Parent(x);
        Finite field of size 7
```

Note that after the coercion we can perform operations on *x* that might return different values or just be nonsensical if attempted prior to the coercion.

```
>   Order(x);
        3
```

Coercion is of fundamental importance in Magma, since it allows the easy transfer of objects from one mathematical structure to a related one where questions of interest may more readily be answered.

1.8 Boolean expressions

Magma has the inbuilt Boolean constants true and false. The logical operators **and**, **or**, **not** and **xor** operate on these in the expected manner. Boolean values may arise in other ways; for instance, the six relational operators **eq**, **ne**, **lt**, **le**, **gt**, and **ge** take two comparable objects and return the truth-value of the appropriate comparison. These actions are summarised in the following table.

Operator	Usage	Meaning
not	*not a*	true if *a* is false, else false
and	*a and b*	true if both *a* and *b* are true, else false
or	*a or b*	true if either *a* or *b* is true, else false
xor	*a xor b*	true if exactly one of *a* and *b* is true, else false
eq	*x eq y*	true if *x* is equal to *y*, else false
ne	*x ne y*	true if *x* is not equal to *y*, else false
lt	*x lt y*	true if *x* is less than *y*, else false
le	*x le y*	true if *x* is less than or equal to *y*, else false
gt	*x gt y*	true if *x* is greater than *y*, else false
ge	*x ge y*	true if *x* is greater than or equal to *y*, else false

Functions may also return Boolean values, of course.

```
>   G<a,b> := Sym(4); // Symmetric group on four elements
>   Order(a) eq 4 or Order(b) eq 4;
        true
>   IsAlternating(G);
        false
>   IsSymmetric(G) and IsSoluble(G);
        true
```

1.9 Conditionals: **if** and **select**

To perform different commands based on some condition, an **if** statement is used.

>   **if** *condition* **then** *statements* **else** *statements* **end if**;

If the condition is true then the statements between **then** and **else** will be executed; otherwise, the statements between **else** and **end if** will be executed.

```
>     p := 341;
>     if 2^(p−1) mod p eq 1 then
>             p, "is a pseudo-prime to base 2";
>     else
>             p, "is definitely composite";
>     end if;
              341 is a pseudo-prime to base 2
```

In this example the expression evaluated to true, so the statements between **then** and **else** were executed.

The **else** section is optional; without such a section no statements will be executed if the condition is false. Also, an adjacent **else** and **if** can be combined together into **elif**; the reason to do this is that then only one **end if** will be required. For long chains of **if**...**else if**... sequences this is highly desirable.

```
>     p := 191;
>     if not IsPrime(p) then
>             p, "is not prime";
>     elif IsPrime(2*p + 1) then
>             p, "is a Germain prime";
>     else
>             p, "is prime (but not a Germain prime)";
>     end if;
              191 is a Germain prime
```

Sometimes it is desirable to have an expression whose value depends on some condition. This is particularly useful when creating recursive sequences (described later), or in **where** clauses, or even just as a shorter form than the corresponding **if** statement would be. The **select** expression is used for this purpose.

>   *condition* **select** *expression*₁ **else** *expression*₂

If the condition is true then the entire expression has the value of *expression*$_1$, otherwise it has the value of *expression*$_2$.

```
>     m := 17;
>     x := 56 mod m;
>     modnegx := x eq 0 select 0 else m − x;
>     modnegx;
```

12

2 Sets and sequences

Sets and sequences are extremely important in Magma. Sequences in particular are the second-most commonly used type of object (integers are the most common), and it is rare to find a significant piece of Magma code that does not use them. An understanding of their use often leads to compact yet expressive and understandable programs.

Sets and sequences are collections of objects belonging to the same structure (this structure is said to be the *universe* of the set or sequence). Sets are unordered collections, and so may not contain duplicated values.[2] In contrast, sequences are ordered collections and duplications are allowed.

Sets and sequences use brackets in their creation and printing; the brackets involved are { } for sets and [ ] for sequences.

2.1 Creation of sets and sequences

The simplest way to create a set or sequence is to explicitly list its elements.

```
>    [ false, true, true, false, true, false, true ];

        [ false, true, true, false, true, false, true ]

>    { 2, 7, 1, 8, 2, 8 };

        { 1, 2, 7, 8 }
```

In many cases, explicit enumeration of elements in this manner is not necessary because one of the special constructors described below can be used. Magma has two special constructors for sets and sequences; the first has the form

$$\{ a..b \text{ by } k \}$$
$$[a..b \text{ by } k]$$

where a, b, and k are integers. This creates the set or sequence containing the values from the arithmetic progression a, $a+k$, $a+2k$, ... and bounded by b. (It is not required that b be part of the progression.) The **by** k part may be omitted, in which case the increment is taken to be 1.

The second special constructor has the following form.

$$\{ \text{ expression in } x : x \text{ in } D \mid \text{ condition on } x \}$$
$$[\text{ expression in } x : x \text{ in } D \mid \text{ condition on } x]$$

Using this form, the result consists of the values of *expression in* x evaluated for each x in D such that the *condition on* x evaluates to true.

```
>    [ p : p in [1..100] | IsPrime(p) ];

        [ 2, 3, 5, 7, 11, 13, 17, 19, 23, 29, 31, 37, 41, 43, 47,
          53, 59, 61, 67, 71, 73, 79, 83, 89, 97 ]
```

---

[2]There is a related *multiset* type in Magma that allows duplicated values; it will be briefly described later.

The condition part of the constructor may be omitted; this is equivalent to using a condition of true.

```
>    K := GF(7);
>    { x² : x in K };
            { 0, 1, 2, 4 }
```

It is possible to iterate with respect to more than one variable in the same constructor.

```
>    V := VectorSpace(K, 2);
>    m := Matrix(K, [ [1, 2], [4, 1] ]);
>    m;
            [1 2]
            [4 1]
>    { v : v in V, a in K | not IsZero(a*v) and v*m eq a*v };
            {
                (4 1),
                (3 6),
                (2 4),
                (6 5),
                (1 2),
                (5 3)
            }
```

There is actually one more part to these special constructors. The universe of the set or sequence may be specified, and elements arising during the construction will then be coerced into this universe.

```
{ universe | rest of constructor }
[ universe | rest of constructor ]
```

This is preferable to using explicit coercion for clarity reasons, and also because it ensures that the universe is correctly set even when the resulting set or sequence is empty.

```
>    T := { RationalField() | x : x in K | x² eq 3 };
>    T;
            {}
>    Universe(T);
            Rational Field
```

In the context of a sequence constructor, the function Self returns the specified element of the sequence that is being created. In conjunction with the **select** expression to handle base cases, Self allows sequences to be created recursively.

```
>    [ n le 2 select 1 else Self(n−1) + Self(n−2) : n in [1..15] ];
        [ 1, 1, 2, 3, 5, 8, 13, 21, 34, 55, 89, 144, 233, 377, 610 ]
```

The set and sequence constructors can clearly be used to convert between the two types by explicit iteration. Another way to achieve the same goal is to use one of the intrinsics SetToSequence or SequenceToSet.

```
>    S := [ Order(x) : x in K | x ne 0 ];
>    S;

        [ 1, 3, 6, 3, 6, 2 ]

>    T := { x : x in S };
>    T;

        { 1, 2, 3, 6 }

>    SequenceToSet(S);

        { 1, 2, 3, 6 }

>    SetToSequence(T);

        [ 1, 2, 3, 6 ]
```

Many mathematical objects can be naturally viewed as including a sequence of elements from some underlying structure. For example, polynomials have their coefficients and geometric points have their coordinates. In such cases the intrinsic ElementToSequence, usually abbreviated to Eltseq, returns this sequence.

```
>    Zx<x> := PolynomialRing(Integers());
>    f := (x+1)²*(x−2);
>    f;

        x^3 - 3*x - 2

>    Eltseq(f);

        [ -2, -3, 0, 1 ]
```

(Note that for polynomials the ensuing sequence starts with the coefficient of the constant term and finishes with the coefficient of the leading term.)

Elements can also be created by coercing such sequences into the parent structure of the original object.

```
>    Zx ! [ 41, 1, 1 ];

        x^2 + x + 41

>    V ! [3, 5];

        (3 5)
```

There are two other set-like types in Magma: *indexed sets* and *multisets*. These can be created like normal sets, except that they use different brackets—indexed sets use {@ @} and multisets use {* *}. Indexed sets are ordered but may not contain duplicates, while multisets are unordered and may contain duplicates (each element has an associated multiplicity). Most operations that apply to normal sets also apply to indexed sets and multisets. Additionally, indexed sets support many operations that apply to sequences.

2.2 Indexing and indexed assignment

Sequences support an indexing operation (using [ ]) to extract particular elements. Valid indices are integers from 1 to the number of elements in the sequence (also called its *length*).

```
>    S := [ "r", "e", "l", "a", "t", "i", "n", "g" ];
>    S[4];

        a
```

It is also possible to index one sequence by another sequence. In this case the result will be a sequence whose elements are the result of indexing the first sequence by the elements of the second.

```
>    indices := [ 6, 7, 5, 2, 8, 1, 4, 3 ];
>    S[indices];

        [ i, n, t, e, g, r, a, l ];
>    S[ [5, 1, 6, 4, 7, 8, 3, 2] ];
        [ t, r, i, a, n, g, l, e ];
```

Indexing may also be used to set specific elements of a sequence.

```
>    S[5] := "x";
>    S;

        [ r, e, l, a, x, i, n, g ]
```

If a sequence contains sequences (or other indexable objects) then multi-level indexing can take place.

```
>    M := [ [1, 2], [3, 4] ];
>    M[2];

        [ 3, 4 ]
>    M[2][1];

        3
```

In such situations there is a valid alternate syntax, in which the indices are flattened into a single list.

```
>    M[2, 1];

        3
```

Note that this is *not* the same as indexing by a sequence.

```
>    M[[2, 1]];

        [
            [ 3, 4 ],
            [ 1, 2 ]
        ]
```

When setting elements of a sequence, the index used may be larger than the current size of the sequence. If so, the sequence is extended to include the new element at the appropriate index.

```
>    M[3] := [ 5, 6 ];
>    M;

     [
          [ 1, 2 ],
          [ 3, 4 ],
          [ 5, 6 ]
     ]
```

2.3 Operators on sets and sequences

There are a few important operators that can be applied to sets and/or sequences. One with more general applicability is #, which in Magma returns the cardinality of an object. In the case of sets or sequences this is the number of elements. These operators are summarised in the following table.

Operator	Usage	Meaning
#	#S	the number of elements in S
in	x *in* S	true if the element x is in S, else false
notin	x *notin* S	true if the element x is not in S, else false
cat	S_1 *cat* S_2	the concatenation of sequences S_1 and S_2
join	S_1 *join* S_2	the union of sets S_1 and S_2
meet	S_1 *meet* S_2	the intersection of sets S_1 and S_2
diff	S_1 *diff* S_2	the set of elements in set S_1 but not in set S_2
sdiff	S_1 *sdiff* S_2	the symmetric difference of sets S_1 and S_2
subset	S_1 *subset* S_2	true if S_1 is a subset of S_2, else false

There is an extremely important class of operators called *reduction operators*. These have the form &*op* and act on a set or sequence to reduce it to a single element by repeated application of the binary operator *op*. Not all binary operators have a corresponding reduction operator; the valid reduction operators are as follows.

&+   &*   &*and*   &*or*   &*meet*   &*join*   &*cat*

If S contains the elements $\alpha_1, \ldots, \alpha_n$ in that order[3] then

$$\&op\ S = (\ldots((\alpha_1\ op\ \alpha_2)\ op\ \alpha_3)\ldots)\ op\ \alpha_n.$$

Note that most of these operators are commutative (at least under common circumstances). The exception is &*cat*, which concatenates the elements of the set/sequence.

---

[3] For sets, the order is the internal iteration order, which corresponds to the order used when printing.

```
>    &+[ k² : k in [1..24] ];
         4900
>    F₁₁ := GF(11);
>    _<t> := PolynomialRing(F₁₁);
>    &*{ t - a : a in F₁₁ | not IsSquare(a) };
         t^5 + 1
>    P<a,b,c> := PolynomialRing(Integers(), 3);
>    sympols := [ &+[ &*S : S in Subsets({a,b,c}, k) ] : k in [0..3] ];
>    sympols;
         [
             1,
             a + b + c,
             a*b + a*c + b*c,
             a*b*c
         ]
>    &and [ IsSymmetric(pol) : pol in sympols ];
         true
>    &cat [ "Hello,", " ", "world!" ];
         Hello, world!
>    &meet [ { n : n in [2..10⁴] | Modexp(x,n,n) eq x } : x in {2,3,5} ]
>              diff { n : n in [2..10⁴] | IsPrime(n) };
         { 561, 1105, 1729, 2465, 2821, 6601, 8911 }
```

When a set or sequence is empty then it may not be obvious what the result of applying a reduction operator to it should be. The guiding principle is that the result should be consistent, in the sense that (&*op S*) *op x* should produce *x* when *S* is empty. Thus &+ produces 0 and &* produces 1, &***and*** produces true and &***or*** produces false, and &***join*** and &***cat*** produce empty objects of the appropriate types. It is not permissible to call &***meet*** on an empty structure.

In the case of &+ and &*, if they are called on an empty set or sequence then the universe must have already been set. This is so that Magma can determine in which structure the result should lie.

2.4 Important set and sequence intrinsics

The commonly used intrinsics for modifying sets and sequences are summarised in the table below. The intrinsics Include and Exclude can be used on both types, while Append, Prune, and Remove apply to sequences only. Each of these intrinsics has both a procedural and a functional form; the usage with tilde (∼) is the procedural version that modifies the set or sequence directly, while the functional version returns a new object and leaves the original unchanged. For more information on procedures, see the later section on creating functions and procedures.

Intrinsic	Usage	Meaning
Include	Include($\sim$S, x) Include(S, x)	Puts x into S
Exclude	Exclude($\sim$S, x) Exclude(S, x)	Removes the first occurrence of x from S
Append	Append($\sim$S, x) Append(S, x)	Appends x to the sequence S
Prune	Prune($\sim$S) Prune(S)	Removes the last element from the sequence S
Remove	Remove($\sim$S, i) Remove(S, i)	Removes the ith element from the sequence S

Another important intrinsic is Index(S, x), which returns the index of the first occurrence of x in the sequence S (or 0 if x is not in S). Thus, assuming that the sequence S contains x, the effects of Exclude(S, x) and Remove(S, Index(S, x)) are the same.

```
>    S := [ 1, 4, 9, 16 ];
>    Index(S, 16);

        4

>    Remove(~S, 4);
>    S;

        [ 1, 4, 9 ]

>    Append(S, 2); // Functional version—does not change S
        [ 1, 4, 9, 2 ]

>    S;

        [ 1, 4, 9 ]

>    T := { "e", "a", "t" };
>    Include(~T, "r");
>    T;

        { t, e, a, r }

>    Exclude(T, "e");

        { t, a, r }
```

Another handy intrinsic is Explode, which returns the elements of a sequence. This may not seem like it accomplishes much, but it is a great convenience when assigning these elements to variables. Without it, each assignment would have to be written explicitly, index and all.

```
>    first, second, third := Explode([ 3, 1, 4 ]);
>    first; second; third;

        3
        1
        4
```

2.5 Quantifiers: **exists** and **forall**

There are a few ways to test whether some or all elements of a set satisfy a certain property. For example, one could construct a sequence of Boolean values indicating whether each element satisfies the property, and then apply one of the reduction operators **&or** or **&and** (as appropriate) to this sequence. A better idea would be to use a loop (described later) over the set.

However, both of these approaches require the set itself to have been fully constructed, which may be needlessly time-consuming. A better method is available in Magma, using one of the quantifiers **exists** or **forall**. These allow the test and the set construction to be aborted early if an element is found that satisfies (**exists**) or does not satisfy (**forall**) the given property. In such cases the exceptional element may be assigned to a variable.[4]

The syntax for quantifiers is quite similar to that of set constructors.

exists(*variable*) { *expression in x* : *x* **in** *D* | *condition on x* }
forall(*variable*) { *expression in x* : *x* **in** *D* | *condition on x* }

In the case of **exists**, each element of *D* is checked; if one is found that satisfies the condition then true is returned (and no more elements of *D* are checked). If this occurs then the variable is assigned the value of the expression for that particular element. If no elements satisfy the condition then false is returned.

```
>   exists(u) { x³ − 13 : x in [1..100] | IsSquare(x³ − 13) };
        true
>   u;
        4900
```

In the case of **forall**, the early stoppage occurs if the condition is *not* satisfied for some *x*. If this occurs then false is returned and the variable is assigned the value of the expression for that particular element. If all elements satisfy the condition then true is returned.

```
>   forall(u) { x : x in [1..50] | Factorial(x−1) mod x in {0, x−1} };
        false
>   u;
        4
```

More than one variable may be given in the quantifier's arguments. In this case, the expression must evaluate to a tuple (see the next section for an explanation of tuples) with the matching number of elements. Each element of the tuple is then assigned to the appropriate variable.

```
>   p := 89;
>   exists(x,y) { <x, y> : x,y in [1..10] | x² + y² eq p };
        true
```

---

[4]The effects of a quantifier can also be achieved by using a suitable loop; this is more work, however.

```
>    x; y; x² + y²;
         5
         8
         89
```

3 Tuples

A *tuple* is an element of some Cartesian product. As such, it has a number of components, each potentially lying in a different structure. This makes tuples handy for keeping related data of different types together. Tuples are created by enclosing the components in angled brackets < >.

< *first component*, ..., *last component* >

Tuples have a certain degree of similarity to sequences. They are indexable in much the same way, for instance (except that indices larger than the number of components may not be used), and they support the intrinsics **Explode** and **Append**, although uses of the latter intrinsic are uncommon.

```
>    tup := < 7, true >;
>    tup[1] +:= 4;
>    tup;
             <11, true>
>    a, b := Explode(tup);
>    a; b;
             11
             true
```

One place where tuples commonly arise is when factoring. In Magma, the intrinsic **Factorisation** (or its synonym **Factorization**) takes a ring element and returns a sequence of tuples containing primes and exponents that give the factorisation of the element (up to units).

```
>    P<t> := PolynomialRing(GF(7));
>    g := t¹⁵ + 6*t¹¹ + 2*t⁸ + t⁷ + 5*t⁴ + 2;
>    F := Factorisation(g);
>    F;
             [
                 <t + 2, 7>,
                 <t^2 + t + 4, 1>,
                 <t^2 + 2*t + 2, 1>,
                 <t^2 + 5*t + 2, 1>,
                 <t^2 + 6*t + 4, 1>
             ]
>    g eq &*[ t[1]^t[2] : t in F ];
             true
```

4 Creating functions and procedures

New functions can be created using the **function** statement.

function *name*(*arguments*) *statements* **end function**;

Here *name* is the name of the function and *arguments* is a comma-separated list of variable names. When the function is called, the statements between the arguments and the **end function** are executed (these statements are called the *function body*). While executing the function body, the argument names are treated as variables whose initial values are those given when the function was called.

At some point during the execution of the function a **return** statement must be executed.

return *values*;

Here *values* is a comma-separated list of values. When such a statement is executed the function immediately finishes and its return values are those specified in the **return** statement.

```
>    function minmax(x, y)
>        if x le y then
>                return x, y;
>        else
>                return y, x;
>        end if;
>    end function;
>    minmax(8, 6);
        6 8
```

When defining a function, the name may be omitted; this turns the statement into an expression whose value is the unnamed function. In Magma, functions are first-class objects, so this function may then be assigned to a variable. The result is indistinguishable from using the statement form, and both methods are commonly employed.

```
>    area := function(a, b, c)
>        s := (a + b + c) / 2;
>        return Sqrt(s*(s − a)*(s − b)*(s − c));
>    end function;
>    area(3, 4, 5);
        6.000000000000000000000000000000
```

There is a convenient shorthand for the common special case of a function with a single return value that is a simple expression in the arguments.

func<*arguments* | *expression*>

For example:

```
>    harmonic_mean := func<a, b | 2 / (1/a + 1/b)>;
>    harmonic_mean(3, 6);
          4
```

A *procedure* is a function that does not return any values. Procedures are created similarly to functions, except that the relevant keyword is **procedure**.

```
>    procedure print_num(n, obj)
>        if n eq 1 then
>            print n, obj;
>        else
>            print n, obj cat "s";
>        end if;
>    end procedure;
>    print_num(39, "step");
          39 steps
```

Procedures may include a **return** statement that causes them to finish immediately (no values are returned, of course).

Procedures have a feature that functions in Magma do not have—the ability to modify their arguments. An argument that is to be modified is indicated by prefixing it with a tilde ($\sim$), both in the definition and when called.

```
>    remove_gcd := procedure(~x, ~y)
>        g := GCD(x, y);
>        x div:= g;
>        y div:= g;
>    end procedure;
>    a := 15;
>    b := 20;
>    remove_gcd(~a, ~b);
>    a, b;
          3 4
```

5 Loops: **for, while,** and **repeat**

Magma has three looping constructs, which are similar to those of other languages. Each of the loops has a group of statements called the *loop body* that is executed repeatedly until the loop ends.

The most commonly used type of loop is the **for** loop. It has two versions; the less used but possibly more familiar one has the following form.

> **for** variable := a **to** b **by** k **do** statements **end for**;

In this form the values a, b, and k must be integers. The loop body is executed for each integer in the specified range; the first time with the variable set to a, the next time to $a+k$, and so on. The increment is allowed to be negative.

```
>    for i := 13 to 1 by −3 do
>        i;
>    end for;
```

```
         13
         10
          7
          4
          1
```

The **by** subclause is optional (and usually omitted) in this form; if it is omitted then the increment is taken to be 1. A common use is when it is desired to perform the same operation a specific number of times. The following example demonstrates that performing eight perfect out-shuffles returns a deck of 52 cards to their original order.

```
>    os_image := [ 1..52 by 2 ] cat [ 2..52 by 2 ];
>    out_shuffle := func<cards | cards[ os_image ]>;
>    cards := [ 1..52 ];
>    for i := 1 to 8 do
>        cards := out_shuffle(cards);
>    end for;
>    cards eq [ 1..52 ];
```

```
         true
```

The second version of the **for** loop allows the variable to take on values other than integers.

for *variable* **in** *domain* **do** *statements* **end for**;

In this form the loop body is executed once for each value in the specified domain, with the variable taking on these values.

```
>    G<a,b> := AbelianGroup([2, 4]); // ℤ₂ × ℤ₄
>    for g in G do
>        printf "%o\thas order %o\n", g, Order(g);
>    end for;
```

```
         0         has order 1
         a         has order 2
         b         has order 4
         a + b     has order 4
         2*b       has order 2
         a + 2*b   has order 2
         3*b       has order 4
         a + 3*b   has order 4
```

The first version of the **for** loop can be simply converted into the second form by using a sequence constructor, changing *variable* := *a* **to** *b* **by** *k* into *variable* **in** [*a*..*b* **by** *k*]. In fact, this is exactly how Magma handles these statements internally.

The other two types of loops are **while** loops and **repeat** loops.

> **while** *condition* **do** *statements* **end while**;
> **repeat** *statements* **until** *condition*;

In a **while** loop, the condition is first checked, and if it evaluates to true then the loop body is executed. This process is continued until the condition becomes false.

```
>    collatz := func<x | IsOdd(x) select 3*x + 1 else x div 2>;
>    x := 13;
>    while x ne 1 do
>        x := collatz(x);
>        print x;
>    end while;
         40
         20
         10
         5
         16
         8
         4
         2
         1
```

In a **repeat** loop, the loop body is executed and then the condition is checked. If it evaluates to false then the process will be repeated. This process is continued until the condition becomes true.

```
>    Zx<x> := PolynomialRing(Integers());
>    newton := func<f,r | r − Evaluate(f, r)/Evaluate(Derivative(f), r)>;
>    g := x² − x − 1;
>    φ := 1.0;
>    repeat
>        oldphi := φ;
>        φ := newton(g, φ);
>    until φ eq oldphi;
>    φ;
         1.6180339887498948482045868343
```

The **break** and **continue** statements can be used to cause the early termination of a loop (**break**) or of a particular execution of the loop body (**continue**).

> **break**;
> **continue**;

When a **break** statement is executed it causes the innermost loop to finish immediately; the flow of execution resumes after the loop, just as if the loop had terminated normally. When a **continue** statement is executed it causes the innermost loop body to finish immediately; the flow of execution resumes at the end of the loop body. For **for** loops this causes the next variable in the iteration to be used; for **while** and **repeat** loops the corresponding condition is checked to see whether the loop should continue or not.

```
>    for x in [0..10] do
>        if IsOdd(x) then continue; end if;
>        print x;
>        if not IsPrime(2^x + 1) then break; end if;
>    end for;
         0
         2
         4
         6
```

Each of these statements has a variant that allows outer **for** loops or loop bodies to be terminated.

> **break** *variable*;
> **continue** *variable*;

In this form, the specified variable must be the loop variable of an enclosing **for** loop. It is the loop or loop body associated with this **for** loop that will be terminated.

```
>    n := 91;
>    for a in [-6..6] do
>        for b in [-6..6] do
>            if a^3 + b^3 eq n then
>                <a, b>;
>                break a;
>            end if;
>        end for;
>    end for;
         <-5, 6>
```

If the *a* were omitted from the statement **break** *a* then only the inner loop would have been terminated, and the other solutions $<3, 4>$, $<4, 3>$ and $<6, -5>$ would have been found.

6 Maps

Another highly important kind of object is the *map*. The map type in Magma corresponds to the mathematical notion of a map from one structure to another. While the application of a map could also be achieved by defining a suitable function, the use of the map type allows structural information to be retained on the map. It also enables special maps such as homomorphisms to be created in a simple fashion. How maps can be created will be explained a little later; some basic uses of maps will be described first.

6.1 Map operations

When an intrinsic computes one structure from another, it is quite common for a map to also be returned that describes how to move from the original structure to the new one. One such intrinsic is MinimalModel.

```
>   E := EllipticCurve([3, 1, 4, 1, 5]);
>   M,φ := MinimalModel(E);
>   φ;
        Elliptic curve isomorphism from: CrvEll: E to CrvEll: M
        Taking (x : y : 1) to (x + 1 : y + x + 1 : 1)
```

The intrinsics Domain and Codomain can be used to extract the appropriate structures from a map.

```
>   Domain(φ);
        Elliptic Curve defined by y^2 + 3*x*y + 4*y = x^3 + x^2 + x
        + 5 over Rational Field
>   Codomain(φ);
        Elliptic Curve defined by y^2 + x*y + y = x^3 + 3*x + 4 over
        Rational Field
```

If the map supports it, the intrinsic Inverse returns the inverse of the map.

```
>   Inverse(φ);
        Elliptic curve isomorphism from: CrvEll: M to CrvEll: E
        Taking (x : y : 1) to (x - 1 : y - x : 1)
```

(The alternative form ϕ^{-1} is equivalent to Inverse(ϕ).)

The image of an element under a map can be obtained in one of two ways, either by treating the map like a function or by applying the @ operator.

> *map(element)*
> *element @ map*

Additionally, if supported by the map, the @@ operator can be used to return a preimage of an element of the codomain.

> *element @@ map*

```
>   P₁ := E ! [ −2, 1 ];
>   P₂ := E ! [ 0, 1 ];
>   φ(P₁);
        (-1 : 0 : 1)
>   P₂ @ φ;
        (1 : 2 : 1)
>   (M ! [ 2, 3 ]) @@ φ;
        (1 : 1 : 1)
```

In fact, both image and preimage calculation can be applied to an entire sequence or set of elements at once.

```
>    [ M | [ −1, 0 ], [ 1, 2 ] ] @@ φ;
         [ (-2 : 1 : 1), (0 : 1 : 1) ]
```

Given maps $f : A \to B$ and $g : B \to C$, the iterated application $g(f(x))$ is legal since the codomain of f is the same as the domain of g. The corresponding composite map $h = g \circ f : A \to C$ may be created as $h := f*g$ (note the order of f and g).

```
>    φ * φ⁻¹;
         Elliptic curve isomorphism from: CrvEll: E to CrvEll: E
         Taking (x : y : 1) to (x : y : 1)
>    φ⁻¹ * φ;
         Elliptic curve isomorphism from: CrvEll: M to CrvEll: M
         Taking (x : y : 1) to (x : y : 1)
```

6.2 Map creation

Maps can be created with either the **map** or **hom** constructor, specifying the domain and codomain of the map together with the information that determines how to find the image of an element of the domain.

map< *domain* → *codomain* | *image specification* >
hom< *domain* → *codomain* | *image specification* >

(The two character sequence −> is represented by → throughout this book; thus a code fragment like $A \to B$ would actually be typed as A−>B.)

There are two ways to define a map, either by specifying the images of each element (an *image map*) or by providing a rule that determines the image of an arbitrary element (a *rule map*). If the **hom** constructor is used with an image map then it is only necessary to provide images for each generator of the domain, since these determine the homomorphism uniquely.

There are two equivalent notations for explicitly specifying the image of an element.

element → *image*
< *element, image* >

An image map specification will contain either or both of these forms. They may be listed directly, or contained within a set or sequence. So in the following example the three map definitions are equivalent.

```
>    K := GF(2);
>    ψ₁ := map< K → Booleans() | 0 → false, 1 → true >;
>    ψ₂ := map< K → Booleans() | <1, true>, <0, false> >;
>    ψ₃ := map< K → Booleans() | [ x → IsOne(x) : x in K ] >;
```

If specifying a homomorphism, a third form may be used whereby only the images are listed. These will be implicitly matched up with the generators of the domain (the first generator yields the first image, and so on).

```
>   C<i> := QuadraticField(−1);
>   conj := hom< C → C | −i >;
>   conj(3 − 4∗i);
        4∗i + 3
```

A rule in a rule map is given by a variable and an expression involving that variable. The following format is used:

$x \mapsto$ expression in x

(The three character sequence :−> is represented by $\mapsto$ throughout this book; thus a code fragment like $x \mapsto x + 1$ would be typed as x :−> $x + 1$.)

When the map is applied to an element, the value returned is the result of evaluating the expression with the variable temporarily assigned the value of the element.

A rule map may have a second rule, which specifies how to find the preimage of elements under the map. Continuing the example from the previous section, it turns out that the Mordell–Weil group of E is generated by the two points P_1 of order 2 and P_2 of infinite order. Thus this group is isomorphic to $\mathbb{Z}_2 \times \mathbb{Z}$. The following is a quick-and-dirty implementation of the map between this abstract group and E itself.

```
>   G<a, b> := AbelianGroup([ 2, 0 ]);
>   GtoE := func<g | t[1]∗P₁ + t[2]∗P₂ where t is Eltseq(g)>;
>   EtoG := function(P)
>        n := Round(Sqrt(Height(P)/Height(P₂)));
>        Q := n∗P₂;
>        m := Q[1] eq P[1] select 0 else 1;
>        Q −:= m∗P₁;
>        if Q[2] ne P[2] then n := −n; end if;
>        return G ! [ m, n ];
>   end function;
>   gmap := map<G → E | x ↦ GtoE(x), y ↦ EtoG(y)>;
>   gmap(a − b);
        (1 : 1 : 1)
>   (P₁ + 3∗P₂) @@ gmap;
        a + 3∗b
```

There may also be special forms of map constructors allowed for particular combinations of domain and codomain. One important case is that of maps between schemes, which may be specified by providing the coordinates of the image as rational functions of the coordinates of the element (and similarly for the inverse, if supplied).

```
>   Q := RationalField();
>   P₁<u, v> := ProjectiveSpace(Q, 1);
>   P₂<x, y, z> := ProjectiveSpace(Q, 2);
>   C := Conic(P₂, x² + y² − z²);
>   ψ := map< C → P₁ | [ y, x + z ], [ v² − u², 2*u*v, v² + u² ] >;
>   ψ(C ! [ 3, 4, 5 ]);
        (1/2 : 1)
>   Inverse(ψ)(P₁ ! [ 2, 3 ]);
        (5/13 : 12/13 : 1)
```

References

1. John Cannon, Wieb Bosma (eds.), *Handbook of Magma Functions*, Version 2.12, Volume 1, Sydney, 2006.

Index